Irrigation and Society in Medieval Valencia

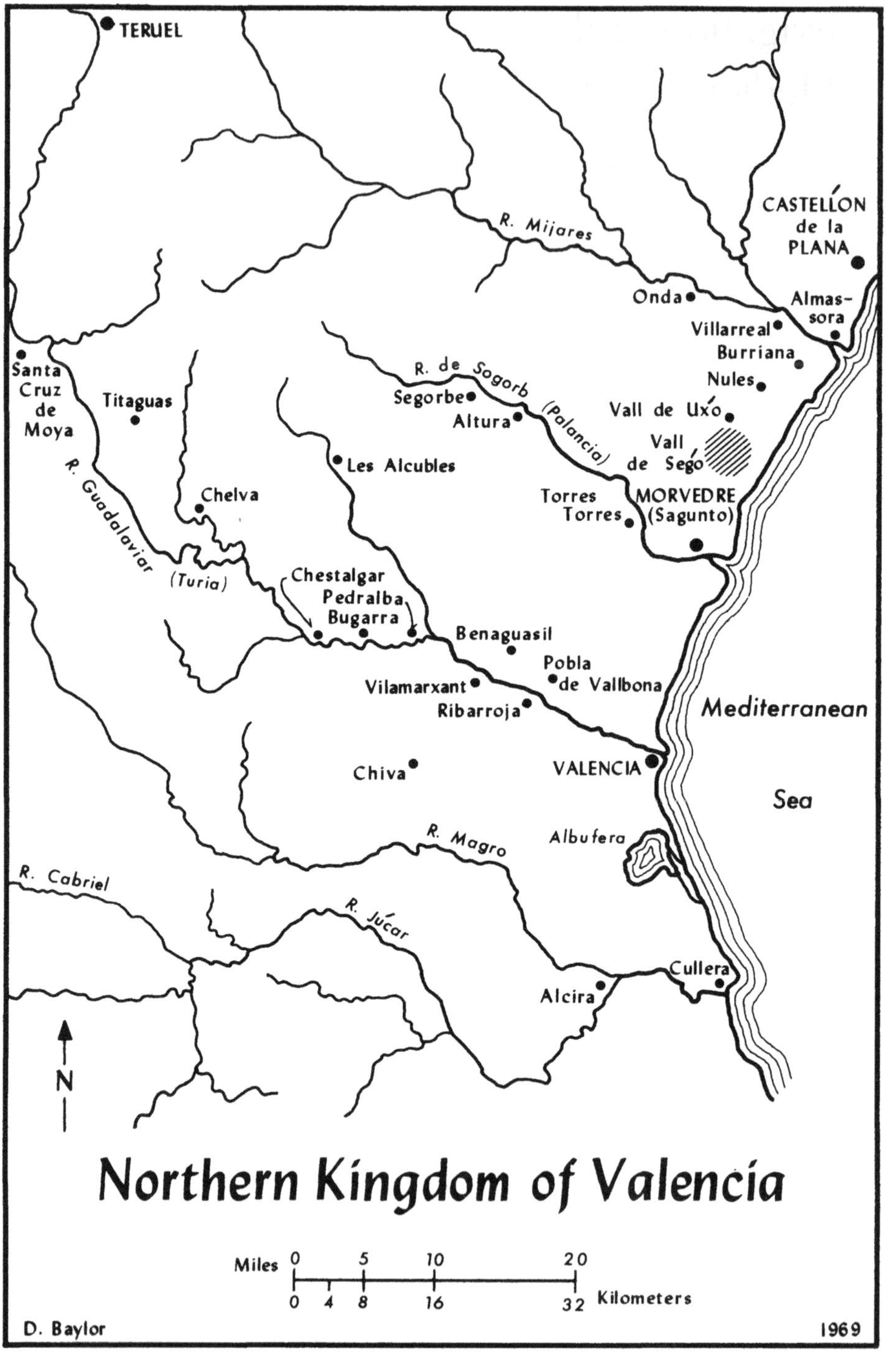
TERUEL
R. Mijares
CASTELLÓN de la PLANA
Onda
Almassora
Villarreal
Burriana
Nules
R. de Sogorb (Palancia)
Segorbe
Altura
Vall de Uxó
Vall de Segó
Santa Cruz de Moya
Titaguas
Les Alcubles
R. Guadalaviar (Turia)
Chelva
Torres Torres
MORVEDRE (Sagunto)
Chestalgar
Pedralba
Bugarra
Benaguasil
Pobla de Vallbona
Vilamarxant
Ribarroja
Mediterranean Sea
Chiva
VALENCIA
R. Magro
Albufera
R. Cabriel
R. Júcar
Cullera
Alcira
N
Northern Kingdom of Valencia
Miles 0 5 10 20
0 4 8 16 32 Kilometers
D. Baylor
1969

Irrigation and Society in Medieval Valencia

Thomas F. Glick

The Belknap Press of
Harvard University Press
Cambridge, Massachusetts
1970

Distributed in Great Britain by Oxford University Press, London

Library of Congress Catalog Card Number 79–95923

SBN 674-46675-6

Printed in the United States of America

TO

Robert T. Follestad
Anthony G. Freeman
Timothy L. Towell

Preface

My interest in the history of irrigation in Valencia began in 1959 when, in answer to an advertisement in the *Harvard Crimson,* I set to work translating the regulations of the Valencian irrigation communities. This was the beginning of a most fruitful friendship with Professor Arthur Maass that led me to the irrigated fields of Valencia and Gandia in 1960–61 and finally in 1965–66 to the archives of Valencia, Castellón, and Alicante on a grant from the Social Science Research Council. Over the years we identified several areas of historical development which seemed challenging and ripe for research. The first, most elusive, most often discussed, and of most immediate interest to me was that of the influence of Islamic culture and practice upon present institutions of irrigation. We searched for documents; there were none. We talked to historians, arabists, lawyers, and civil servants; few had anything to offer but conjecture.

Discussion with some of the few social scientists who are familiar with irrigation in Islam convinced me of the utility of the comparative method. Even in the absence of historical documentation, the main outlines of irrigation practice in Al-Andalus could be unearthed by working carefully backward from the present traditional systems of Spain and by comparing these systems to Islamic ones. This paper is by no means that comparative study; but I have tried in Part Two to block out some of the major areas where comparative study might yield worthwhile results. My own comparative data, largely from Yemen, North Africa, and Syria, have been selected (or, rather, preselected, for studies of these irrigation systems are rare) more on the basis of intuition and general knowledge of Islamic settlement patterns in Al-Andalus than on a rigorous search of all the possible Islamic models for Valencian irrigation. Nonetheless, it has always been known that the areas under consideration were settled principally by Syrians and Yemenites, and on this basis I have limited most of my attempts at comparison to the homelands of those two peoples.

Although my attention was initially centered on the problem of cultural continuity, in time I became interested in the institutions for their own sake. When I set out for Spain I had originally planned to study social control in medieval irrigation, the agrarian institutions of the *huertas,* and, as far as possible, the technological and economic ramifications of irrigation agriculture. Unhappily — and, perhaps, luckily — the sources did not permit such a broad inquiry.

During a year in the Valencian archives I was able to study a vast number of documents dealing solely with irrigation problems. Though these documents shed some light on certain technological problems of irrigation and on economic problems internal to the irrigation systems, generally they are relevant to the areas of social and administrative control. I had to abandon for the present any hope of understanding the broader economic ramifications of irrigation agriculture in Valencia and, indeed, I remain largely ignorant of the actual farming methods, since the interest of the water documents typically is in the distribution of water from the river up to the farmer's "turn-out," but no further. Mentions of crops, for example, are incidental, and data about plowing and other agricultural techniques are absent in my documents.

On the other hand, the political dimension of water control — of which I had been largely unaware when I began my search of the archives — provided an unexpected challenge.

The major limitations of the documents on which this study is based can be summarized as follows:

1) Data concerning actual day-to-day operations of the medieval irrigation systems are very scarce. Most of the documents (irrigation ordinances, royal privileges, and so on) describe a generalized or idealized situation. Litigation documents fill in some gaps by describing specific situations in great detail, but because

these documents are the product of stress or conflict they do not describe the daily routine either.

2) Almost all of the documents are administrative or legal in nature. They tell much, therefore, about social control, the nature of conflicts and their resolution, but little about the agricultural regime or economy. The litigation material, however, is composed largely of accounts of the governor's court, which heard only important cases. Most instances of trivial, local conflict were adjudicated orally by the local irrigation officers; hence almost no records of minor offenses exist.

3) Reliance on the records of municipal archives, which are particularly rich in irrigation material, tends to exaggerate the towns' control over irrigation affairs and to de-emphasize the role of the communities of irrigators. The council books of Valencia (extant from the early fourteenth century to the present in an almost unbroken series), Castellón, Alcira, and Elche contain an inexhaustible wealth of detail about medieval irrigation.

4) The records of the irrigation communities themselves are mostly modern (that is, from the seventeenth century to the present), with only sporadic holdings from the late fifteenth and sixteenth centuries. But similar material from the Middle Ages, the study of which would go far in rectifying any bias resulting from over-reliance on municipal records, does survive in notarial series (*notals* and *protocols*). These sources have hardly been searched at all for irrigation material, yet as my unsystematic and largely fortuitous findings attest, their study is of the utmost importance in filling gaps left by more specialized series of documents.

5) The time span of this study is naturally dependent upon the available documentation, most of which falls between the years 1375 and 1460. Considerable documentation, largely royal privileges, does exist for the thirteenth century, however, and in some cases, where medieval concerns were continued into the sixteenth century or later I have found it worthwhile to carry through into the postmedieval period. Furthermore, in traditional systems such as those of eastern Spain, where custom dies hard, modern practices contain many medieval elements. Where medieval documents are lacking, I do not hesitate to draw on

more recent material. The regulations of the Valencian communities, most of which were codified in the eighteenth century, contain many provisions of medieval origin.

The spelling of Valencian place names presents numerous problems, owing to the existence of many variant spellings — both Catalan and Castilian — for the same place. No one criterion is completely satisfactory. For major place names I have used the familiar spellings found in standard maps and atlases. For minor toponyms I have followed, when possible, Catalan orthography according to M. Sanchis Guarner's *Contribució al nomenclàtor geogràfic del País Valencià* (Barcelona, 1966), which does not cover all the names in this study. In certain cases I have preferred an archaic or Castilian orthography when it more closely accords with English pronunciation (Benacher rather than Benatger, for example; Chiva rather than Xiva). All common variants are included in the index.

Professors Arthur Maass and Giles Constable read the original draft of this study and offered many insightful comments, the former from the point of view of a political scientist and irrigation specialist, the latter from the vantage point of the medievalist. Robert I. Burns read a later draft and suggested numerous emendations and corrections based on his profound knowledge of medieval Valencia. Harold A. Thomas helped me to interpret the hydraulic background of specific situations discussed in Chapter III. The portions of Chapter IX dealing with the diffusion of hydraulic techniques were presented as a paper at the December 1968 meeting of the Society for the History of Technology, in Dallas. As a result I have benefited from the suggestions and criticisms of Paul W. English, Eugene S. Ferguson, and Chris Field who were able, I hope, to redress somewhat my lack of technical expertise. Portions of Chapters III and XI have appeared in *Technology and Culture* (© 1968, 1969 by The University of Chicago and the Society for the History of Technology).

Many persons were of great help to me in Spain. In Valencia, the assistance of Rosa Rodríguez de Tormo, archivist of the Archivo del Reino, and her staff was invaluable. My two good friends Vicente Giner Boira, assessor of the Tribunal of Waters, and José M. Cueco Adrián, a distinguished historian of Spanish irrigation, gave generously of their time and interest. Historians of Castellón de la Plana are most fortunate in that the archivist, Don Eduardo Codina Armengol, is also mayor and strives to make foreign scholars feel at home; he and the historian José Sánchez Adell made my work there both enjoyable and fruitful. Lastly, Lluís V. Aracil made many strong and compelling suggestions toward solving my orthographical dilemmas. That I did not follow his advice in all cases reflects only my own timorousness. The final draft benefited greatly from the work of my wife, Betty, which has spared the reader numerous long-winded digressions of which medievalists are overly fond, and of Mrs. Margaret Ritchie; my debt to both is great.

Thomas F. Glick

Austin, Texas
June 1969

Contents

Maps

drawn by David W. Baylor

Illustrations

Tables

Glossary

Ar. = Arabic Cast. = Castilian
Val. = Medieval Valencian

açut (Val.): diversion dam
alamí (Val.): see amīn
Al-Andalus (Ar.): Islamic Iberia
Albufera (Val.): fresh-water lake south of the city of Valencia
almenara (Val.): see return ditch
alter (Val.): field at so high an elevation that it could be irrigated only when the canal ran full or was backed up by checks
amīn (Ar.): irrigation or guild functionary
boquera (Val.): field turn-out
cafizate: square measure equal to six fanecates
canal check: obstruction placed in a channel to raise or slow down flow
cequia (Val.): irrigation canal
cequia maior (Val.): main canal, usually of a town
cequia mare (Val.): "mother canal," that is, main canal
cequiatge (Val.): canal maintenance tax
cequier (Val.): irrigation officer
crida (Val.): "cry," an official municipal communication as announced publicly by the town crier
divisor: structure dividing flow of an irrigation channel into two or more parts
fanecate: square measure equal to 831 square meters
fila (Val.): 1) proportional measure of water; 2) secondary irrigation channel
fueros (Cast.): laws of the realm
hereter (Val.): member of an irrigation community
huerta (Cast.): intensively irrigated alluvial region

jovate: square measure equal to six cafizates or thirty-six fanecates
jurate: municipal executive officer
justiciar: municipal judge (there were two, a civil justiciar and a criminal justiciar)
Libros de Repartimiento (Cast.): registers of land newly distributed after the conquest from the Muslims
llaurador (Val.): cultivator
marjal (Val.): marshy, swampy area (when capitalized, the marshland to the south of the city of Valencia, between Russafa and the Albufera)
mulnar (Val.): water measure (amount of water that could move one millstone)
mustasaf (Val.): municipal public health and market officer
partidor (Val.): 1) divisor; 2) minor official of the irrigation community
pecunia comuna (Val.): general fund of the municipal treasury in which fines were deposited
porter (Val.): enforcement officer on staff of governor
Pueblos Castillos (Cast.): several villages upstream of the Valencian huerta which drew water from the Guadalaviar River
qaḍā' (Ar.): judgeship, the office of qāḍi
rech (Val.): service area of a canal
regadiu (Val.): irrigated land
return ditch: channel carrying water from an irrigation canal back to the river or from a secondary channel to the main canal
roll (Val.): canal turn-out
ṣāḥib al-sāqiya (Ar.): "master of the canal" (irrigation official)
secá (Val.): unirrigated land
syndic: 1) elected deputy of any group; 2) municipal financial and administrative officer subordinate to the jurates
tanda (Val.): irrigation turn
turn-out: channel or orifice leading from a main canal to a secondary channel, or from a canal to an individual field
vega (Cast.): alluvial valley floor, usually irrigated

Irrigation and Society in Medieval Valencia

Abbreviations

ACCV	*Anales del Centro de Cultura Valenciana*
ACV	Archivo de la Catedral de Valencia
AHN	Archivo Histórico Nacional (Madrid)
AMC	Archivo Municipal de Castellón
AMV	Archivo Municipal de Valencia
ARV	Archivo del Reino de Valencia
BSCC	*Boletín de la Sociedad Castellonense de Cultura*
CSIC	Consejo Superior de Investigaciones Científicas
RDTP	*Revista de dialectología y tradiciones populares* (Madrid)

Introduction

The social creativity of medieval Spanish townsmen and villagers was expressed in a variety of communal institutions which have long attracted the interest of historians and social scientists. Municipal councils, herding and fishing collectives, and irrigation communities were characteristic of a society in which feudal customs had taken but shallow root. Of all these institutions the irrigation communities of eastern and southern Spain were possibly the most important, certainly the longest lived.

In the nineteenth century, foreign engineers, agronomists, and social scientists, drawn by centuries of experience in communal water use and control, made the irrigated districts — the *huertas* — of Valencia and Murcia the object of intensive study. Englishmen particularly were eager to apply the Spanish experience to the irrigation problems of British India. Sir Clements Markham, Sir Colin Scott Moncrieff, and J. P. Roberts all had personal acquaintance with Indian irrigation systems and hydraulic problems.[1] The French engineer Maurice Aymard was interested in irrigation possibilities in Algeria, and the agronomist Jaubert de Passa was intimately familiar with the irrigation economy of southern France.[2] The American William H. Hall, who described the irrigation systems of Valencia, Murcia, and Granada in the 1870's, was the state engineer of California, a region which underwent extensive irrigation development in the latter part of the last century.[3] All of these men hoped that some practical benefit to their countries might result from their studies.

In Spain itself a body of local historical and hydraulic studies steadily accumulated, and around the turn of the century Spanish and foreign scholars began at last to consider these irrigation customs in a scientific context. The French geographer Jean

Brunhes attempted to delineate the relation of irrigation institutions to geographical conditions, and the Spanish social thinker Joaquín Costa discussed the irrigation communities within the general framework of traditional agrarian collectivism in Spain.[4]

The historians and the social scientists had different aims. The historians were concerned with discovering — largely by conjecture — the origins of the irrigation canals and with describing the growth of a corpus of customary irrigation law. The problem of origins, so typical of nineteenth-century historiography, was elaborated polemically by partisans of a Roman origin theory and by those who emphasized Islamic influences.[5] The social scientists, on the other hand, were only marginally interested in the past of the institutions they studied. The divergence in approaches and concerns of the different disciplines accounts for the lack of any systematic study of the social structure of medieval Spanish irrigation, a void which the present study attempts to fill.

The weakness of the nineteenth-century historical approach was owing to the lack of documentation from the Islamic period and an inability to use comparative material. A prime historiographical problem, then, is to reconstruct the irrigation institutions of Islamic Spain from indirect evidence, by working backward from the copious documentation of the Christian Middle Ages and by applying data drawn from the study of modern Islamic systems. Not only do Valencian irrigation institutions come onto the historical stage full-blown in the thirteenth century, but, further adding to the historian's dilemma, they hardly undergo any palpable change during the medieval centuries. The basic objectives of the system of distribution remained constant throughout, notwithstanding frequent — and usually minor — institutional and political adjustments necessary to the vitality and efficiency of the system.

Medieval patterns of water use, agricultural and irrigation techniques, the settlement patterns of the irrigated areas, and traditional modes of social control in irrigation affairs all survived practically intact into modern times. Profound and far-reaching changes in the old patterns of irrigation agriculture belong largely to the nineteenth century, not (as it is all too frequently claimed in popular accounts of Spanish history) to the time of the Reconquest or the expulsion of the Moriscos. Land-use patterns of the irrigated districts manifested remarkable stability until recent times; it was not until the mid-nineteenth century that the walls of Valencia were torn down and the city began to encroach upon the irrigated area.[6] The economic contours of huerta life showed considerable continuity as well, the most significant break with the past coming with the large-scale introduction of new crops and modern agricultural methods in the last century. The commitment of more and more land to the growing of orange trees and the heightened productivity of such staple crops as rice made possible by the introduction of guano diverted the economic orientation of the huerta increasingly to export markets.[7]

Traditional patterns of huerta social life lasted well into the present century, as the novels of Blasco Ibañez attest.[8] However, elements of the old institutions of water control lost their effectiveness as the conditions which they were designed to serve were altered. The completion of the Generalísimo Reservoir in 1950, by assuring the Valencian huerta of the permanent availability of irrigation water, initiated the decline of the venerable Tribunal of Waters, the most salient component of the traditional system of control, whose purpose was to resolve conflicts in time of water shortage.

The long-term stability of irrigation agriculture in the Valencian Middle Ages does not by any means suggest that its institutions were stagnant or lifeless. The historian's task is to describe a dynamic society acting within an extremely stable institutional framework. Without understanding how the system functioned in the vigorous centuries following the Reconquest, it would not only be impossible to understand the relation of that system to the Islamic past, but it would also be difficult to appreciate the significance of subsequent changes.

My investigation of irrigation society describes several levels of social organization. The basic water-use unit in medieval Valencia was the community of irrigators comprising all who drew water from a single canal system. The social function of these communities, whether autonomous or municipally controlled, was similar throughout southeastern Spain and was the common denominator among local systems that differed in other aspects of irrigation practice and law. The communities' objectives, expressed in their regulations and in the principle of proportional division of water, were to apportion water justly and fairly to each user and to prevent conflict among the irrigators. In spite of the fact that the regulatory system was designed as a control over all disruptions, no matter what the cause, the communal authorities were not always able to fulfill their statutory roles. Large communities often split into a number of political factions which brought pressure to bear upon the communal authorities in order to prevent the enforcement of certain regulations. Especially in disputes involving large numbers of people, the community's internal mechanisms of control sometimes proved ineffective, and the intervention of higher authorities was sought.

The irrigation communities are seen in another dimension when considered in light of their relations with higher levels of local and regional authority: the towns, the Church, and the king. Relations between local water-use communities and political rivalries within the watershed reveal a fluctuating situation in which the outcome was likely to be determined more by political pressure than by statute, in contrast to the general stability of community institutions of control.

The irrigation policies of higher authorities raise questions concerning the nature of irrigation societies. The long-recognized association of irrigation agriculture with the development of the great civilizations of antiquity (China, Egypt, the empires of the Fertile Crescent, Inca Peru, for example) has led in the past few years to the formulation of general theories of hydraulic civilizations. The best-known hypothesis, and that of most interest to historians, is that of Karl A. Wittfogel.[9] The Wittfogel theory is derived from the observation that water by its very nature imposes certain economic and social necessities upon those who would control it. Large-scale water projects require capital and mass labor which can only be supplied by a powerful, centralized authority. But recent studies, made primarily by anthropologists, of irrigation systems differing widely both in structure and in level of social and economic development, suggest that political centralization in irrigation is an independent variable. Successful systems are just as likely to be cellular and decentralized as centralized.[10]

The Valencian evidence confirms the findings of anthropologists: in the huerta of Valencia, social control has traditionally rested in the collective will and institutions of the irrigators themselves. The autonomy of local irrigation communities was accepted by all levels of the power structure and enshrined in customary law; throughout the Middle Ages, even as the system expanded, no attempt was made by the city to lessen the autonomy of the communities. But in other areas of the kingdom, such as La Plana, irrigation systems were typically controlled by the municipality. There is evidence both of towns acquiring control of previously autonomous communities and also of the opposite process — the devolution of control from the town to the community of irrigators. Even if one can accept the general proposition that regional hydraulic development is best achieved (as modern experience would indicate) under the aegis of a central power endowed with adequate financial and planning resources, it is fruitful to consider the particular geographical and social factors which ran counter to and effectively prevented such a development in medieval Valencia.

The focal point of this study is conflict, not only because conflict stimulates community activity and molds the agencies of social control, but also because the dynamics of medieval irrigation communities are most visible to the historian in records of litigation.[11] It has become a commonplace, especially in anthropological literature, that irrigation necessitates cooperative action.[12] But the fabric of relations inherent in any system of water distribution must also have an implicit potential for disruption. As some students of Mexican irrigation have described it: "An irrigation system establishes relationships among individuals and groups which are simultaneously relationships of interdependence and potential antagonism, stemming equally from the need for and the constraint imposed by cooperation and coordination. The necessity for an ordered, predictable system of water allocation has divisive as well as cohesive potentialities." [13] All irrigation systems experience similar forms of conflict: for example, upstream versus downstream users. The principles underlying the distribution system and the values by which the society seeks to ensure its efficient and just functioning are what distinguish one style of irrigating from another.

In the present discussion the role of weather and the relation of climatic factors to institutional structure is of special significance. Recent studies have sought to distinguish conflicts arising from strains within a given social system from those produced by stresses external to the system.[14] The case of the highly developed administrative and legal system regulating Valencian irrigation practices does not confirm the validity of this dichotomy. No stress can be considered effectively *external* to the system: no climatic, hydrological, or hydraulic crisis could find the irrigators institutionally unprepared to meet it, inasmuch as the very rationale of the regulatory system was to provide solutions for the broadest possible range of problems. Climatic change and variation, however extraordinary, were accounted for in the system of control. Even the effects of such an "external" disruptive force as invasion from abroad were taken into account by the regulatory apparatus, which included procedures to be undertaken in the event of physical damage to the water system that such a disruption might cause.

The first part of this study is concerned with the dynamics of both irrigation communities and regional water-use units. The stability of local institutions allowed the practice of irrigation agriculture to proceed with little interference from the vagaries of daily political life — in contrast to the more visible rate of change in the regional arena.

The second part is devoted to the diffusion of Islamic irrigation technology in general and, more specifically, to the problem of cultural and institutional change during the transition from Islamic to Christian rule in Spain. Having first described medieval Valencian irrigation in the epoch of its fullest documentation (1238–1500), I shall examine its antecedents on the basis of Islamic survivals in medieval Christian institutions and of comparative data from other Islamic irrigation systems. This investigation raises questions of acculturation and cultural transition that extend beyond the restricted geographical and temporal bounds of the present study. For one thing, irrigation was only one of the aspects of Islamic life which the conquering Christians found worthy of emulation. For another, the experience gained by thirteenth-century Spaniards in adapting to their own life style the institutions of an alien, conquered people served as a model and an example which Spaniards of the sixteenth century put to good use when confronted with similar institutions in the New World. It was not merely coincidental that the Law of the Indies stipulated norms for administering native American irrigation systems in terms almost identical to those found in the dispositions of James I of Aragon and Alfonso the Wise of Castile in Valencia and Murcia. Both on the peninsula and in the New World the history of Spanish irrigation is but one example of the administrative creativity and genius for cultural synthesis which characterized Iberian culture at the dawn of the modern age.

Part One | The Irrigation Communities of Medieval Valencia

I | The Huerta Environment

The place name Valencia has three different geographical applications. It refers to the city, to the huerta of some fifty square miles of rich, alluvial fields surrounding the city, and to the kingdom, an area extending north to the River Cenia and south to Orihuela where, prior to the eighteenth century, the *fueros* of Valencia were in force.

Within the kingdom were two radically different agro-ecological systems: dry farming (*secá* in Valencian, *secano* in Castilian) in the mountainous hinterland and irrigation agriculture (Valencian, *regadiu;* Castilian, *regadío*) in the rich alluvial lands along the Mediterranean coast. In the northern part of the kingdom much of the irrigation water was supplied by five river systems, each of which created near its mouth an alluvial core area that became the highly cultivated, densely populated huertas of the Middle Ages.[1] Each huerta supported an urban commercial center. The Mijares watered the huerta known as La Plana, whose chief town was Castellón, and to the south were the Riu de Sogorb (now called the Palancia), serving the huerta of Morvedre (now Sagunto); the Guadalaviar (now the Turia), which was the river of Valencia; the Júcar, which irrigated the fertile area around Cullera; and the Serpis, which entered the sea near Gandia (see Frontispiece).

Along the margins of the huertas, stretching north and south along the seacoast, were low-lying, swampy lands — the *marjals* — typically given over in modern times to rice cultivation.[2] The reclamation of these swamps, a constant preoccupation of medieval townsmen and rulers, has been man's contribution to the development of the huerta just as alluviation has been nature's. Although the lowness of the marjals exacerbated problems of drainage and made their cultivation more difficult than that of the huerta proper, still, the most crucial distinction between huerta and marjal was one of development. The two formed a continuum, the former being reclaimed from and extended at the expense of the latter, the swamp encroaching on the cultivated areas and gaining control when vigilance was relaxed. This relation was pronounced during the Middle Ages, and it lent a characteristic rhythm to irrigation development.

Numerous social and economic factors combined to draw the town, the huerta, and the marjals into close interdependence. First, the Valencians of the littoral, descended from Catalan and southern French settlers, continued the Catalan tradition of the city-state. Most of the huerta was part of the municipal bounds (*terme*) of the city, and the cultivators were thus considered townsmen; also, many townsmen owned extramural farms. Second, as the towns developed as agricultural and commercial centers their growth and that of the surrounding huertas went hand in hand, the huerta supplying the town with food requirements and with some raw materials for industry. Third, the irrigation networks themselves served as physical links between town and huerta, since the canals served both agricultural and urban needs. Finally, the needs of the urban population determined the rate and extent of expansion of the irrigated farmland, and indeed the financial and administrative resources of the city were often required for the turning of barren marjal into productive huerta.[3]

Farther inland and upstream from the main huerta-town complexes were numerous smaller huertas, watered either from the rivers or from the region's many springs that have accounted for dozens of village sites.[4] In the south of the kingdom especially, where there are no large rivers, the small, oasis-like huerta was typical. These southern huertas where water was relatively scarce — Alicante, Novelda, Elche — were characterized by a water tenure system different from that prevailing in the large river huertas. In Valencia and Castellón water and land were inseparable, and water rights could not be alienated from the irrigated parcel. Water was distributed proportionally and free to all whose land had rights to it. In Alicante, Elche, and other southern huertas water rights were detachable from the land and could be sold. Most scholars who have attempted to systematize Spanish irrigation systems have done so on the basis of alienability of water rights and have insisted on a rigid dichotomy between the two systems.[5] This is an entirely valid criterion, which has not only a geographical basis (as Brunhes has shown) but a cultural one as well (see Part Two, below). Nevertheless, excessive contrasting of the two systems has obscured the basic homogeneity of administrative and social organization in all the huertas of the kingdom. This study is largely concerned with the Valencia model, which was more important economically and better documented; but regionwide similarities will be pointed out and the two systems will not merely be contrasted.

IRRIGATED AND UNIRRIGATED FIELDS

In the medieval huerta, the right to water inhered not in the landowner, but in the land itself. The donations of the Repartimiento specified what kind of land the grantee had received, as in 1248 when A. Boschet was granted three jovates of irrigated land in Gandia and two of unirrigated land ("III jo. terre in regadivo et II in seccano").[6] The terms for irrigated land (regadiu) and unirrigated land (secá) indicated the legal status of the land in question, a status which was most difficult to alter. Regadiu implicitly had the right to the water of a specific source,[7] from which the owner was entitled to take an amount of water proportional to the area of regadiu he held.[8] Accordingly, specific water allotments were not included in medieval land grants, since the grantee receiving land *in regadivo* was understood to be entitled to a quantity of canal water in proportion to the size of his parcel.[9] According to a dictate of Roman law universally observed in eastern Spain, nonuse did not void a right. (The principle of *rega o pert* — irrigate or lose — applied only to the irrigation turn, not to the underlying right.)

Changing the legal status of irrigated land to unirrigated, and vice versa, was a complicated procedure. One of the fueros of James I indicates that in the initial moment of Christian settlement an option existed concerning the legal status of a donation:

> The vineyards and estates which are able to irrigate pay *cequiatge* [canal maintenance tax] unless the lords of those vineyards or those estates shall desire not to take the water for irrigating; if anyone possesses some places which have not, up to now, been used to irrigate, he can take the water with which those places irrigate according to the custom of the place . . . and he can irrigate those places, without anyone impeding him, and pay *cequiatge* according to the manner of the other residents on that canal.[10]

The settlers coming in the wake of the conquerors seem to have had an option: in occupying land previously irrigated they could elect not to irrigate and thus be freed from having to pay *cequiatge.* They could also take previously unirrigated land and arrange to join an existing community of irrigators, so long as they agreed to pay the cequiatge and abide by the rules of the commons of that canal. In practice, however, land which once had been irrigated could not be converted to secá merely on the whim of its owner, whether he used the water to which his land was entitled or not.

In 1421 the Muslims of Chiva petitioned the governor to change the status of their regadiu land, which they had irrigated from the spring of Chiva, to secá status. The Muslims stated that the land in question had never been secá, and that because it was huerta and regadiu they had paid fifty pounds to the king each year. Now, however, the land had de facto become secá because of a long drought, and they wished that their taxes could be made equal to those of Christians holding unirrigated land.[11] The request had been made repeatedly before and, even though the Moorish quarter (*morería*) was fast becoming depopulated, the probabilities of its being granted were small. Irrigated lands were taxed at a much higher rate than unirrigated and no magnate would easily consent to a change in status.[12] The opposite case, changing secá to regadiu, was another way of saying "irrigating without rights." In the fine-books of the *cequiers* of Castellón, irrigators were frequently cited and fined for "irrigating in a place where there is no regadiu," or in a place where they "have no right." [13]

Grants to the same settler of both irrigated and unirrigated lands did not necessarily mean that the two were contiguous; the holdings might have been located in different parts of the district. The frequency of unirrigated fields in a medieval huerta area was also a function of development; as existing canals were extended or new ones dug, new regadiu land came into existence.

SIZE AND SHAPE OF FIELDS

Fields were measured in fanecates, cafizates, and jovates (six fanecates to the cafizate, and six cafizates to the jovate).[14] Irrigated fields are typically smaller than unirrigated parcels because they are much more productive. In 1963, 72.2 percent of all parcels in the province of Valencia measured less than 0.5 hectares[15] (equivalent in medieval measures to somewhat less than one cafizate), reflecting not so much a "minifundia" problem as the typically minute parcelation of irrigated land.

Average plots in the huerta of Valencia in the fifteenth century measured from three to ten fanecates. For example, a woman's will in 1442 disposing of *terra campa* that she held (apparently in Burjassot) on the Alborg Canal listed irrigated parcels of nine, six, and eight fanecates plus three fanecates of vineyard [16] — an average of six and a half fanecates. The estate of Johan Monflorit in the huerta of Massamagrell (Moncada Canal) included parcels of *terra campa* and vineyards of five, four, ten, and five fanecates,[17] an average of six. A list of parcels in the Mislata Canal system in 1462 included some fields with larger areas: two cafizates of field and vineyard, five and nine fanecates of vineyard, and other fields of three and a half, eight, eleven, and thirteen fanecates, and one-half and one cafizate (that is, three and six fanecates).[18]

Irrigated fields were most often quadrilateral, with straight border lines to facilitate the uniform flow of water.[19] Figure 1,

Figure 1. Irrigated Fields on the Favara Canal, Valencia

Legend

1. Land taken for straightening the canal and for construction and reinforcement of its banks
2. Former channel of Favara Canal
3. Land belonging to the Cathedral chapter, with house
4. Canal which crosses Favara in order to irrigate lower lands
5. Road to Mislata
6. Roll (canal turn-out) which conducted water from the old channel
7. Roll of the new channel
8. Hemp tank (*balza de cáñamo*)
9. Drainage ditch (*escorrentia*) which runs from the Mislata Canal through the Brazo de los Moros and into Favara

an eighteenth-century drawing, shows fields bordering the Favara Canal in the huerta of Valencia. These fields are trapezoidal, or nearly square, and the middle two are approximately three fanecates in area. Figure 2 is an eighteenth-century plan of fields bordering the Sierra del Puig in Játiva and shows the typical

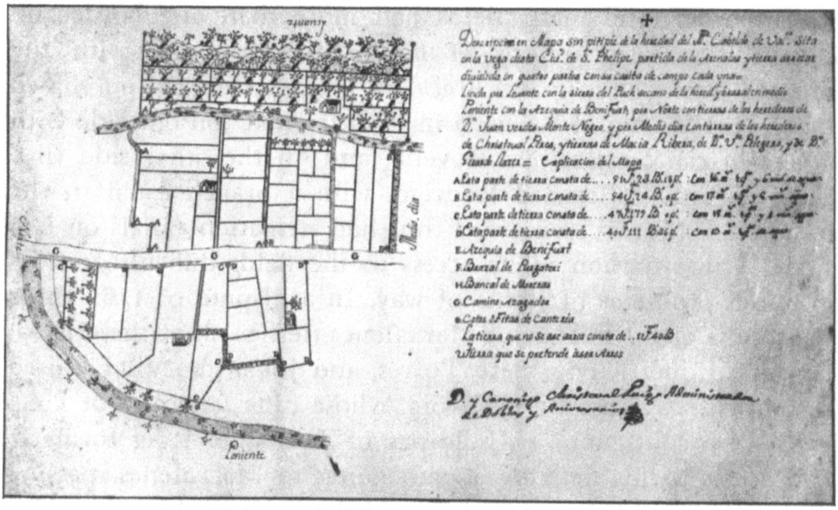

Figure 2. Irrigated Fields on the Benifurt Canal, Játiva

Legend

ABCD	Estate subdivision boundaries
E	Benifurt Canal
F	Purgatori branch canal
G	Cattle path
H	Mulberry plot

pattern of rectangular fields and straight ditches. The plan represents an estate owned by the chapter of the Valencia cathedral. Each of the four sections of the estate has a *casita de campo* (presumably a farmhouse inhabited by tenant laborers). The majority of medieval *llauradors* lived either in the villages and towns skirting the huerta or in *alqueries,* small hamlets dispersed throughout the huerta. The *barraca* (a thatched, whitewashed adobe building), the typical huerta house of modern times (eighteenth century to the present), is rarely cited in medieval documents.[20]

CANALS AS FIELD BOUNDARIES

A commonplace of medieval Valencian field perambulations is the medial canal. All irrigated fields had at least one channel for a border, and many fields had more than one border on canals. Thus descriptions of landholdings abound with the phrases *cequia en mig* and *braçal en mig*. A vineyard purchased by Martin Morata of Valencia in 1427 fronted "on one side with a certain canal called Na Rovella, and on the other side with the vineyard of Bernard Rodrigo, with a branch canal in the middle." [21] Morata's vineyard thus had irrigation canals on two sides. This situation made access to the fields difficult and aggravated problems of rights of way. In a dispute of 1462 three llauradors of the district of Marxalenes (left bank of the Guadalaviar), Miquel Ruvio, Pere Torres, and Johan del Vilar, raised a complaint against Bernat Roig, whose nine fanecates of vineyard were contiguous to holdings of each one. Roig formerly had access to his field by a path going to Marxalenes, passing near Vilar's field, but now he wanted to change his access. Instead of using the former route, he made a new path through the fields of Ruvio and Torres and across a canal dividing them in order to reach the road going to Benicalapet. Not only did he destroy trees and crops to do this, but the new access crossed and, indeed, broke down the medial canal.[22] Such complications were quite frequent and are noteworthy as illustrations of the rather rigid structuring which irrigation canals lend to any system of communications.

In customary law, canals running between estates were presumed legally medial (like a wall medial between two houses, responsibility for which was shared by both neighbors) if there were no title to the contrary. If silt from cleaning the canal were piled on only one side of the canal, however, this was taken as proof of sole ownership of the channel by the man on whose side the pile was; it had to be his, the reasoning went, if he assumed sole responsibility for maintenance.[23]

The huertas, with their complicated canal networks, proved formidable obstacles in the paths of would-be invaders. James I, during the Valencian campaign, worried lest the battlefield be an irrigated one and some of his men fall into canals: "So at night we sent out brave men to see if the fields were irrigated or not, and if they were not irrigated, the men were to return to us and we could then deem it well that the attack be made." [24] The strategic difficulties of facing an enemy across canals is self-evident. Ramon Muntaner recounted the Murcian foray of Prince Peter of Aragon against the Muslims: "Surely, had it not been for the canals which stood between the two hosts, the prince would have attacked them. But the canals and the water were so great in between them, that they were unable to do so." [25]

ROADS AND BRIDGES

Obviously the huerta — much more than unirrigated lands — depended greatly on the good condition of its roads and paths for communication. The roads had to be kept clean and passable to travelers at all times. James I decreed that if a landholder "whether irrigating or not, or in any other manner shall throw water on highways or roads, he must pay five *sous* and make restitution to those who have suffered harm owing to the splashing of those waters." [26] The fathers of Murcia went so far as to declare that flooding the road was especially contemptible because it made passers-by blaspheme.[27]

Roads passed through a variety of jurisdictions and so responsibility for their maintenance fell to more than one administrative organ. In the huerta itself the irrigation communities and their officers assumed this task. The ordinances of the communities set fines for watering the road: "*Item,* if any one [through] irrigating or in another way should flood fields or public paths, let him pay for each contravention, ten *sous,* payable one-half to the *cequier* and the other half to the accuser, and beyond this, let him be held [responsible] for all the harm which results from the said flooding, not only to neighbors, but to others as well." [28]

The burden of enforcement fell to the irrigation officials.[29] In the fifteenth-century fine-books of the cequier of Castellón, watering the road (*rega cami*) was the third most cited offense (113 of 910 fines, or 12.5 percent), the usual fine being one penny (1/12 of a sou) or two. But the problem was so continually aggravating and vexatious that the cequiers and their assistants, who had in any case to fulfill their primary function of overseeing the distribution of water, could not alone adequately enforce the cleanliness of the roads. Especially near or within the city, other jurisdictions became involved with road problems.

Royal roads[30] within the municipal bounds of the city of Valencia and the ditches draining them (*escorredors*) were under the care of the municipal office of walls and sewers (*fabrica de murs i valls*). Their maintenance therefore fell within the jurisdiction of the town council,[31] which, on February 12, 1417, set a fine of twenty sous for causing water to flow onto these roads, considering that "the said roads be destroyed and the drainage ditches broken down by the water which irrigators fronting on, or who have possessions near to, the roads" let flow thereon. To enforce this ruling the council further suggested appointing an inspector (*visitador dels camins*) to make the round of the royal roads once or twice a month.[32]

The ordinary jurisdiction of the fabrica (and its chief official, the *sotsobrer*) did not extend to the small roads of the huerta, and the council frequently took note of the poor state of the roads and of the difficulty in getting anyone to patrol them and enforce the regulations. The problem was debated in full by the council at the meeting of August 14, 1396 (Appendix 1), at which it was decided that since neither cequiers, *mustasaf* (the urban public health officer), or *sotsobrer* were keeping the roads in good repair, another officer would have to be appointed to do so. Castellón had similar problems, and its council noted in 1408 that harm had come to the city because of the fouling of the road to the sea—the fault of poorly cleaned canals. The council agreed that either the mustasaf or the cequiers ought to clean and repair the canals in such a way that the road not be dirtied.[33]

It can be appreciated that within the municipal bounds, which included the whole huerta in both Valencia and Castellón, the problem of keeping the roads dry fell untidily between jurisdictions. The mustasaf, who was really an urban and commercially oriented official, could hardly be expected to spend time walking through the fields. The cequiers were concerned only secondarily with what was an unwelcome by-product of irrigation. The continued dispositions on the subject make it clear that the problem was endemic to the irrigation environment and that alleviation by decree was close to impossible.

Failure to comply with road regulations was the subject of a communication from Prince John of Navarre to the bailiff general of Valencia in 1450. The jurates (municipal administrative officials) and councillors had informed the Prince that the roads of the huerta were totally ruined as a result of the great abuses daily committed by the residents, especially in flooding the roads. Moreover, the cequiers had ceased applying the due punishment because, the town fathers asserted, "they are themselves landholders and cultivators participating in the abuse." The petitioners suggested that the treasurer (*racional*) of the city be given charge of enforcing the road regulations.[34]

A basic physical reason for the frequent involvement of roads with the irrigation network was that roads often ran along the banks of canals. Canals were considered public, as were the banks (*caxers*), which had to be left open and unplanted for a short space on either side of the canal to allow men and animals to pass and to leave unencumbered the access to the canal.[35] The tendency for public banks of canals to become natural arteries of communication was entirely logical. The road system of the huerta was institutionally homologous to the canal network. Secondary roads, for example, were called *filloles* (daughters), an expression borrowed from irrigation terminology.[36] Just as all the members of the commons of an irrigation canal were known as the *hereters* of that canal, those sharing the same road or path were called its hereters.[37] Those hereters whose property fronted on the road formed a discrete community — more informally constituted, to be sure, than the community of irrigators — with certain common responsibilities regarding the maintenance of the public way like those of members of irrigation communities in the maintenance of the canals.

The interweaving network of roads and canals passing over and under each other by means of bridges, aqueducts, and siphons was a striking aspect of the medieval huerta. Bridges were particularly important links in the huerta communication system: to the irrigator the primary purpose of bridges was to protect the sides of the canals against damage by fording men and animals. The convenience of the traveler was only a secondary consideration.

In Valencia bridges on royal roads were the responsibility of the fabrica,[38] which was also in charge of building and repairing all the bridges in the Marjals, an area whose development was the special charge of the city.[39] But even when royal roads or special projects were not involved, the city had an interest in assuring that even small bridges be maintained, for the common good. The foreman (*sobrestant*) of the fabrica had to see "that bridges be made in all places which pass over *cequiols* [small channels] in order to reach the fields. Because, since individuals excuse themselves from making these bridges, the *cequiols* cave in, and the water cannot flow, whence a great harm ensues. . . ."[40] Bridges, like the roads themselves, were public property and when crossing a canal had to be supported on public space on either side. The council of Castellón agreed in 1414 that the bridge called En Roca ought to be torn down and rebuilt anew of a good breadth at its base from bank to bank, and that the said bases not exceed the limits of the banks of the canal."[41]

THE PROBLEM OF CATTLE

Although there may be a basic association of domestic animals (for plowing and manure) with irrigation agriculture,[42] clearly there must be a severe limitation on numbers and movement of beasts if the system is to be kept free from the threat of physical destruction. Medieval records are filled with instances of damage caused by animals to canals, and repeated attempts to legislate against this circumstance are indicative of the relative lack of effectiveness that the regulations had.[43]

Sheep, goats, and any cattle not for purely domestic use were forbidden to pasture in the huerta. For this reason rather strict bounds were defined, and special areas — unirrigated — were set off outside the limits of the huerta as pasture (*bovalar*).[44] The council of Valencia defined the bounds of the huerta, within which no herds were permitted (see Map 1):

> Now hear that the jurates and councillors of the city inform you that according to a regulation made . . . on July 30, 1332 . . . no one . . . may keep sheep or goats in the *huerta* of the city, that is from this side of the Gully of Catarroja, going down towards the sea, and then along the canals of Quart and Moncada, down to the cross set up in the road to Morvedre, near Meliana, and down to the sea.
>
> No one ought to have any sheep or goats within the *huerta* of the city, according to the limitations as declared above, that is to say, in vineyards, or fields of alfalfa, cereals or rice, by night or by day, in any way.
>
> Nor may anyone have large cattle, such as oxen, mares, mules, asses, hogs or sows . . . except on his own property alone.[45]

No flocks could be grazed within these bounds without a special license from the council. On August 20, 1400, for instance, the council, looking to the provisioning (*refrescament*) of the city gave such permission to Domingo Maçana of Valencia, "to keep within the huerta and pasture therein twenty goats and one ram, so long as he causes no destruction or harm." [46] Such a license took the form of a letter from the jurates to the guards of the huerta — *guardians de la orta* — whose job it was to see that the cattle restrictions were obeyed. The appointment of two such guards, one for either side of the river, was authorized on November 16, 1340, after repeated complaints of damage to vineyards and fields by shepherds of flocks belonging to meat dealers.[47]

As indicated before, the problem was aggravated by the fact that the cows did not merely get into the corn, they also broke down canals, the most common natural field boundary. This problem was especially acute in the less well developed swampy areas, where communications were hampered by poor drainage and an insufficient number of bridges. The bridges were crucial,

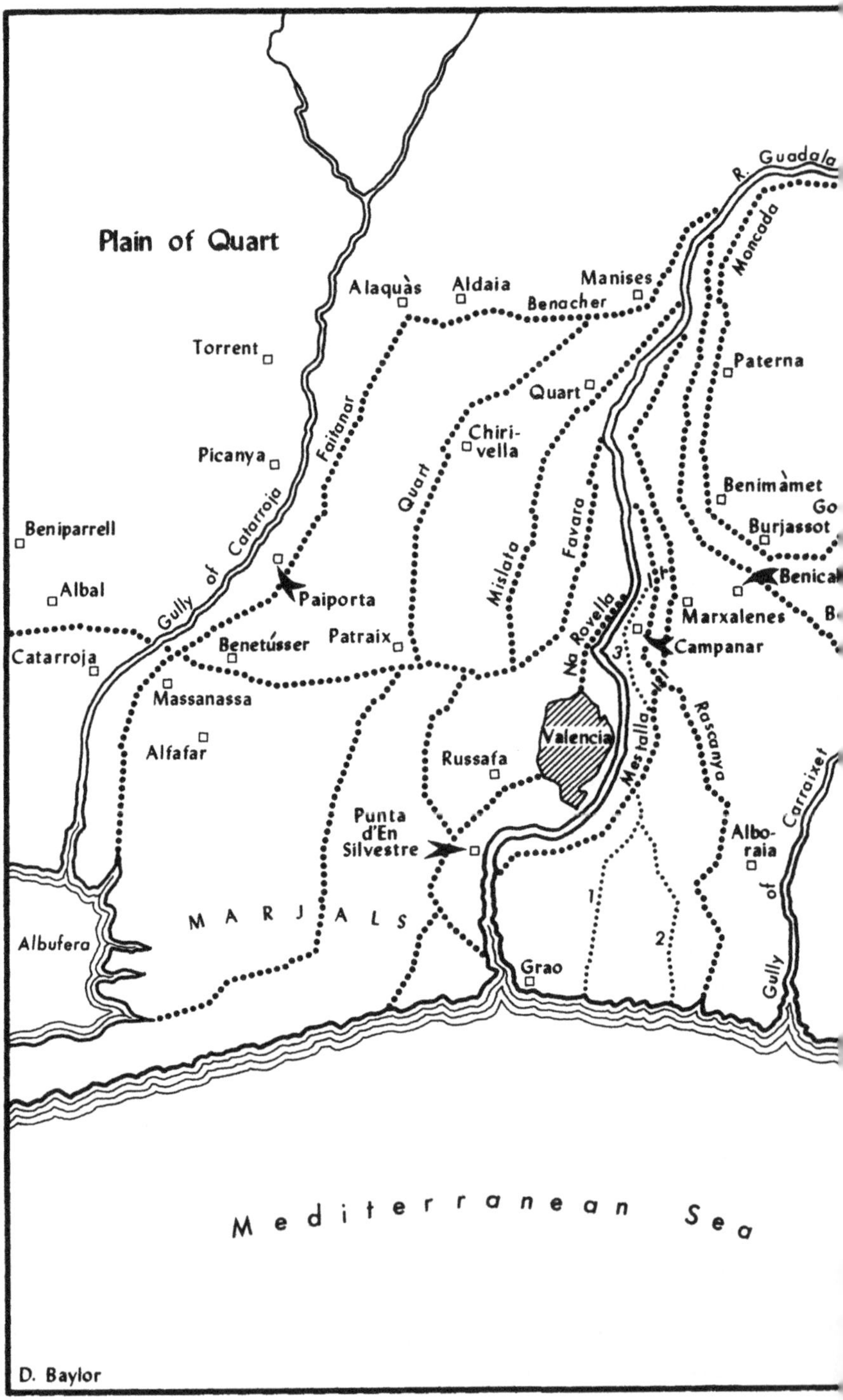

Plain of Quart
R. Guadala
Moncada
Alaquàs
Aldaia
Manises
Benacher
Torrent
Paterna
Quart
Chiri-
vella
Faitanar
Picanya
Benimàmet
Quart
Gully of Catarroja
Beniparrell
Burjassot
Mislata
Favara
Albal
Paiporta
Benical
Marxalenes
Catarroja
Benetússer
Patraix
Na Rovella
Campanar
Massanassa
Valencia
Rascanya
Alfafar
Russafa
Mestalla
Punta
d'En
Silvestre
Albo-
raia
Gully of Carraixet
M A R J A L S
Albufera
Grao
Mediterranean Sea
D. Baylor

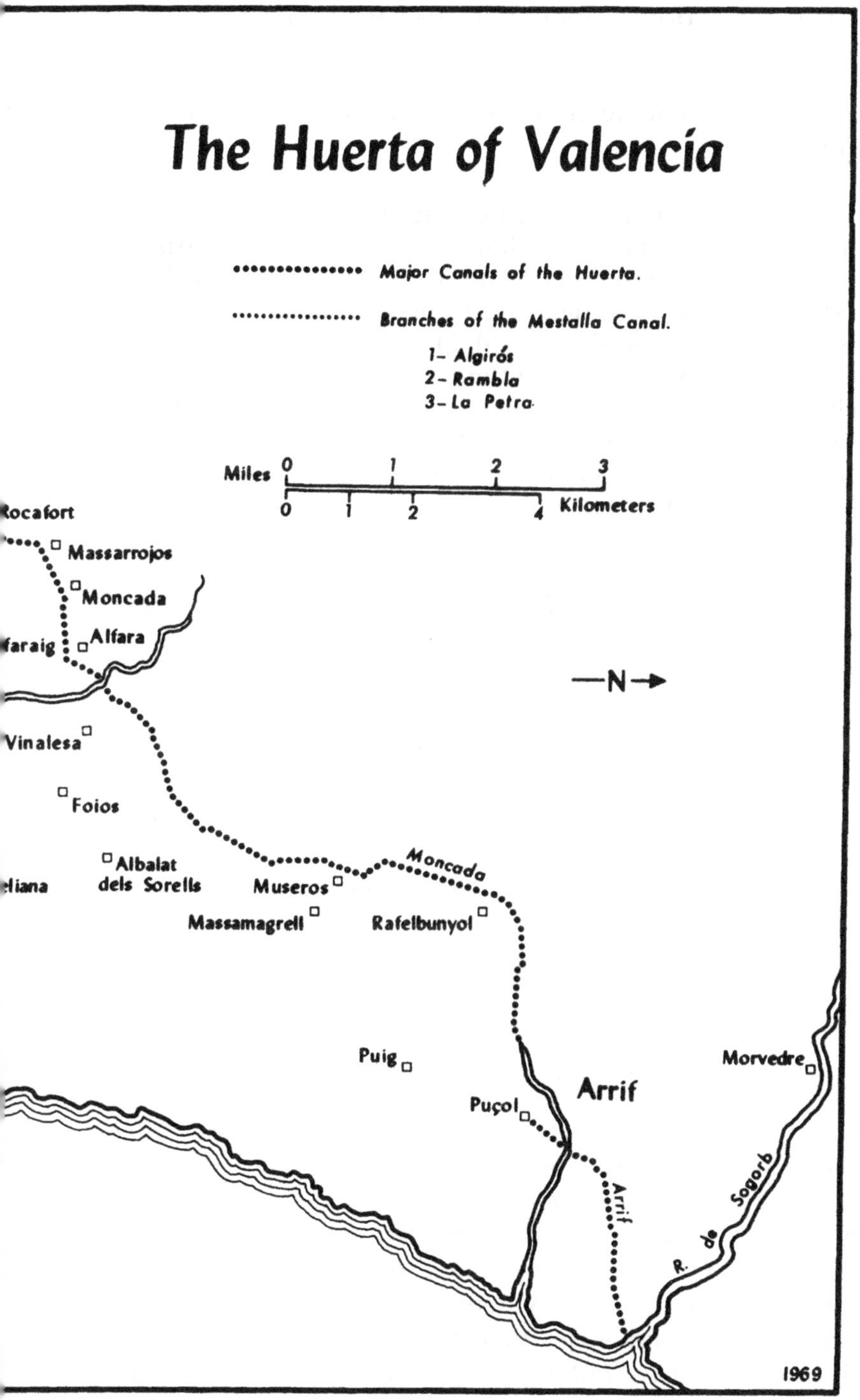

The Huerta of Valencia
Major Canals of the Huerta.
Branches of the Mestalla Canal.
1- Algirós
2- Rambla
3- La Petra
Miles
0
1
2
3
0
1
2
4
Kilometers
Rocafort
Massarrojos
Moncada
Alfara
Vinalesa
Foios
Albalat dels Sorells
Museros
Massamagrell
Rafelbunyol
Moncada
Puig
Puçol
Arrif
Arrif
Morvedre
R. de Sogorb
1969

as the complaint of some residents of the Marjals of Valencia in 1396 attests:

> Some people having possessions in the Marjals . . . in the places of Russafa and Alfafar . . . suffer great harm, notably owing to the indolence and delay on the part of the city, through ceasing or delaying to make bridges and pontoons over the canals of the said Marjals, through lack of which bridges the oxen and other beasts, large and small, which pasture in the Marjals, have crossed over through the said canals and branch canals and ruined and destroyed them, blocking up the channel and the water so that it flows back on their properties and makes them return to their former swampy state.[48]

A report on the New Canal of the Marjals in 1406 revealed that the channel in question had been "bridged" (*pontada dalguns ponts*) only in part and that at the present time it was "ruined and broken down by beasts." [49] And again in 1445, when another canal was built in the Marjals at Castelló d'En Arruffat, it was noted by the council that bridges would be necessary over which oxen, cows, and other beasts could pass without ruining the canal.[50]

The peril to which uncontrolled cattle subjected the huerta cannot be overemphasized. The huerta was largely a cattle-free zone, and this heightened even more its ecological difference from unirrigated areas.

CROPS AND CROP CYCLES

The medieval huerta presented the following paradox: in a kingdom rich in animal resources it deliberately eschewed the numerous herds as a source of fertilizer in order to better maintain the physical condition of the canal network. Human fertilizer gathered from the city was to some extent used, but generally the high yields of medieval Valencian agriculture can be attributed more to plain water than to the advantageous use of animal waste. Extensive fallowing — just as on unirrigated land

—and, complimentarily, the alkaline properties of the soil (making it conducive to the growth of nitrogen-fixing plants) also contributed to high productivity. Medieval agricultural use of water was radically different from what it is today. Irrigation was not generally used for growing exotic plants which could not otherwise be grown in the Valencian climate; it was used, rather, to increase the yield of ordinary crops grown generally throughout medieval Europe without the benefit of irrigation and which, indeed, are not today irrigated in eastern Spain. Thus cereal grain, grapes, and even carobs were commonly irrigated in medieval Valencia.

It is a striking, if vexatious, characteristic of medieval irrigation documents that they are primarily concerned with delivery of water from the source to the farmer's turn-out, but not beyond. Information about what the farmer did with the water is sparse. Thin though the data may be, they still enable some comment on medieval crops and occasion some contrasts with modern practices.

Table 1 is a summary of the crops mentioned specifically in the fine-books of the cequiers of Castellón (1443 and 1486). The

Table 1

Crops mentioned in fine-books (frequency of citation)

Crop	1443[a]	1486[b]
1. *Aylls* (garlic)	2	–
2. *Blat* (wheat)	3	2
3. *Faves* (broad beans)	1	1
4. *Fezols* (string beans)	6	–
5. *Forment* (golden wheat)	3	–
6. *Maloll* (new vine)	1	–
7. *Ordi* (barley)	1	–
8. *Pomar* (apple trees)	1	–
9. *Sexa* (high-quality wheat)	–	2
10. *Vinya* (grape vine)	3	10
Totals	21	15

[a] ARV, Real Patrimonio (Maestre Racional), leg. 482, no. 9852.
[b] ARV, Varia, libro no. 402.

fines are for "overirrigating" neighboring fields with these crops in them; therefore, the evidence is indirect and we cannot be sure that all the fields were irrigated, although the supposition is that they were since all were in the jurisdiction of the cequier of the *cequia maior.* In Table 2 the dates of the fines are indicated by month, in hopes of defining the seasons in which specific crops were watered and grown.

Table 2

Crops in fine-books[a] by month

Crop	Jan.	Feb.	Mar.	Apr.	May	June	July	Aug.	Sept.	Oct.	Nov.	Dec.
A. Wintercorn												
1. *blat*	o	o	x	x	–	–	–	–	–	–	o	–
2. *forment*	–	–	o	–	o	–	–	–	–	–	–	–
3. *ordi*	–	–	–	–	–	–	–	–	–	–	–	o
4. *sexa*	–	–	–	–	x	–	–	–	–	–	–	–
B. Springcorn and garden vegetables												
1. *aylls*	–	–	–	–	–	oo	–	–	–	–	–	–
2. *faves*	–	–	–	ox	–	–	–	–	–	–	–	–
3. *fezols*	–	–	–	–	–	oo	ooo	o	–	–	–	–
C. Grapes												
1. *maloll*	–	–	–	o	–	–	–	–	–	–	–	–
2. *vinya*	o	–	x	xxxx	x	ox	o	xxx	–	–	–	–
D. Fruit trees												
1. *pomar*	–	–	–	–	–	o	–	–	–	–	–	–

[a] Symbols:
o = 1443 book
x = 1486 book

Although the grain consumed by the city of Valencia was primarily imported, wheat and barley were nonetheless widely grown in the huerta for fodder and also for human consumption. Winter crops mentioned are *alfalç*[51] (alfalfa, surely used for fodder), several varieties of wheat (*blat* and *forment,* the most common two, *triga,*[52] and *sexa,* a high-quality wheat for white bread), and *ordi* (barley, apparently grown as wintercorn).[53] Winter wheat was sown in September or October (see Table 2, wherein grains are recorded in the ground from November through May) and harvested in late spring, when it was most susceptible to the effects of drought.[54] Spring crops, of which broad and string beans (*faves* and *fezols*) were the most typical representatives, were planted in April or May and grew throughout the summer.

Grapes were either irrigated or not, but most often they were.[55] The Castellón evidence indicates that they were watered as they ripened throughout the spring and summer. In 1438 one Bernat Ebri, llaurador of Chirivella, brought suit against the family of Miquel Giner in order to receive payment for work done in caring for Giner's vineyards (the account is given in Appendix 2). One of the vineyards in question was not irrigated, but four others were. Indeed, one of Ebri's witnesses was Pere Garcia, cequier of Chirivella, who testified that Ebri had irrigated the four vineyards mentioned. When asked how he knew this, Garcia replied that "he himself had irrigated them and was paid for the said irrigation." [56]

Certain trees, such as carobs and figs, were also found both in regadiu and secá;[57] but oranges were always irrigated. Most references indicate that the orange was a luxury crop and that it was also grown ornamentally in such places as the Ort dels Tarongers del Real of the city of Valencia[58] and the Taronger del Palau of Castellón,[59] orange groves associated with public buildings. Other fifteenth-century documents refer to oranges in Chirivella (Canal of Benacher and Faitanar),[60] in the Campanar-Marxalenes area on the left bank of the Guadalaviar (Rascanya system)[61] and near the Albufera.[62] There must have been considerable production of the fruit if Ramon Muntaner could report the shipment of fifty cartloads from the kingdom at one time.[63]

In times of drought the canals could not supply all of the irrigators and priorities had to be established. In July of the drought year 1376 the Valencia council decreed that in all the canals of the huerta cereals and vegetables were to be watered first, then vineyards, then fallow fields "about to be sown," and finally rice,[64] the cultivation of which was generally prohibited in huertas associated with large towns because of the danger of disease. (The text of the council's decree is in Appendix 3.)

The actual schedule of crop rotation is not clear from the

documents I have examined. Although more than two harvests a year would have been possible—as the practice today[65]—there is no concrete evidence that this was taken advantage of. Tables 2 and 3 indicate a three-course rotation of winter wheat,

Table 3

Fallow and stubble in fine-books[a] by month

	Jan.	Feb.	Mar.	Apr.	May	June	July	Aug.	Sept.	Oct.	Nov.	Dec.
Rastoll (stubble)	–	–	–	–	–	ooox	x	xxx	oxx	–	–	–
Guareyt (fallow)	ooo	o	oooo oooo ooxx xx	oooo xxxx	ooo xxx	ooox x	ooox	o	oxxx xxx	ooxx xx	o	–

[a] Symbols:
o = 1443 book
x = 1486 book

spring cereals and vegetables, and considerable fallowing. Table 3, based on fines for "overirrigating" stubble and fallow, indicates that once the winter wheat was harvested the fields were left in stubble, probably for limited grazing, rather than sown immediately with vegetables. Fallow fields are documented at all seasons, somewhat more frequently in the spring.

II | Irrigation Communities and Their Administration

The basic unit of Valencian irrigation society was the community of irrigators, the commons (*comu* or *comuna*) of those irrigating from a single main canal. The sum of all the regadiu holdings irrigated by a single canal system constituted the *rech,* or service area, of that canal. In the highly developed huertas one man's land might be bounded by canals belonging to two different systems. But he could draw water from only one system and be a member of only one irrigation community (for that particular parcel), to which he had to pay certain dues and perform certain services and by whose regulations he was governed in irrigating.[1]

The concept of having rights in a particular canal is illustrated by the accusation of one irrigator in Castellón against another: "He irrigated by the canal [in] which he had no right" — a formula complementary to another typical complaint: "He irrigated in a place where he had no right." [2] These two rights, the right to irrigate a parcel and the right to do so from a specific canal and no other, were necessarily interdependent. Only when the course of a canal was changed could a parcel change rechs.[3]

In the case of communally owned and administered canals, such as the eight main canals of the huerta of Valencia, the service area was the prime administrative and jurisdictional unit in irrigation matters, taking precedence over the village jurisdictions through which the canals passed. Although village lines often determined the constituencies of members of the governing body of the communes, the administrative districts were most typically natural subdivisions of the canal system itself, such as branch channels.

Many towns throughout the kingdom, in contrast to Valencia, irrigated from a single main canal, the cequia maior; these canals, as in Castellón, Burriana, and Orihuela, were owned and administered by the town. In such instances the town assumed the role and carried out the same functions as did the autonomous communities of the Valencian huerta: the rech lost its jurisdictional discreteness and became subordinate to the municipal jurisdiction.[4]

GENERAL AND EXTRAORDINARY MEETINGS OF THE COMMONS

The self-governing irrigation communes of the medieval Valencian huerta were those of the four upper canals of Moncada, Quart (including the Benacher and Faitanar branches), Tormos, and Mislata and the four lower canals of Mestalla, Favara, Na Rovella, and Rascanya[5] (see Map 1). All persons having land in the service area of the canal were members of the commons.[6] At a general meeting held in the cloister of the monastery of St. Francis in Valencia on May 29, 1435, eighty-four men of Benacher and Faitanar gathered to draw up and approve regulations; they were described as "landowners and those having lands and possessions which are irrigated by the waters flowing through the canals of Benacher and Faitanar, originating in the Guadalaviar River." [7] Present at such meetings were not only owners of cultivated lands, but also millers and other industrial water users who were considered as equal hereters of the canal and members of its commons.[8] In the parlance of general meetings of the fifteenth century the phrase "millers and landowners" [9] meant all water users.

Meetings were held at traditional places of assembly, usually ecclesiastical buildings. In the fifteenth century both the commons of Benacher-Faitanar and that of Na Rovella met in the Franciscan monastery of the city of Valencia.[10] The commons of Rascanya met at the Monastery of the Holy Trinity in the 1420's and 1430's.[11]

Such meetings were held under license from the king's governor[12] at fixed intervals, for the purpose of electing officers:

> In the huerta of the city of Valencia there is a canal among the others called Tormos, the water of which continually serves to irrigate certain quarters of the said huerta . . . and each year at a certain time the landholders and those having lands there meet together and elect certain persons necessary for the administration of the waters of the said canal. They are accustomed to elect deputies, a treasurer, a cequier or administrator of the canal, and a collector, giving to each certain duties. . . .[13]

The annual meeting, therefore, was the occasion for transacting the regular business of canal and commons. But meetings could be convened at any other time upon petition of some of the members; frequently they were held to discuss construction and financing of works projects meriting the immediate attention of the hereters. In 1436, for instance, the hereters of Rascanya met at the Holy Trinity to discuss the feasibility of moving the diversion dam higher up the river in order to increase the canal's intake capacity.[14] On October 5, 1446, the deputies and others of the Favara Canal petitioned the lieutenant governor to order a meeting of the commons for the following Sunday in order to provide for some expensive and urgent repair work.[15] For any new work which an individual irrigator might wish to make in the canal system the "will and consent" of all the hereters was needed.[16]

Meetings sometimes failed because of low attendance. Not always did fourscore irrigators turn out, as at the meeting of Benacher and Faitanar mentioned above. In November 1430, at the instigation of some residents of Quart, a general meeting was called in the name of the governor and, accordingly, public notice (that is, a cry, or *crida*) was given along the canals of Quart, Benacher, and Faitanar. "With a public cry the irrigators

. . . were convoked in order to gather at the monastery of the Franciscan brothers, there where the said canals are accustomed to meet for many necessary acts." The meeting was a distinct failure, however, because only eight or ten irrigators showed up in addition to those who had called the meeting in the first place;[17] so another meeting was called and held in the same place on December 3. This one was a success, and a reconnaissance of needed works on canals and dams was authorized. The record of the second meeting states that, among others residing on the canals in question, the important magnates were present (the Commander of Torrent, the Mayoral of Quart, and the lords of Alaquàs and Manises[18]); this suggests that the first meeting's lack of success was not due to poor attendance per se, but rather to the absence of those whose interest and support was necessary for the successful outcome of any important project.

The primary business of the commons as a whole was to enact regulations for the distribution of water and maintenance of the canal system and then to elect the executive and administrative officers to whom authority for the day-to-day running of the canal's normal affairs was delegated. Ordinances such as the *Capitols* (Chapters) of Benacher and Faitanar (1435) established the duties of the cequier and his assistants, set fines for various misdemeanors, and stipulated obligations of the hereters regarding observance of turns, maintenance of the canal, and contribution of dues.

DELEGATED AUTHORITY

The offices of the Tormos Canal, cited above, were standard. The "deputies" (*diputats, elets,* or *sindichs*) formed a kind of governing council and were selected by varying criteria. The most usual form of representation, especially on the longer canals, was by village. Thus the administration of the Moncada Canal has traditionally been entrusted to twelve syndics, one each elected from Paterna, Burjassot, Moncada, Alfara, Vinalesa, Meliana, Foios, Albalat dels Sorells, Massamagrell, Museros, Puig, and Puçol.[19]

Sometimes class as well as geographical interests were represented. In 1414 two syndics of the Moncada Canal, one representing the "knights and townsmen" (*cavallers e ciutadans*), the other representing the peasantry (*pagessia*) requested a general meeting of the commons.[20] The status of these representatives and their actual functions are not clearly delineated in the documents. The syndic or deputy functioned primarily as spokesman for a particular group within the commons and sometimes as representative of the whole commons (as when petitioning the town council or pleading before the governor's court). The medieval syndics of the Moncada community apparently did not possess the administrative authority which in modern times came to be concentrated in their hands.[21] (The present-day syndic who sits on the Tribunal of Waters is not the same as the medieval syndic. He is, rather, the medieval cequier; the change of name occurred gradually and became standardized in the nineteenth century.)

The powers of the deputies were quite indefinite and their position within the commons often unstable, as indicated by challenges to their power. In 1446 Gilabert Canoguera, "one of the deputies of the canal and commons of Favara [said] that some hereters of the said canal had attempted to take power along with the said Gilabert, on behalf of [all] the commons, in order to carry out certain acts and elections in the said commune of Favara, among others [that of] electing and creating other deputies and syndics." [22] Deputies did not go unchallenged when they attempted to extend their constituency arbitrarily.

The chief irrigation officers of the kingdom were the *sobrecequiers* (or *çabacequies*) and cequiers. The *sobrecequier* was a municipal officer; a cequier usually had jurisdiction over just one canal.[23] The *sobrecequier* was the typical official in towns where irrigation was municipally controlled, as in Orihuela and Burriana. The title was not uniformly applied, however, for in Castellón the municipal irrigation officer was called, simply, cequier.

The canals of the Valencian huerta were administered by a cequier delegated by each community of irrigators at the general meeting. Huerta-wide jurisdiction in irrigation matters was exercised by the town executive officers, the jurates.[24] The distinction between the role of the huerta cequiers, who were officials of the autonomous communes, and the jurates *qua sobrecequier* is revealed in litigation between the canals of Favara and Mislata in 1404:

> According to the fueros and privileges of the kingdom each canal of the district of the city has its cequier, and he has jurisdiction and decides the questions which arise between hereter and hereter of [each] canal. [He is also in charge of] cleaning the said canal and doing other things pertaining to it.
>
> [But, on the other hand] all the canals of the district of the city ought to have a *çabacequies* who [ought to have jurisdiction] over the waters and canals of the whole district of the city; and these are the jurates [of the] city. . . .[25]

Nevertheless, at some time prior to 1283 the city did in fact have a *sobrecequier,* to whom the other cequiers were subordinate, as indicated by a privilege of Peter I outlawing the office: "We order that the *sobrecequier* be banished from his office in perpetuity, and that each one of the cequiers should go and take charge according to what was the ancient custom." [26]

There is some evidence that a huerta-wide jurisdiction did survive in a single office, with authority delegated by the jurates, as late as the fifteenth century. But the office seems to have been ad hoc rather than permanent, and dictated by extraordinary circumstances. The jurates addressed a letter in July 1415 to "the honorable Berenguer Rochafort, cequier of the waters of Valencia," who appears to have been in charge of turn arrangements with the upstream villages known as the Pueblos Castillos.[27] The jurates often appointed cequiers or ex-cequiers of individual huerta canals to serve as their representatives in particular irrigation affairs. Rochafort was most likely a temporary appointee with specific enforcement duties upstream during a time of crisis.

The duties of cequiers, whether municipal or community,

were described in general terms in the fueros.[28] They had to see "that no one dare steal water, nor disturb the canals, nor cut off the water of any canal, nor divert it through another, nor break down the main canals or branches." Once a year they were to clean the main canals from head to tail, remove vegetation from them, and not return the water to the canals until the cleaning had been accomplished in the prescribed manner. They were entrusted with the maintenance of divisors and bridges and had to repair the diversion dam whenever it was breached. (In the latter eventuality the cequier had ten days in which to act if the break occurred in winter, eight if in summer.) If a hereter failed to clean his prescribed section on the day assigned by the cequier, the latter could undertake the work himself and charge the delinquent double the expense. The cequier could impose a fine of five sous for flooding the road and sixty for stealing water. He could furthermore impose these fines either with a hearing or summarily, as he saw fit, and could use force to execute them.[29] Besides his enforcement activities (described in Chapter III), the cequier oversaw the normal distribution of water. He watched the diversion dam and controlled the amount of water diverted from the river into the canal.[30] He had broad discretionary powers in the setting of turns among the hereters and as a result was frequently involved in disputes with them.

The daily activities of the irrigation community were necessarily directed by someone who knew the topography, regulations, and local traditions intimately. Custom thus decreed that cequiers of huerta canals had to be cultivators — llauradors — themselves, and members of the commons of the canal.[31] This meant that the cequiers were sometimes inclined to sacrifice administrative efficiency to the natural sympathy of one irrigator for the problems of all, as in their reluctance to enforce the penalties against road flooding. They suffered too from other failings common to llauradors of the huerta. In a dispute between several villages on the Favara Canal, the complaint was raised that "the cequier of Favara and his inspectors . . . are peasants and gross men and don't know how to read." [32] Thus the cequier's capacity to decide the point in question was held in doubt. A privilege of James II was designed to overcome just this difficulty by stipulating that a cequier might appoint, ex officio, a jurist to help him in his decisions.[33]

The method of electing cequiers differed from community to community. The usual methods were election from a slate or competitive bidding to see who could get the job done at the least expense. A syndic of the Tormos Canal described the latter method: "When anyone is made cequier of Tormos or of another canal of the said city of Valencia, it is customary that the said canal be bid for publicly before the commons of its hereters who are gathered together . . . and he who gets the canal . . . at the lowest price, administers the said canal according to its regulations . . . and is held to observe and comply with the regulations of the canal."[34] The cequier of Tormos, in this case Jacme de la Torre, served for three years.[35] The assigning of the office of the cequier was referred to as the sale (*venda*) or lease (*arrendament*) of the canal.[36] The total bid — in the bidder's estimation, the amount necessary for the administration of the canal during the stipulated term — would then be divided into units according to the cafizates of land in the service area. This pro rata sum was the cequiatge rate or the rate of assessment that each hereter had to pay as his share of the upkeep of the canal system.[37]

The cequier was assisted in his duties by a variety of subordinate officials: lieutenants, inspectors, guards, and water-dividers. These terms had great fluidity, and the same man might have been called by various titles depending on the particular task he was executing at the time. The cequier himself was sometimes called *partidor de les aygues,* and Jacme Marti, cequier of Mislata in 1430, was also called *açuter,* indicating the importance attributed to maintenance of the dam (*açut*).[38] No matter what the title, all the assistants performed administrative and police tasks as delegates of the cequier.

The cequier usually had a lieutenant who was empowered to act in his stead when circumstances required. In the case of the Rascanya Canal, the lieutenant was elected by the commons.[39] It may have been that in other communities the office was appointive, or only nominally elective. The cequier tended to have as his chief assistant a man whom he knew intimately, even a member of his family. In 1413 Gil Roiç of Castellblanch was lieutenant to the cequier of Moncada, Johan Roiç of Castellblanch — either his father or his brother.[40] The council of Castellón de la Plana proposed in 1472 that the cequier, J. Marti, have two aides (*moços*) one of whom ought to be "a familiar of the cequier." [41] Whether "familiar" here connotes a dependent or servant or only a close acquaintance, the cequier's principal aides were close associates who enjoyed his full confidence.

In the thirteenth century the term *çabacequies* or *çabacequier* was a synonym of *sobrecequier*. But in the fifteenth-century huerta the *çabacequier* was commonly an officer subordinate to the cequier.[42] Jurisdiction over the canals of Benacher and Faitanar was exercised by a cequier and *çabacequiers;* and a fifteenth-century document relates that "the cequiers and *çabacequiers* of the Moncada Canal have responsibility . . . for administering and guarding the waters and of notifying the chief cequier of the misdemeanors which they find." [43] It appears from these references that there was considerable flexibility in the appointment of the cequier's subordinates and no standard disposition of offices.

In the 1435 ordinances of Benacher and Faitanar there is no mention of *çabacequiers,* although it is specified that the cequier may have a lieutenant and shall be aided by inspectors, guards, and water-dividers. *Guardians* and *partidors* are synonyms in this text, although a document of 1413 for the same canal indicates a special jurisdiction, the *guardiatge de la aygua,* filled by two men who were to guard the canals of Benacher and Quart from Alaquàs to Manises, day and night, so that no one would steal water.[44] In any case the two titles stress the administrative and police functions of the same office: as partidor the official divided the water, diverting it to specified channels in the normal course of the turn, as *guardian* he prevented the unlawful diversion of water by anyone else.[45]

The Benacher-Faitanar ordinances make it clear that partidors had local jurisdiction and did not officiate along the entire length of the canal. There was to be one divider for the upstream irrigators and one for the downstream; the partidors were the officials who actually saw to the proper execution of the order of turns.[46] An earlier compilation of the ordinances further stipulated that the canal turn-out called Palmar (located at the point where Benacher and Faitanar met) must have as partidor one of its own hereters.[47] In this way authority was subdivided and subdelegated until a local man was held responsible (to the cequier) for seeing that the water of his branch canal was apportioned according to the regulations of the community and to the rights of the individual irrigators. The partidor had to be present at a specific place of diversion (divisor or turn-out) during the turn and was responsible for diverting the water into the proper channels at the appointed times. Jacme Ilari, of Patraix, "divider of the water" of the Mislata Canal, testified in a dispute of 1414 that on a night in question, at the first hour of the morning, "he divided the water and gave it to Jacme Fuster" by placing a check in the canal.[48]

Each irrigation community also had two or more inspectors (*veedors*), officials typical of medieval Valencian corporations,[49] who assisted the cequier in a variety of tasks but were not subordinate to him. Rather, their purpose was to represent the rights of the commons. Frequently the cequier counseled with the inspectors before rendering a decision,[50] and thus the inspectors provided an added guarantee for the observance of customary practice and for the probity of the cequier. And finally, the communities had financial officers, such as the collector of the cequiatge and the treasurer (*clavari*).[51]

Valencian cequiers did not and were not allowed to succeed themselves in office; there is no evidence that any tried to make the office a career. But subordinate officials often had lengthy tenures: Nadal Tarrega held the office of collector of the cequiatge of the Tormos commons for more than ten years in the second quarter of the fifteenth century.[52] Likewise, the çabacequiers who assisted the cequier of Moncada in their places of residence served for long terms. Guillem Cabot, llaurador of Puçol, stated in 1438 that "for thirty years up to the present year . . . [he had been] *çavacequier* [sic] on the Moncada Canal"; and Johan Aparici, of Rafelbunyol, testified that he had served in the same capacity for sixteen years.[53] Only in canals under municipal control does there seem to have been a tendency to make the more important posts into standing offices. In the Valencian huerta, where the basis of authority was the consensus of all the irrigators, every hereter could be cequier.

MUNICIPAL CONTROL OF IRRIGATION

Administrative procedure on municipally controlled canals can be described in greater detail because more records have survived. In towns where its administration was a subdivision of municipal administration, irrigation is alluded to frequently in the series of general municipal records, notably the minutes of the council. The council of Castellón took special precautions to ensure that irrigation records were kept. In 1405 it ordered Johan d'Arenys, the cequier, to enter in his book the ordinances and judicial decisions which it lacked and "with which the cequier guides himself in order to do justice and equality to the people according to the said decisions and ordinances of the canal [the cequia maior of Castellón]." [54] Similarly, on August 5, 1414, the council ordered that all documents pertaining to the disputes (largely concerned with water matters) between Castellón and the other towns of La Plana — Burriana, Almassora, and Villarreal — be put into "un memorial o libre." [55]

The council of Castellón made repeated attempts to bring order into the registers of the cequiatge owed by the irrigators. On July 12, 1416, it ordered that a book be drawn up "of all those who hold lands in the regadiu of the cequia maior, in order that from now on the cequiatge be clear." A public cry was ordered stipulating that the hereters had to bring their certificates of ownership (*cedules*) to the jurates and the scribe of the council.[56] The councillors continued to discuss the "book of the cequiatge" [57] until, in 1432, they complained to the jurates that the book was "confused and unclear" and had many omissions, to the detriment of the town.[58] In addition, the cequier kept a record of the fines which he and his assistants imposed [59] and an account of the yearly cleaning of the canal.[60]

Doubtless the autonomous communes of the Valencian huerta also kept records. All of them now have archives with documents extending back several centuries, and their regulations often indicate elaborate bookkeeping procedures.[61] Nonetheless, records comparable to those of Castellón have not survived from the medieval period, probably because the irrigation communities lacked the archival facilities available to medieval municipalities.

When towns built new canals, the municipal authorities established their jurisdiction in clear terms. In a dispute between Valencia and Alcira over the right to draw water from the Júcar through the Canal of the Siphon (Cequia del Canó), Alcira asserted that the water of this new channel was only to irrigate lands within its bounds. When the new work had been completed, Jacme Garcia, syndic of Alcira, personally supervised measurement of the channel's width and depth along its entire course; then, seeing that all of the canal was within Alcira's municipal limits, he performed a symbolic act to establish the town's jurisdiction.

> And in token of the proclamation of a new work and in order to verify the possession of the town of Alcira . . . he took three stones in his hand and threw them into the new canal, and throwing them—one on this side, one on the other—in the presence of me, the notary, and of the below-signed witnesses . . . he said: "I, here announcing a new work in your presence, throw these three stones, one after another, in order to preserve the possession of the said town and its citizens, in and by that form, manner, and condition which suits best, according to the fuero and good reason." [62]

In such a case the town would administer the canal and the irrigators would pay cequiatge to the town and look to it for resolution of conflicts.

The actual administration of municipal canals such as the

cequia maior of Castellón differed little from that of the communal canals of the huerta of Valencia. The cequier had the same duties; the town council filled the role of both deputies and plenary meetings of the commons alike. The council was a readily accessible forum for irrigation matters; indeed, it probably heard too many minor complaints which in an autonomous community would have been handled at a lower level.

The traditional date for election of the cequier of Castellón was December 26. For example, on that day in 1385, on the jurates' recommendation (they and the justiciar customarily elected the best man for the job), the council agreed to ask Berthomeu del Mas to serve.[63] But by the latter part of the fifteenth century the method of election had changed. On December 26, 1472, a slate of six men was submitted and the nominee chosen by lot — by the common procedure of drawing wax balls, *redolins,* out of a vessel. The man selected, Jacme Marti, took the oath of office on December 27.[64] He then petitioned the council that "he have an aide [*moço*] for the canal, that is Nadal Galent . . . and that it please the council to give him the salary of an assistant."[65] The council provided that the cequier have two assistants, Galent and some "familiar" of the cequier. Galent was serving as *moço* as early as 1472 and still played a vital role in the system in 1487. In that year (see Table 6) he initiated action in 214 of 492 (43.5 percent) of the fines entered in the cequier's book, while the cequier himself was responsible for only 28 accusations (1.2 percent). At least in policing the canal, the *moço* was the effective officer at this time.

In other towns of the kingdom where irrigation was administered directly by the jurates and council, a cequier or *sobrecequier* functioned much as did the Castellón official. In Elche the *sobrecequier* was elected yearly by the council at the same time as other town officials; similarly, the jurates of Orihuela chose a townsman as *sobrecequier* to patrol the canals, divide the water, and hear irrigation cases.[66]

Where the towns controlled irrigation administration the council functioned as a general forum for irrigation matters. Municipal control did not necessarily reduce the voice of the irrigators in governing their own affairs; however, the potentiality for abuse was greater. Jurates in particular had the power to subvert the system in order to obtain more water for themselves. In 1625 the Bishop of Orihuela discovered in the course of a pastoral visitation that the water of the town reservoir was unevenly distributed among the landowners because of collusion of the powerful with the jurates and the *sobrecequier* and the fact that "each one strives to take advantage of his office during the year he is jurate in order to irrigate his field." As a result, the normal turn of three weeks was extended by some ten to twenty days.[67] In the Valencian huerta, decentralization of authority among the autonomous communities virtually prevented any meddling by town officials whether they were water users or not. Indeed, there is no evidence that any such meddling took place during the medieval centuries. Within the communities a powerful individual might seek to thwart the will of the commons or cequier, but the success of such a maneuver depended purely on the resources of the individual and could have no institutional support.

ADMINISTRATION OF THE MARJALS OF VALENCIA

The area known in the Middle Ages as the Marjals of the city of Valencia was the region to the south and east of the city bounded by a line running from Russafa to Alfafar on the west,[68] the Gully of Catarroja and the Albufera on the south, and the sea to the east, and on the north by the river from the Punta d'En Silvestre to the sea. It therefore included the lower portion of the Rovella and Quart systems. The city, although it did not intervene directly in the internal administration of the huerta canals, did in the course of the fifteenth century undertake a sustained effort in irrigation administration in the Marjals. In attempting to administer the canals of the Marjals (for whose improvement and reclamation it had spent vast amounts of money), the city demonstrated the problems arising from a situation in which irrigators do not have full control over the selection of their administrative officials.

As part of the agreement drawn up in 1386 between the city and the Cathedral chapter concerning improvement of the irrigation network of the Marjals, the city was responsible for naming a cequier with authority over all the canals of the Marjals. The cequier was to serve for a two- or three-year term and was to be assisted by two inspectors[69] with whose counsel he would have full jurisdiction over canals and branch canals, drainage ditches, bridges, and their hereters, "just as, by fuero and privilege, each cequier [that is, of the huerta canals] has over his canal and its hereters." In spite of the fact that the city appointed the cequier, a way was left open for his recall by the hereters of the Marjals. "If anyone should feel himself aggrieved by a judgment or act of the cequier . . . let him recur to the [civil] justiciar and the jurates [of Valencia] and to no one else," the agreement stipulated.[70]

The post was hard to fill. In 1433 the jurates provided that, since at the present time there was no cequier in the Marjals, Johan Marroma, subsyndic of the city and syndic of the Marjals, ought to take charge of removing the vegetation from the Marjals canals. In addition, he was empowered to collect the cequiatge from the hereters.[71] On June 30, 1435, the council authorized the election of a cequier, and on November 5 the same Johan Marroma was named to the post.[72] Several years later, on February 10, 1439, Marti de Mariana, a furrier (*pellicer*) and citizen of Valencia, was elected cequier of the Marjals.[73] Mariana continued in office, on and off, until at least 1470. From the time of his appointment until 1442, he carried out his duties in the Marjals and was also in charge of works on the canals and drainage ditches of the spring of the port of Valencia (Font del Grau).[74]

On May 8, 1442, the first hint of Mariana's difficulties with the irrigators is revealed. One of the jurates told the council that Mariana, "because of his office, had been denounced by some hereters who had brought suit against him when the queen was in the city, for which reason he had incurred many expenses in defending himself." Mariana sought some remuneration from the city.[75] In spite of his difficulties he was at work in the port in autumn 1442 and spring of the following year.[76]

In November 1445, Johan Marroma, syndic of the city, and not Mariana, was in charge of keeping the New Canal of the Marjals in repair,[77] and the council's action in August 1447 makes it clear that the office had been vacated. The jurates, "mindful that in the Marjals of the city there was no cequier nor any guard," recommended that Marti de Mariana be appointed as "cequier and guard of the Marjals, with the salary and regulations pleasing to the jurates." [78] Back on the job, Mariana worked on the Font del Grau in the autumn, received his salary of 15 pounds (300 sous) for one year's tenure as cequier, and worked steadily and without incident both in the Marjals and the port until January 1452.[79] At that time complaints were again raised against him: "Continually there have been infinite complaints from the hereters of the Marjals owing to the great . . . blunders (*senrahons*) which he [Mariana] frequently made, in such a way that they can in no way put up with him any longer." [80] On March 1, Mariana was paid 50 pounds for cleaning the cequia maior of the Marjals, but on May 15 the council acted on the issues raised in January and removed him from office.[81] The same day Bernat Mojoli, llaurador and resident of Russafa, was named in his place.

Mojoli continued in office until October 1454 (indicating a three-year term, in accordance with the original statute), at which time he was removed—"without infamy," however—and replaced by Marti de Mariana again.[82] Mojoli returned to office on October 8, 1460, only to cede once more to Mariana just six days later.[83] What then happened is not clear: Mariana worked in 1461;[84] Mojoli returned in 1463;[85] but in September 1465, one Johan Alonso was elected administrator of the Cequia del Riu, in the Marjals. Moreover Alonso was elected by the deputies of all the hereters of the canal.[86] This incident clearly represents a step toward the "regularization" of canal administration in the Marjals, whereby a rech jurisdiction along the lines of those of the canals of the huerta proper was carved out of the general, municipally-controlled administration of the Marjals. Mojoli was still cequier at the time of his death in February 1466, whereupon Mariana again regained the office.[87] The last reference I have to the veteran cequier is from May 19, 1470, on which day the city paid him 100 sous to replace his horse, which had died in the Marjals, "whither he was going to exercise the office of cequier."[88]

Throughout the Middle Ages the Marjals were a demographically and economically unstable area. It was difficult to get people to settle there; the dangers from disease were great and the battle with nature to reclaim the swamps was constant. This instability was reflected in the administrative history recounted above. The city had financed the reclamation of the area in the 1390's and continued supporting it financially throughout the fifteenth century (see Chapter IV); it was reasonable that the city should want to exercise some control. On the other hand, there is no doubt it would have gladly surrendered control to the hereters themselves (as in the case of the Cequia del Riu) if settlement had been dense enough to ensure that the community itself would be able to keep the irrigation system in repair. The principal reason for the city's presence in the Marjals was to care for those stretches of canals — and there were many — that had no *frontalers* responsible for their maintenance. The need for municipal control in the Marjals, in other words, was a function of the low population density.

The faults of running the Marjals from the Cambra Secreta of the council were many. First, the cequier came to be considered a municipal irrigation officer, and his energies were diverted to other areas, such as the Grau, far from the scene of his original charge. This development was in part responsible for the second weakness of the arrangement: lack of communication with the hereters, which appeared to undermine Marti de Mariana's authority and for a while at least made his tenure impossible. Why did he keep returning? Probably because he was willing to make a career of what must have been a most undesirable job and because, after his years of service, no one could be found with comparable knowledge of the area and its problems.

RESPONSIBILITIES FOR CANAL MAINTENANCE

Constant vigilance in maintenance (*condret*) and repair (*adobs*) of canals and dams was a vital necessity. Especially near the tails of canals, where the land bordered on and had been reclaimed from marjals, there was danger of the land regressing to its former swampy condition if the channels were not kept clear.[89]

In arrangements elaborated for canal maintenance, a balance was struck between the responsibilities of the cequier (or of the town) on the one hand and those of the individual irrigators on the other. The usual formula was that the hereters were collectively responsible for cleaning and maintaining the branch canals, while the cequier directed the cleaning of the main canal. The overall arrangements and the assurance that hereters carried out their duties were within the cequier's purview. Canal cleaning was his task par excellence. In a dispute over a mill on the Favara Canal in 1426, the miller asked testily that the cequier stay away from his turn-out and concentrate on cleaning the canal, because "that is the job of a cequier!" [90]

The customary time for cleaning canals was during the month of April, that is, before the spring crops were planted: "It is the custom in the present city of Valencia and among the cequiers of all its canals that each year in the month of April . . . they shut off the water and order the cleaning of all the canals, and during the rest of the year it is not customary to clean main canals." [91] The Benacher–Faitanar ordinances stipulated that when the cleaning was done and before the water was let back into the canal the cequier was to show the cleaned canals to the syndic, deputies, and inspectors. The job had to be completed in a week's time, and if it was not the cequier was barred from receiving any of that year's cequiatge — that is, he would have to finish the job out of his own pocket.[92] Generally the cleaning lasted for one week,[93] work continuing on holidays if need be;[94]

if more than a week was required the furor raised was great.[95] The council of Elche took note in June 1525 that the canals and branch canals of the town were "right poorly cleaned": only five men had come each day to work and ten days had passed without the job being finished. The council decreed that lazy workers (*peons flachs*) were to be docked.[96]

The most important tasks of canal maintenance were clearing away vegetation from the sides of the canal (*erbejament*) and scouring the channel (*escurament*), that is, digging out the silt (*tarquim*) and mud (*fanch*). The *erbejament* was done with scythes (*dalls*) and sickles (*corbelles*),[97] and sometimes the growth was burned away.[98] These implements, along with the tools for digging, were usually rented; the *sotsobrer* of Valencia expended 7 sous 6 diners in 1390 for rental of three scythes and two sickles for nine days of cutting on the banks of the En Fluvia Canal in the Marjals.[99]

Diggers, called *palers* because they worked with the *pala* (shovel),[100] were aided by *glevers,* who carried away the lumps of earth in rush baskets (*cabaçets*).[101] Holes in the channel were plugged with a mortar of quicklime, gravel, and sand;[102] stakes were inserted in the banks of the canal to hold back vegetation and support the sides;[103] and, finally, the silt which had been heaped on the banks was leveled down.[104] The workers were guided in their digging by esparto cords with which they checked the width of the canal.[105]

On the Marjals project of 1390–1392 *palers* were paid 3s. daily, *glevers* 2s. 6d., and young boys (*fadrins*) 2s. for doing odd jobs. The *erbejadors* who cut the undergrowth while standing on land were paid 2 sous each, but "reapers who were standing in the water" received 2s. 6d.[106] It is possible to formulate some idea of the cost of cleaning an entire canal because estimates were sometimes quoted per unit of length (in this case, the *braça,* or ell, measuring about 2 meters). In 1449 the Cequia Morisca in the Marjals had been measured as 503 ells long and the cleaning cost estimated at ten diners per ell, which amounts to 5,030 diners, or 20 pounds 19 sous 2 diners.[107] This figure can be compared with expenditures in the cleaning of the same canal in

July 1390, when more than 50 men worked for 4 days for total wages of 27 pounds 1 sou 6 diners.[108] Assuming that the canal was the same length in 1390 as when it was measured in 1449 and taking into account that wages were higher in the latter year than in the former,[109] the estimate given the jurates in 1449 must have been based on the assumption that the work could be completed with fewer man-days. This would be logical if the canal were in reasonably good condition, bearing in mind that the work of 1390 was more an overhauling than a simple annual cleaning.

Thus far work on main canals directed by the cequier or by municipal authorities has been considered. But on secondary channels the irrigator himself was responsible for maintaining the stretch of canal on which his property fronted (hence the term *frontaler*). At the same time as the city made plans for its share of the work in the Marjals, the *sotsobrer* moved to inform the *frontalers* of their obligations: "I gave to En Artus, the town crier, [the sum of 13s. 6d.] for himself and his companions, that is trumpeters and drummers, for a cry which he made throughout the city . . . that everyone who fronts with the canal called En Fluvia within fifteen days must have cleaned his front on the said canal, making clear that they who dare not do so will pay double." [110] One of the cequier's duties was to see that the hereters cleaned the canals for which they were responsible. If the landowner was recalcitrant, the usual formula was for the cequier to hire men to do the work and charge the owner double the fee.

The responsibility of millers was somewhat different: they were required to clean not only the stretch of canal in front of their mills, but the canal up to the first diversion of water below the mill. In a dispute of 1444 a miller declared that not he, but the *frontalers,* were responsible for cleaning a certain length of canal near his mill. In the course of the litigation many witnesses, including several millers from the Favara and Mislata canals, gave testimony, and all agreed that the miller was wrong and that millers must clean up to the first diversion downstream. Pere Roger, a miller of Valencia, testified that in the twelve years he had owned his mill he had cleaned the canal only up to the first divisor, "and this he had seen practiced in several mills of the district and huerta of Valencia: that they cleaned only below the mill up to the first canal check or the first divisor." [111]

In the Marjals, too, it was the duty of the *frontalers* to maintain the secondary canals, but in many places there were stretches of canal with no *frontalers.* In these cases the city had to appropriate funds for the task.[112] Even in the huerta communities there were lengths of canal the responsibility for the cleaning of which was argued by the cequier and the *frontalers.* A dispute of 1414 involved the technical point of what constituted a main canal. If the section in question was main (*mare*), then the cequier had to clean it; if not, the *frontalers* were responsible. After an investigation the governor declared that Rascanya, the canal in question, was main up to the divisors of Alboraia and Almenara, whereas the cequier had claimed some of this section to be a branch deriving from the main canal.[113]

III | Intracommunity Conflict and Its Resolution

The cequier was much more than an administrative officer. The Valencian poet Jaume Roig characterized him as the supreme arbiter of irrigation in the *huerta:*

> Ab orde, pauses,
> conte, mesura,
> pes e dretura,
> laygua abandona,
> ffrancha la dona
> lalt çequier
> al ereter.[1]

(With order, calm, deliberation, measure, judgement, and rectitude he apportions the water; the lofty cequier gives it free to the hereter.)

The fairness with which the cequier carried out distribution of the canal's water was the test of his authority in the settlement of conflict. If he was not fair his position might be challenged by the community.

In fulfillment of his role as enforcement officer of the canal, the cequier made regular inspections. The regulations of Benacher and Faitanar stipulated that the cequier or his lieutenant "patrol the said canals at least once each month, with one or more of the inspectors and the syndic of the commons."[2] In practice the canals were constantly patrolled, if not by the cequier then by his guards or by officials in charge of the division of water in the secondary channels. The patrols had a police function. Gil Roiç of Castellblanch recounted in March 1413 that "as lieutenant of the cequier of the [Moncada] Canal, he was walking along the said canal the past Wednesday evening with six men whom he included in his company, in order to prevent the dishonest acts which might be committed with the water of the said canal." [3]

These dishonest acts (*fraus*) were defined in the regulations of the communities or municipalities (stealing water, flooding the road, irrigating without right, for example). The cequier would "denounce" such acts in the commission and the fine was exacted later. The cequier could also settle summarily such disputes as he might encounter; the decision, though ad hoc, was binding on both parties.[4] The power to initiate a judicial process without one of the disputants actually seeking a decision from him was crucial to the cequier's authority.[5] Justice must be swift lest valuable water be lost. This summary justice was a completely oral process; no records were made of the judicial act. Thus, in the case of the canals of the Valencian huerta, little is known of the ordinary justice of the medieval cequiers or of the fines they exacted for simple infractions of the regulations.

An oral decision of the cequier might later be written down, however, if the parties wished to obtain a legal record of it. One such notarial record describes the judicial process of the cequier of the Favara Canal in 1418. A conflict had arisen between a certain Francesch Armengol and some irrigators living upstream from him in the suburb of Russafa. Armengol claimed that the others were letting drainage water into the canal illegally. The cequier, Bonanat Prats, with the counsel of four inspectors, "summarily, and without any writing" ("de una paraula e sens scriptura alcuna") decided in favor of Armengol and ordered that the channel in question be closed "with a good canal check at the head of Armengol's land." If the upstream irrigators continued to drain into that channel, they would incur the penalty of 60 sous. Normally such a decision would not have been recorded, but in this case Armengol requested that the sentence "be reduced to writing and a public instrument be given him." [6]

By a stroke of good fortune, two account books of the fines

imposed by the cequiers of Castellón on irrigators of the cequia maior in the years 1443 and 1486 have survived.[7] They record 898 separate actions and give a picture of daily irrigation problems and the methods used to deal with them. In most respects the picture should be applicable to the Valencian huerta as well. It may well be that similar accounts were kept by Valencian cequiers; but if only two books survived from Castellón, it is understandable that none should have survived in Valencia.

The Castellón fine-books describe two kinds of situation (see Tables 4 and 5). In the first the cequier, his assistant, the guard

Table 4

Misdemeanors by category

Misdemeanor	1443	1486	Total	Percent
1. Forbidden water (*aygua vedada*)	84	180	264	29.0
2. Wasting water (*lançar l'aygua a perdicio*)	79	83	162	17.3
3. Flooding the road (*rega cami*)	76	37	113	12.5
4. Flooding fallow field (*rega guareyt*)	40	41	81	9.0
5. Stealing water (*levar aygua*)	17	60	77	8.5
6. Washing in the canal (*lavar en la cequia*)	26	19	45	5.0
7. Flooding a crop (*sorregar*)	21	15	36	3.9
8. Draining water in wrong place (*escorrer*)	14	9	23	2.6
9. Installing or undoing canal check illegally (*fer parada, desfer parada*)	12	9	21	2.3
10. Taking water by force (*levar l'aygua forcivolment*)	6	8	14	1.6
11. Irrigating without right (*regar on no tenie anpriu*)	5	4	9	1.0
12. Miscellaneous	31	34	65	7.3
Total	411[a]	499[b]	910	100.0

[a] 1443: 406 entries, 10 double fines, 5 no charge specified.
[b] 1486: 492 entries, 10 double fines, 3 no charge specified.

(*moço*), or some other official punished irrigators found committing infractions. The charges most frequently brought by the officials were (1) irrigating with "forbidden water" (*aygua vedada*), that is, taking water out of turn; (2) wasting water (*lançar l'aygua a perdicio*), which might have included a number of offenses such as failing to take a scheduled turn, having small checks placed improperly in furrows so that the water ran back into the canal, or putting water on a field that did not need it; (3) washing clothing, sheets, or wool in the canal — an offense committed only by women. In the second type of situation one irrigator accused another of breaking the regulations to his detriment, denouncing the infractor to the canal officials. In this category the most frequent offenses were flooding someone's fallow field (*guareyt*) or crop, or stealing his water. Flooding the road (*sorregar el cami*) was a charge brought by officials and by

Table 5

Accuser and charges

Accuser	Washing	Forbidden water	Wasting water	Flooding road	Flooding fallow	Flooding crop	Theft of water
			1443				
Irrigator versus irrigator	0	11	1	15	40	21	17
Unspecified[a]	26	65	68	47	0	0	0
			1486				
Irrigator versus irrigator	0	1[b]	0	7	41	15	60[c]
Officials:							
1. Nadal Galent	7	102	77	13	0	0	0
2. Mateu Barofet	2	57	5	3	0	0	0
3. cequier	0	17	1	5	0	0	0
4. unspecified[a]	10	3	0	9	0	0	0

[a] Assumed to be an official.
[b] Confused reading: "li leva laygua vedada."
[c] Actually only 58, but in 2 entries officials charge one irrigator with stealing water from another, named irrigator.

other irrigators alike. Table 5 shows the incidence of participation of both officials and irrigators in the most common accusations.

The proportion of charges brought by community officials to those brought by irrigators was approximately two to one in both books (see Table 6). In neither year did the cequier himself impose many fines (only 1.2 percent in 1443 and 5.7 percent in

1486); rather, the guards assumed the major role in enforcing the regulations. The 1443 book contains 217 entries (53.5 percent) in which the accuser is not specified (the form of the charge is "X was found . . ." or "Y was accused . . ." — *fou atrobat, fou acusat*). Most likely, the accusations were made by guards. In 1486 the two guards Nadal Galent and Mateu Barofet participated in 58.5 percent of all actions recorded.

Table 6

Initiators of actions

1. Cequier, guards, and other officials

	Cequier		*Guards* Galent		*Guards* Barofet		Promens		Unspecified	
Year	No.	%	No.	%	No.	%	No.	%	No.	%
1443	5	1.2	–	–	–	–	37	9.1	217	53.5
1486	28	5.7	214	43.5	74	15.0	2	0.4	22	4.5
Totals	33	3.7		288	32.1		39	4.3	239	26.6

2. Officials and irrigators

	All officials		Irrigators		Total	
Year	No.	%	No.	%	No.	%
1443	259	63.8	147	36.2	406	100.0
1486	340	69.1	152	30.9	492	100.0
Totals	599	66.7	299	33.3	898	100.0

In irrigation regulations, whether of autonomous communities or towns, fines were usually set high. The mandatory fine for erecting an illegal canal check was 20s., both for the cequia maior of Castellón and for the Benacher-Faitanar system of Valencia.[8] One-third of the fine went to the accuser and two-thirds to the cequier. But paying the full statutory fine was exceptional. In practice the fines levied were very low (a few pennies at the most) and also variable, depending on the gravity of the offense, on general economic conditions, and probably on the individual's ability to pay. Erecting a check illicitly was one of the more serious infractions, but in 1443 those who did it were charged 2d. (in five instances), 1d. (three), 4d. (two), and 6d. (only once).[9] In 1486 six infractors paid 1d., rather than 20s., and three paid 2d.; fines of 2d. for all offenses predominated in 1443, whereas fines of 1d. were the rule in 1486 (see Table 7). This difference

Table 7

Value of irrigation fines

Year	1d. No.	1d. %	2d. No.	2d. %	3d. No.	3d. %	4d. No.	4d. %	5d. No.	5d. %	6d. No.	6d. %	Higher No.	Higher %
1443	85	21.0	242	59.6	17	4.2	40	9.9	1	0.2	14	3.4	4	1.0
1486	355	72.1	73	14.9	0	0	2	0.4	0	0	0	0	0	0

Year	Charge withdrawn (*remes*) No.	Charge withdrawn (*remes*) %	Value not specified No.	Value not specified %	Total fines No.	Total fines %
1443	0	0	3	0.7	406	100
1486	36	7.3	26	5.3	492	100

is in accord with higher commodity prices and lower money wages in 1486 in comparison with the earlier year.[10] Apparently the full allowable fine was only sought when the malefactor was recalcitrant and would not submit to summary judgment.

A low irrigation fine is characteristic of traditional irrigation societies. Robert Gray, in a study of such a society in Africa, found data similar to mine: relatively low fines varying according to the seriousness of the crime.[11] Breaking the rules is worth the nominal penalty if one's crops are badly in need of water. Gray suggests that theft of water might be considered an institututionalized form of conflict and concludes that society's acceptance that some irrigators "may break the law to steal water with only a small risk renders the irrigation system more flexible and averts more serious conflicts if some of the men should lose their crops because of failure to obtain water legally." [12]

Evidence suggests that the way in which regulations were enforced in Castellón (for example, the preponderance of *aygua vedada* fines over those for stealing water) did dissipate somewhat the aggressive energies of the irrigators, in that the commons tended to make itself the subject of abuses which otherwise might

Table 8

a) Repeaters in misdemeanors

Irrigators participating in more than one action per year				
Year	All fines	Total number of individuals	All repeaters	Percent of individuals repeating
1443	406	299	114	38.1
1486	492	317	118	37.2
Totals	898	616	232	37.6

b) Frequency of repeats per year

Year	All repeaters	2 actions		3 actions		4 actions		5 actions		More than 5	
		No.	%	No.	%	No.	%	No.	%	No.	%
1443	114	55	48.2	25	21.9	14	12.3	8	7.6	12	10.5
1486	118	41	34.7	32	27.1	20	16.9	9	7.6	16	13.7
Totals	232	96	41.4	57	24.5	34	14.5	17	7.6	28	12.0

c) Irrigators participating in at least one action as accuser and one as accused in the same year

Year	No. of individuals	Percent of total repeaters
1443	31	10.4
1486	38	12.0
Totals	69	11.2

d) Irrigators participating in multiple actions as accuser

Year	No. of individuals	In 2 actions	In 3 actions	In more than 3 actions
1443	27	14	9	4
1486	33	19	6	8
Totals	60	33	15	12

Repeaters in misdemeanors (continued)

e) Irrigators participating in multiple actions as accused

Year	No. of individuals	In 2 actions	In 3 actions	In more than 3 actions
1443	78	39	15	24
1486	91	37	25	29
Totals	169	76	40	53

have been directed at individuals. The incidence of theft by force was a low 1.6 percent.[13] The fact that two-thirds of the actions were initiated by officials suggests that they were vigilant enough — the guards especially — to discover and stop the infraction before the malefactor came into direct conflict with another irrigator.

The involvement of the average irrigator in Castellón in conflict with either the officials or a neighbor was minimal. Only one-third of all individuals mentioned in the fine-books participated in more than one action in the course of the year (Table 8a). Of these repeaters, 41.4 percent participated in only two actions, 24.5 percent in three, 14.5 percent in four, and 7.6 per-

Table 9

Fines against Francis Agramunt (1443)

Number	Date	Charge brought by	Charge	Fine
1.	10 Jan.	[a]	*rega cami*	2d.
2.	10 Jan.	Gilabert	*rega cami*	2d.
3.	11 Apr.	[a]	*rega cami*	2d.
4.	27 Apr.	[a]	*regave ab aygua vedada*	2d.
5.	28 May	[a]	*lançave l'aygua a perdicio*	2d.
6.	12 June	[a]	*lançave l'aygua a perdicio*	4d.
7.	13 Sept.	[a]	*lançave l'aygua a perdicio*	2d.
8.	8 Oct.	[a]	*regave ab aygua vedada*	2d.
9.	14 Dec.	Johan Rodrigo	*rega cami*	2d.
		Total fines paid 1443		20d.

[a] "Fou atrobat" (charge assumed to have been made by an official).

cent in five, but 12 percent in more than five (Table 8b). Most, therefore, were fined but once or twice a year and could easily afford the penny or two assessed them. Repeating accusers were relatively few in number (Table 8c and d); chronic fining was the result of official vigilance. Of the nine accusations against Francis Agramunt (Table 9) only two were raised by other irrigators and these were for flooding the road; Pere Mut (Table 10) was ac-

Table 10

Fines against Pere Mut (1443)

Number	Date	Charge brought by	Charge	Fine
1.	6 Feb.	[a]	*rega cami*	2d.
2.	9 Feb.	[a]	*rega cami*	2d.
3.	9 Feb.	Johan [?]	*li ha regada la sua marjal arrabaçada*	2d.
4.	9 Feb.	Pere Modia	*li ha regada la sua marjal arrabaçada*	2d.
5.	21 Mar.[b]	[a]	*lançave l'aygua a perdicio*	2d.
6.	28 May	[a]	*lançave l'aygua a perdicio*	2d.
7.	5 June	[a]	*lançave l'aygua a perdicio*	1d.
			Total fines 1443	13d.

[a] "Fou atrobat."
[b] On 21 March Pere Mut accused En Forner regarding "lo postell que era mal tancat." Forner was fined 2d.

Table 11

Fines against Pere Veciano (1443)

Number	Date	Charge brought by	Charge	Fine
1.	11 Apr.	Domingo Gasquo	*a omplit lo forment*	2d.
2.	5 June	[a]	*lançave l'aygua a perdicio*	1d.
3.	12 June	[a]	*lançave l'aygua a perdicio*	6d.
4.	21 June	Domingo Gasquo	*li ha regat lo rostoll*	6d.
5.	6 July	[a]	*lançave l'aygua a perdicio*	1d.
6.	15 July	[a]	*lançave l'aygua a perdicio*	2d.
7.	13 Sept.	[a]	*lançave l'aygua a perdicio*	4d.
			Total fines paid 1443	22d.

[a] "Fou atrobat."

cused by other irrigators twice out of seven times; and similar patterns are shown in the records of other chronic offenders (Tables 11 through 16).

Table 12

Fines against Jacme Galceran (1443)

Number	Date	Charge brought by	Charge	Fine
1.	14 June	[a]	*lançave l'aygua a perdicio*	4d.
2.	26 July	Johan Rodrigo[b]	*li levave l'aygua e la mete en la scorentia*	2d.
3.	29 July	[a]	*regave ab aygua vedada*	2d.
4.	29 July	[a]	*regave ab aygua vedada*	2d.
5.	29 July[c]	[a]	*rega cami*	2d.
6.	4 Sept.[c]	[a]	*lavava la roscada en lo cequiol*	2d.
7.	13 Sept.	[a]	*lançave l'aygua a perdicio*	2d.
			Total fines paid 1443	16d.

[a] "Fou atrobat."
[b] See Table 9, no. 9.
[c] Charge against Galceran's wife.

Table 13

Fines against Asmet Bocaxo, carter (1486)

Number	Date	Charge brought by	Charge	Fine
1.	7 Mar.	Johan Figuerola	*ha regat la troç*	1d.
2.	7 Mar.	cequier	*regave cami*	1d.
3.	16 June[a]	cequier	*regave ab aygua vedada*	1d.
4.	26 June	Anthoni Vicent	*ha feta parada en la sua cequia*	2d.
5.	18 July	Nadal Galent	*lançave l'aygua a perdicio*	2d.
6.	27 July	Nadal Galent	*lançave l'aygua a perdicio*	1d.
7.	12 Aug.	Mateu Barofet	*regave ab aygua vedada*	1d.
8.	21 Aug.	Mateu Barofet	*regave ab aygua vedada*	2d.
9.	22 Sept.	Mateu Barofet	*regave ab aygua vedada*	2d.
10.	14 Oct.[b]	–	*lavave en la cequia*	4d.
			Total fines paid 1486	17d.

[a] Also on 16 June the cequier fined one Exemeno 1d. for "shutting the *fila* of Asmet Bocaxo while he was irrigating [from it]" ("tanquave la fila asmet boquaxo que regave").
[b] Charge against Bocaxo's wife.

Table 14

a) Charges brought by Pere Johan (1486)

Number	Date	Charge brought against	Charge	Fine
1.	12 Mar.[a]	Pere Veciano	*li levave l'aygua*	1d.
2.	17 Apr.[b]	Johan Veciano, *menor*	*li levave l'aygua*	1d.
3.	17 Apr.	Pere Borat	*li levave l'aygua*	1d.
4.	19 May[c]	Pere Veciano	*li levave l'aygua*	1d.
5.	26 June	Jaume Roquamarti	*li levave l'aygua forcivolment*	1d.

[a] Compare Table 14 b) no. 3.
[b] Compare Table 14 b) no. 4.
[c] Compare Table 14 b) nos. 7–9.

b) Fines against Pere Johan (1486)

Number	Date	Charge brought by	Charge	Fine
1.	7 Mar.	Nadal Galent	*lançave l'aygua a perdicio*	1d.
2.	7 Mar.	cequier	*lançave l'aygua a perdicio*	1d.
3.	12 Mar.	Pere Veciano	*li levave l'aygua*	1d.
4.	17 Apr.	Johan Veciano, *menor*	*li ha levat l'aygua*	1d.
5.	17 Apr.	Pere Veciano	*troba aygua en la sua escorredor*	–
6.	17 Apr.	Pere Veciano	*li levave l'aygua*	1d.
7.	19 May	Pere Veciano	*li levave l'aygua*	1d.
8.	19 May	Pere Veciano	*li levave l'aygua*	1d.
9.	19 May	Pere Veciano	*li levave l'aygua*	1d.
10.	19 May	Nadal Galent	*regave ab aygua vedada*	1d.
11.	26 June	Andreu Dolo	*ha levat l'aygua de la sua escorentia*	1d.
12.	7 Aug.	Nadal Galent	*regave ab aygua vedada*	1d.
13.	31 Oct.	Johan Valenti, *notari*	*li ha sorregat la sua eretat*	–
			Total fines paid 1486	11d.

Gray suggests that disadvantaged irrigators tend to break the rules, but these chronic offenders as a group do not seem to have had any distinguishing social features setting them apart from the other irrigators. Some were of more than average means, suggesting that ability to pay made them less hesitant to break the rules. Lorenç Gasquo (Table 15) was a notary; Pere Veciano (Table 11) was a member of one of Castellón's most prominent families. Asmet Bocaxo, a *traginer* (carter) (Table 13), was one

Table 15

Fines against Lorenç Gasquo, notary (1486)

Number	Date	Charge brought by	Charge	Fine
1.	7 Mar.	Nadal Galent	*lançave l'aygua a perdicio*	1d.
2.	7 Mar.	Nadal Galent	*regave ab aygua vedada*	1d.
3.	12 Mar.	Guabarel Noz, *corder*	*per la broça que ell [ha] del seu enfront lançada en la cequia a feta parada en la cequia, se resguarda la sua vinya e regar lo cami de la mar*	1d.
4.	17 Apr.	Nadal Galent	*regave ab aygua vedada*	1d.
5.	19 May	Nadal Galent	*lançave l'aygua a perdicio*	1d.
6.	11 July	Nadal Galent	*lançave l'aygua a perdicio*	1d.
7.	27 July	Nadal Galent	*lançave l'aygua a perdicio*	1d.
8.	7 Aug.	Nadal Galent	*regave ab aygua vedada*	1d.
9.	12 Aug.	Nadal Galent	*regave ab aygua vedada*	1d.
10.	21 Aug.	Nadal Galent	*regave ab aygua vedada*	1d.
11.	29 Aug.	Nadal Galent	*regave ab aygua vedada*	1d.
12.	5 Sept.	Mateu Barofet	*regave ab aygua vedada*	1d.
		Total fines paid 1486		12d.

Table 16

Fines against Berthomeu Loqular (1486)

Number	Date	Charge brought by	Charge	Fine
1.	17 Apr.	Nadal Galent	*regave ab aygua vedada*	1d.
2.	17 Apr.	En Noç	*ha regat la vinya*	1d.
3.	16 June	Nadal Galent	*lançave l'aygua a perdicio*	1d.
4.	16 June	Nadal Galent	*lançave l'aygua a perdicio*	1d.
5.	16 June	Arnau Ferer	*ha regat la vinya*	1d.
6.	26 June	Nadal Galent	*regave ab talponera*[a] *y lançave l'aygua a perdicio*	1d.
7.	2 Aug.	Nadal Galent	*regave ab aygua vedada*	1d.
8.	29 Aug.	Nadal Galent	*lançave l'aygua a perdicio*	1d.
		Total fines paid 1486		8d.

[a] *Talponera:* burrow or tunnel of a mole; here, euphemism for a secret and illicit irrigating conduit. See Sarthou Carreres, *Provincia de Castellón*, p. 70; compare Jaume Roig, *Spill*, ll. 14801–14803: the hereter "opens intake gates, shuts mole-tunnels of his bank" — *obre boqueres/clou talponeres/del seu ribatge.*

of the few Muslim irrigators on the cequia maior in the fifteenth century, and his fines ran somewhat higher than average. The Bocaxos owned a substantial amount of land in the parish of St. Nicolas, and in 1462 Asmet himself owned at least eight fanecates of huerta land and three of unirrigated vineyard near the Jewish cemetery.[14] His cultural distinctiveness may have made him a target for relatively severe treatment by the guards.

The record of one particularly fractious individual, Pere Johan (Table 14), is a rare example of an irrigator in chronic conflict with his neighbors. Between March and June 1486 he made five accusations of theft of water, and all five accusees were fined 1d. He himself was accused thirteen times between March and October; two accusations were disallowed, and he paid a total of 11d. in fines for the year. Moreover, ten of the eighteen actions were conflicts with one family, the Vecianos. Twice Johan accused Pere Veciano (probably the son of the Pere Veciano mentioned in the 1443 book, Table 11) of stealing water from him and once he made the same charge against Johan Veciano the Younger. In turn, the Vecianos accused Pere Johan six times of stealing water and once of a drainage offense. It is inconceivable that relations between the Vecianos and Pere Johan were anything but hostile. It would be interesting to know the physical disposition of the fields owned by Pere Johan and the Vecianos. A close study of the "Libres de Values" — massive registers of landholdings in medieval Castellón[15] — could yield valuable data on the morphology of intracanal conflict. (During the short time I spent in the Castellón archives, I was unable to connect Veciano holdings with any of Johan's; but Pere Johan's land definitely did border that of the notary Johan Valenti, who charged him with flooding his field in October 1486.[16]) The fact that chronic conflict between individual irrigators was rare is a tribute to both the efficiency of the distribution system and the vigilance of the guards.

THE TRIBUNAL OF WATERS OF VALENCIA

Although the Valencian evidence does not yield as complete a picture of the cequier's role as we have for Castellón, still it is possible to confirm the practical identity of procedure in those two towns as in the others where irrigation was conducted according to the dictates of the fueros of the kingdom. In modern times, though, the Valencian cequiers have participated in a weekly joint session called the Tribunal of Waters, an institution whose medieval roots are shrouded in mystery.[17]

When a huerta guard finds an irrigator committing an infraction today, the irrigator is ordered to appear before the Tribunal of Waters, at the Gate of the Apostles of the Cathedral of Valencia, on the following Thursday. There the syndics of seven canals[18] — inheritors of the dignity of the medieval cequiers — meet to pass judgment on all the infractions committed during the previous week. The infractor is "denounced" by the guard of his own canal, whose syndic then questions him and hears testimony. Then the syndic stands aside, and the merits of the arguments are weighed by the syndics of the canals of the opposite bank. If the accused is found guilty as charged, the president of the Tribunal so announces.

Valencians of today unanimously believe that the Tribunal antedates the Reconquest.[19] The fact that its traditional meeting place is at the site of the city's former mosque is deemed especially crucial by several authors. There is ample precedent in the Islamic world for public judicial functions that took place in the mosque;[20] and Vicente Giner Boira suggests that, when the mosque was consecrated a Christian church and Muslims were forbidden to enter, the court was moved outside so that the Muslim irrigators could continue to attend.[21]

The arabist Julián Ribera took exception to the hypothesis of the Tribunal's Islamic antecedent, stating that in Islamic Law judicial authority is "unipersonal," that is, that there can be only one competent authority in any case.[22] But Ribera misinterpreted the true nature of the Tribunal. There is no question of any collegiate jurisdiction. On the contrary, the syndics today only state whether, in their opinion, a man is guilty or not; they function not as a panel of judges, but as a jury. The Tribunal can neither stipulate nor execute the punishment: the syndic of each canal applies the punishment and fine in accordance with the regulations of that canal.[23] The syndic does not intervene in the verdict of his co-syndics (just as the judge does not sit with the jury); he only executes the verdict. Each syndic, therefore, retains his discrete, "unipersonal" jurisdiction.

The members of the Tribunal today do act collegiately in nonjudicial matters, however. Since the eighteenth century, at least, they have jointly exercised the right to petition Madrid on behalf of the irrigation rights of the entire huerta.[24] Under the terms of article 295 of the Law of Waters of 1866 the Tribunal was finally accorded official status in public law as a special tribunal.[25] In 1943 its President, on behalf of his colleagues, addressed a memorial to the Minister of Public Works, urging that the turns between the huerta and the Pueblos Castillos established in 1321 by privileges of James II remain in force.[26] In so doing, the members of the Tribunal stressed that they were moved by the desire to defend the rights of the canals of the entire vega.[27]

But in the fourteenth and fifteenth centuries the jurates and council of the city of Valencia, not the cequiers, were the protectors of irrigation rights of the huerta as a whole.[28] This is not to say that the medieval cequiers were lacking in group definition. On occasion the city did exercise its authority in irrigation matters through the cequiers. In 1345 the council ordered two jurates and some cequiers to proceed upstream to the diversion dam of Benaguasil in order to inspect the water and to see that the city and huerta had a sufficient supply.[29] In August 1414, in the course of a great drought, the jurates, with the permission of the governor, summoned "all the cequiers" to a Saturday morning meeting in the council room in order to discuss "affairs of the canals and repair of their intakes." [30] In 1457 the jurates elected as administrators of the water of the lake of Tortajada[31] (which was to be diverted through the Alfambra River into the Guadalaviar) Bernat d'Almenara, cequier of Moncada, Guillen Lorenç and Johan Perez, cequiers of Mestalla, and Pere Marti, cequier of Favara.[32]

It seems logical for the city to have consulted the cequiers as a group from time to time when their experience and knowledge were needed in matters affecting the entire huerta. These ad hoc meetings and commissions of cequiers demonstrate that the huerta cequiers did have some group definition; but such definition falls far short of indicating the existence of any formal and permanent body such as the modern Tribunal. Two recently published documents verify meetings of the cequiers in the traditional seat of the Tribunal in the early fifteenth century.[33] First, a decision of the cequier of Favara made in situ in settlement of a conflict in September 1418 [34] was committed to writing the following March 30 — a Thursday — in the morning (*hora prime*), "in the Plaça de la Seu of the city of Valencia, in front of the Gate of the Apostles" in the presence of the cequier, the inspectors, and the interested parties. Even more significant are the Benacher-Faitanar regulations of 1435 which stipulated that the "fines assessed by the cequier against delinquents should be exacted and judgment of them should be made in the Plaça de la Seu, and in no other place." [35] This evidence shows that in the early fifteenth century the cequiers were holding court and exacting fines in the traditional assembly place of the Tribunal. But it is doubtful that this proves the existence at the time of any *formal* Tribunal.

The Valencian historian Francisco Javier Borrull wondered why James I and his successors failed to assign the Tribunal a special building in which to meet, for they had provided for the court of the justiciar, the bailiff, and the Consulate of the Sea. He reasoned that, had the Muslims themselves had a special building, James I would have reserved its continued use for the Tribunal; he therefore concluded that the meetings must have taken place outdoors, in the atrium of the Great Mosque, where the meetings continued to be held after the mosque became the Cathedral. Borrull regarded the Tribunal as a judicial corporation, assumed it had always taken such a form, and thus could not explain the absence of documentation concerning this judicial body which was every bit as important (in his view) as the justiciar's office and the Consulate.[36]

This insistence upon regarding the Tribunal as a collegiate panel of judges rather than as a jury, and as a necessarily formal rather than originally informal institution, is at the heart of the general inability to uncover its origins. I believe that the Tribunal could not possibly have been the invention of some king or caliph, but that it was a customary, informal institution of long evolution. Records of its existence in the Middle Ages are sparse, because it had no status in public law and as a judicial body was nonessential,[37] because the fueros provided for irrigation justice in the office of cequier and in channels of appeal to the governor's court. One can, therefore, only hypothesize concerning the ultimate origins of the Tribunal. It could have originated in the Islamic period as a subjurisdiction of the judgeship or *qaḍā'*;[38] it could have more ancient roots in customary meetings of elders to discuss irrigation matters;[39] it has obvious analogies with the law merchant.[40] Possibly all of these elements are present in the Tribunal's history.

Here we can only consider what the function of the Tribunal, or proto-Tribunal, might have been in the Middle Ages, long before it acquired formal status. What advantage was there in splitting the cequier's judicial and executive procedure into two phases, the denunciation in situ and the formal act of sentencing and fining? What advantage did a joint meeting of all the cequiers in public have over the individual action of each cequier? It seems fair to suggest that the primary function, historically and practically, of the Tribunal and its informal antecedent of the Middle Ages was not judgment, but consultation among those who best knew the custom of the huerta. Part of the cequier's authority was thus brought under the scrutiny and consideration of all the irrigators of the huerta, through their delegated officials and through the presence of irrigators at large,[41] just as nonjudicial aspects of irrigation such as arrangements for turns were also regarded as the responsibility of all. The convening of the cequiers was an embodiment of the basic precept of consensual authority in irrigation.

PATTERNS OF CONFLICT

One March evening in the drought year of 1413, the lieutenant cequier of the Moncada Canal, Gil Roiç of Castellblanch, was patrolling the canal in order to prevent acts contravening the regulations.[42] When he came to the mill of En Codinachs in Benimàmet he found the gates of the mill's return ditch closed; as a result water was backing up and spilling over the canal banks, to the detriment of the irrigators. The next day Roiç returned with a certain Aparici Romeu of Puig. He brought with him a hook with which to raise the gates of the ditch in question, in order to lower the level of water in the main canal so that it would not spill over. At this point the mother[43] of the brothers Codinachs told him not to touch anything; Roiç replied that she did ill to waste the water of the canal, which "was of profit to other places as well as her own." The woman answered that if water was spilling over it was because the bank was broken down and that Roiç ought to repair it. While she argued with Roiç, one of her sons appeared, told him to touch nothing, and asked to be shown the "fraud" in question. "And since the lieutenant wished to show him the place where the fraud was, he went to a bridge above the mill." The scene which ensued is indicative of the high tensions conspicuous in irrigated areas when water is short; the slightest misunderstanding can set off a fight:

> There came then Master Ausias, Master Manuel, and another brother (only it wasn't Master Pere) and they came with squires and with others of their vassals, bearing arms. And as he was showing them the said fraud and saying to them, "Look, here it is!" one of them came over towards the man called Aparici Romeu who was in the company of the lieutenant, saying, "Halt, Sir Humpback, that I might undo your hump!"
>
> Aparici turned, and when he had turned, they gave him a lance-blow. Recoiling backwards from the blow, he fell into the canal. Another of the brothers—he didn't know which — jumped into the canal with his dagger drawn, and here the two men locked arms. As Aparici placed his hand on the dagger of the other so that he would not harm him . . . they gave him a lance-blow on the back so hard that were it not for the hauberk which he wore, they would have wounded him.

While Aparici was maneuvering in the canal, placing one brother between him and the others so that their lances could not reach him, the lieutenant said to one of the Codinachs that they wanted to kill his friend without reason, and "that is not gallantry" ("aço no es jentillessa"). To this Codinachs replied that Aparici had dishonored his mother, but Roiç said that, speaking with due honor, he had done her no villainy but that, not knowing her, had called her "Madona." [44] The other brother (Roiç didn't know his name but said he could recognize his face and that he stuttered a bit) climbed out of the canal with his sword drawn and said three times to the lieutenant, swearing to God, "Were I not a gentleman, I would slash your face." The official was thus violently prevented from carrying out his task of enforcing the regulations and had to seek action against the Codinachs in the governor's court.

In spite of the elaborate mechanisms of control whose very basis was the consensus of the irrigators themselves, violence lay ever just below the surface, ready to erupt, especially against officials, if provocation should be given. The medieval irrigators were extremely excitable about their rights being "disturbed," and millers and cultivators alike were willing to fight in an instant if they felt that their water supply was jeopardized in any way.

The balance of rivalry between millers and agricultural irrigators was always precarious; the millers, outnumbered, disliked, and feeling that the cequier — as the farmers' man — was naturally inimical to their interests and needs, were ready to fly off the handle at the slightest challenge. Cultivators, too, turned their hostility against officials. Here the provocation was usually associated with turns — with installing and removing canal checks, and with opening and shutting channels and divisors. For example, Jacme Ilari, divider of the water of Mislata in the district of Patraix, was officiating at a turn one night in August 1414, a drought year. "He divided the water and gave it to Jacme Fuster," and after he had done so Pere Tortosa came riding up with three men, bearing "lances, swords, and lashes," menacing Ilari and Fuster both with his lance. Ilari threatened Tortosa with a fine (raising it by ten sous with each threat). Finally one of the armed men got into the canal and Tortosa ordered him to undo the check. With that Ilari, "seeing the furor or fury" of Tortosa, retired and took his claims to the governor's court.[45]

Violence against officials was not the sole province of hotheads such as Pere Tortosa and the Codinachs. It was sometimes the expression of mass resistance as well. During a dispute between upstream and downstream irrigators on the Favara Canal in 1456, the mechanisms of autonomous control broke down completely. The rights of the downstream folk were to have been protected during their turn by the posting of guards upstream to see that no water was diverted there. But the upstream guards had been menaced with cudgels and were so intimidated that they abandoned their posts, and the downstream men felt constrained to ask the governor to send an enforcing agent (*porter*).[46]

In the above examples, violence was associated with individual or mass noncompliance with the orders of irrigation officials. In the first instance, the cequier or guard's discretion was questioned; in the second, the basis of consensual authority was invalidated by mass disobedience. The cequier had considerable discretionary authority in the administration of the turns of his own canal. The less water available, the more the cequier's power assumed critical importance and, consequently, the more likely he was to come into conflict with the hereters. In times of drought control became more stringent; this was the case on both autonomous and municipal canals. In the course of the drought year of 1413 the commons of Benacher, Faitanar, and Quart petitioned the governor for a letter ordering all the residents of those canals to obey the two guards, Marti Darmelles of Valencia and Miquel Ametler of Quart, who were charged with guarding the canals of Benacher and Quart from Alaquàs up to the diversion dam, "guarding night and day, so that no person might take the water of the said canal."[47] The city of Castellon took similar measures to ensure the success of the turn in March 1473. The jurates proposed that since many people wished to irrigate both "in their turns and outside of their turns," and since many were losing their crops because they could not irrigate, it seemed best to elect special officials for the overseeing of turns who would be empowered "to give water to those who had need of it." Therefore the council named Guillem Castell and Bernat Pelegri for the turn of Wednesday and Thursday, Anthoni Lorenç and Jacme Marc for Friday and Saturday, Johan Vilaroig and Pere Angles for Sunday and Monday, and the officials of Canet (the tail region of the canal) for Tuesday, "giving them all the power necessary for giving water to whosoever has need and for making all the necessary orders and rights of access." [48]

Table 17

a) Distribution of fines by month

	1443		1486	
Month	Number	Percent	Number	Percent
January	44	10.9	–	0
February	25	6.2	–	0
March	64	15.6	74	15.0
April	68	16.6	101	20.5
May	23	5.7	38	7.8
June	44	10.9	29	6.0
July	42	10.4	51	10.4
August	38	9.3	112	22.8
September	29	7.2	57	11.5
October	21	5.2	30	6.0
November	2	0.5	–	0
December	6	1.5	–	0
Total fines	406	100	492	100

b) Distribution of fines by quarters

	1443		1486	
Quarter	Number	Percent	Number	Percent
January–March	133	32.7	74	15.0
April–June	135	33.2	168	34.3
July–September	109	26.9	220	44.7
October–December	29	7.2	30	6.0
Total fines	406	100	492	100

Tightening of control in time of drought is illustrated by the monthly and seasonal distributions of irrigation fines levied by the cequiers of Castellón in 1443 and 1486 (Table 17). In 1443, lack of rain led to more fines collected in each of the first two quarters than in the summer quarter.[49] 1486 was a more "normal" year in that the fines in the first quarter were relatively infrequent (15 percent), doubling in the second quarter (34.3 percent), and increasing to 44.7 percent in the summer quarter — of which more than half (22.8 percent) were levied in August, characteristically the hottest and driest month of the year. In both years actions tapered off in autumn, usually a rainy season.

The shorter the water, therefore, the greater the challenge to the cequier's authority. His discretion was challenged on other grounds as well. In most of eastern Spain the principle which guided the order of turns was that the man farthest upstream go first.[50] But the cequier, in times of water shortage, could use his discretion to alter the order according to some other priority; such alterations could easily lead him into conflict with individual irrigators. Miquel Climent owned a vineyard in Almudafer, a suburb of Morvedre, which was irrigated from the cequia maior of Morvedre. Several times Miquel had asked Matheu Folquera, the *cequier mayor* of Morvedre, and his lieutenant to give him the place in the turn which was his by right for the irrigation of his vineyard, according to the customary practice in time of drought. Not only had the cequier not given him water, but what was, in Miquel's view, even worse, he had given it "to the hereters situated, in order, below Miquel's vineyard, and never had he wished to give the water to Miquel." The result was the total ruin of the vineyard; Miquel now sought penalties from the governor (an appeal to the jurates of Morvedre having failed because the cequiers argued that they, not the jurates, had jurisdiction in the affair), and that he be allowed to regain his just place in the turn. Miquel thought he had every reason to expect that, by custom, the top field — namely his — ought to drink first. In this case the cequier appears to have assigned the water to downstream irrigators, perhaps those growing bread grains, in preference to Miquel and his grapes.[51]

Other disputes seem based on the nature of irrigation systems themselves. The location of a man's turn-out determined in large part his relations with the irrigators above and below him. A primary function of an equitable system of distribution is to compensate the downstream irrigators for their unfavorable location by exacting concessions from landowners near the head of the canal. The degree of regulation and the efficacy with which regulations are enforced determine in what way spatial factors make themselves felt.

To be sure, all irrigation disputes are "upstream-downstream" disputes. But methods of control vary in accordance with the distance separating the disputants and the number of people involved. The farther apart the disputants and the greater the number, the less were the communal officers able to deal with the dispute. When disputes took on a political dimension, as frequently happened in the Valencian huerta where the canals were long enough to include several coalitions of villages, intervention of higher authority was often required.[52] The villages near the head of the canal opposed those near the tail; and, on very long canals, the villages in the center section formed a subgroup which sided now with the head villages, now with the tail, depending on how great its need for water was. The huerta canals tailed off into swamp areas, and those villages upstream and nearer the urban nucleus were more highly developed and had more economic leverage than those downstream. The tail villages, as a result, had to find support either from the king — by suing in his court — or from the city fathers.

UPSTREAM-DOWNSTREAM RIVALRY IN THE VALENCIAN HUERTA

Upstream-downstream rivalries on three huerta canals in the fifteenth century, and the related problem of inter-rech disputes, can be examined as specimen cases of political polarity in local irrigation.

THE MONCADA CANAL

The Moncada Canal was the longest in the huerta, entering the sea past Puçol, about twenty kilometers from its diversion dam near Paterna. The flow of water was regulated by four control gates (*tables*) and by a series of seven return ditches (*almenares*), the first of which was near the head of the canal.[53] About halfway along its course the Moncada Canal passed through a siphon under the Gully of Carraixet after a sharp turn seaward at Massarrojos; its course continued in a northeasterly direction after the Gully until it reached Rafelbunyol, where it turned sharply and flowed due east to the sea past Puig and Puçol near the bounds of the municipal district of Morvedre. In the Middle Ages, the villages along the canal formed discrete political groupings: the upstream group (*sobirans*) included all the villages from the head to the Gully (Paterna, Benimàmet, Burjassot, Godella, Rocafort, Massarrojos, Moncada, and Alfara — see Map 1); Puig and Puçol were the keystones of the downstream group (*jusans*); and the intervening villages, the *migans*, had shifting and ambivalent loyalties.

Throughout the course of the fifteenth century Puig and Puçol brought numerous suits in the governor's court against the commons and officials of Moncada. The first such suit, brought by Puçol in 1415, resulted in the Governor's nullification of an order by the cequier to close the canal for cleaning. The cequier, Guillem Campmar of Meliana, had ordered the water shut off on the Monday following April 3. The men of Puçol, pointing out the danger to the crops (already in dire condition in that drought year), argued that cutting off the water was not necessary and could as well be done "at the end of the present month of April, as is customary to do each year, or at another time when the crops would be fuller grown and stronger than they were at present." [54] Finally they appealed, through their procurator, Bernat Steve, to their rights to the water of the canal as vouchsafed by the Pragmatic Sanction of King Peter IV.

Throughout the century, this privilege of the King of Aragon of August 5, 1374,[55] was the mainstay of the argument of Puig and Puçol, many times reiterated, against the cequier and the upstream villages of the Moncada Canal. It provided for the disposition of Moncada's water when the river was in tanda. By provisions in a privilege of James II in 1321, the Pueblos Castillos were to receive all the water of the Guadalaviar in time of drought for four successive days; the huerta canals, Moncada included, received all the water for the four following days.[56] In addition, in times of extreme need it was provided that the four upper canals of the huerta, Moncada, Quart, Tormos, and Mislata, would let some of their water flow to the four lower canals.[57]

Peter's privilege of 1374 provided that, for the aid of its crops, the city had the right in time of drought to take one-fourth of Moncada's water.[58] The other three-fourths remained in the canal and was used for milling and irrigating fields up to the Gully of Carraixet. Moncada's water was to flow through almenares[59] into the river, joining that of the Pueblos Castillos and flowing with it to the city for its four days of the river tanda. On these days the water remaining in the Moncada Canal was to be used by all the hereters up to the Gully of Carraixet; afterward Moncada ran full again, but no water could be taken by the hereters above the Gully. The water in the canal at this time (that is, during the four days on which the Pueblos Castillos had the tanda) could be used only by the hereters below the Gully: three days for Vinalesa, Foios, Albalat, Museros, Massamagrell, and Rafelbunyol, one day for Puig and Puçol. On that final one day the water was to flow to Puig and Puçol to mill and to irrigate crops and fields "at the discretion of the cequier of the said canal, and on that day all the lateral branches of the said canal up to the districts of Puig and Puçol are to be closed and no one might dare take the water or withhold it from the residents of the said places of Puig and Puçol, under penalty of sixty sous." [60]

The discretion of the cequier was critical because he had power not only to arrange details of the distribution but also, as in the above case, to deny the water completely. The revocation of Guillem Campmar's order at the beginning of April 1415 was only a stopgap action. Three weeks later Puçol was back in the governor's court, this time with the support of the Bishop of Valencia.[61] The bishop's procurators and Bernat Steve, who represented Puig, stated that by virtue of the Pragmatic Sanction of Peter IV the residents of Puig and Puçol were entitled to one day of water per week, but that the cequier was abusing this right, causing "the total destruction of the said places . . . which by reason of not having the water which is due them are being turned [from regadiu] into secá, and, for the same reason, are soon to come to depopulation and ruin" if the governor did not provide redress.[62]

Because of their disadvantageous position at the tail of the canal, Puig and Puçol were obliged, both in 1374 and in 1415, to accept subordination to a magnate in return for support. Only with the help of the King in the earlier instance and of the Bishop of Valencia, Hugh of Lupia, in the latter, could the two weak villages countervail the will of the city and its satellite villages upstream.

In June 1416 the *sobirans* fought back. At that time representatives of Burjassot, Godella, Rocafort, Massarrojos, Benifaraig, Carpesa, Borbotó, Moncada, and Alfara — that is, all the places upstream of the Gully of Carraixet — argued before the Governor that according to James I's donation of the Moncada Canal to all the irrigators[63] they had unimpeded use of the water. Recently it had come to their attention that "some of those who irrigate from the water of the said canal below the Gully," following the Bishop of Valencia and the residents of Puig and Puçol, sought to prevent the cequier from administering the canal according to its own regulations and to the fueros; they sought a surety (*ferma de dret*[64]), and this was embodied in a letter from the Governor to Hugh.[65] On July 11, the procurator of the bishop and of the *hereters della lo riu sech* — all those downstream from the Gully — replied that the ordinances of the canal were to be followed in times of abundance of water but not in times of shortage, when the provisions of Peter IV should be put into effect.[66]

By the time of the next suit, in 1424, a political switch had taken place and the city of Valencia was associated with the bishop as co-lord of Puig.[67] The plaintiffs argued that the water of Moncada was customarily divided by jovates and fanecates, and that Puig and Puçol, like the other villages along the canal, should receive water in proportion to the amount of land they had (Puig had sixty jovates, Puçol eighty). But now the cequier of Moncada, Jacme de Almenara, constrained them to pay cequiatge on the basis of number of jovates even though they were not receiving a just share of the water. Puig and Puçol therefore refused to recognize Almenara as cequier but regarded him as a "private person." Almenara, for his part, argued that when there was enough water in the Guadalaviar for all Puig and Puçol received their shares according to the number of jovates. But when there was a shortage a different procedure went into effect: two deputies were elected who, along with the syndics and inspectors of the canal, were to divide the water, taking the scarcity into account, according to their best judgment. Almenara added that he had been elected by a majority of the syndics and therefore had jurisdiction in spite of the fact that he had been opposed by Steve and Pascual Cases, syndics of Puig and Puçol. He therefore justly had "ordinary and contentious jurisdiction" and could indeed force the hereters to pay the cequiatge.[68]

The rebuttal of Puig and Puçol is most significant: they alleged that the upstream irrigators had indirectly stolen their water by conspiring to use the canal regulations against them so they could not obtain justice. The villages upstream of Puig and Puçol controlled the election of the cequier (ten votes to two), whose judgment must be accepted by all. As a result there was no recourse to any other official or judge but those elected by the upstream irrigators, and they were for the most part upstream men themselves.[69] Thus the residents of Puig and Puçol, when denied their day in the Moncada Canal turn, were forced to beg justice from the *hereters sobirans*. The *jusans* could never — they felt — obtain justice from judges who themselves had participated in cutting off their water.[70] Such was their rationale for recourse to the governor's court and for their answer to the other side's contention that only the cequier of Moncada and not the governor was competent.

On occasion the *jusans* had to obtain the physical intervention of the governor. Upstream from Puig and Puçol were numerous fields, called *alters,* which were so high that they could only be irrigated when the canal ran full or by backing up water with checks. In 1438 Johan Ponç, cequier of Moncada, argued that it had been the practice for these alters to be irrigated on Puig and Puçol's day of the tanda.[71] Puig and Puçol claimed that this practice contravened their rights, and they secured from the Governor the assignment of a porter to patrol the entire canal on the day of their turn to enforce the Pragmatic Sanction of Peter IV. On July 13 the porter, Anthoni Perez, together with a notary and the syndics of Puig and Puçol, went personally upstream along the canal. Coming to the turn-out of Massamagrell,[72] a village with many alter fields, Perez ordered Johan Aparici, the *çabacequier,* to close the gate; Aparici complied. A little later Felip Boyl, lord of Massamagrell, came personally to the gate; the porter ordered him not to open it but in spite of the order "the said noble took some hooks and opened the gate." The porter then ordered Aparici to close the gate under pain of 500 florins and Aparici obeyed. Seeing Boyl move to reopen it, Perez ordered him not to do so under pain of first 1,000, then 2,000 florins; when Boyl put the hooks in the water the fine was raised to 3,000. Aparici shut the gate again on orders from Perez and Boyl opened it again, saying that he would always do so because it was his right.[73] On the two succeeding turns of Puig and Puçol, similar action was taken against those who had opened their turn-outs illegally.

On July 18 of the same year Boyl began a countersuit against Puig and Puçol, claiming that for one hundred years or more it had been the custom to open the turn-out of Massamagrell on the days when Puig and Puçol were irrigating.[74] Testimony, given by the present cequier against Puig and Puçol and by past cequiers and çabacequiers, was generally inconclusive. The most telling testimony was that of Vicent Pasqual, who had been cequier in 1435, 1436, and 1437, and who said that during his tenure he had repeatedly sought to have the turn-out of Massamagrell shut on Puig and Puçol's day but had never been successful in enforcing this.[75]

If the situation of Puig and Puçol was precarious, the plight of those whose rights to water were less secure was severe indeed. Between Puçol and Morvedre was a swampy region known as Arrif.[76] James I had given Morvedre the right to take water from the Moncada Canal to irrigate this district from the point where it crossed the Valencia road.[77] Such a donation, especially since Arrif was outside the bounds of the Valencian huerta, must have met with immediate and continuing resistance from the irrigators of Moncada. In 1301 James II took note of Arrif's right to Moncada water, and of the malicious practice on the part of some residents of that canal of refusing to clean the canal and of widening their turn-outs so that water which should have been used to improve the lands of Arrif was instead flowing uselessly into the swamps.[78]

The passage of years did nothing to improve the situation of Arrif relative to the Moncada Canal, and in 1449 a suit was pending between the irrigators of the Arrif (or Riff) Canal as plaintiffs and the irrigators and cequier of Moncada as defendants. The men of Arrif claimed they had the right to a turn in the Moncada rotation. There had been no definitive judgment in the case, and as a result grain and other crops were drying up "through the great fault of the cequier and upstream hereters of Moncada who have permitted, and today still permit, the water of that canal (Moncada) to be lost in the swamps." [79] The lands of Arrif, the argument continued, "could be well irrigated owing to the great richness and fertility of the water of the Guadalaviar and the Moncada Canal." They asked the Governor to order the *sobirans* to return the water to the main canal when they finished irrigating so that it could be diverted into the canal of Arrif.

The defendants pleaded that "owing to the water shortage and dryness of the fields for the past three months and at the present time, the first or upstream irrigators of Moncada had many times had compassion for the irrigators of Puig and Puçol, who are the downstream and last irrigators of the canal and those who have the last turn." And, seeing that the crops of Puig and Puçol were perishing of drought, the *sobirans* "with compassion and reverence for God" had given the *jusans* one or two free days ("hun jorn o dos de gracia") beyond their turn. It was not true that there had been abundant water and that the excess went to waste in the swamps; the irrigators of Moncada had been accustomed to giving some turns to Arrif of their own free will — that is, not through any right of the latter — causing shortage of water to all the hereters of Moncada, "axi als sobirans com als migans e jusans." [80]

Indeed, the tenor of Moncada's argument was that the whole system was under attack and that, though the good upstream irrigators might, when they were able, have compassion for others less fortunate, they still must abide by the prescribed rules and procedures that had provided the most efficient and just way of distributing the available water, especially in times of shortage. No new rights could be admitted, nor could old priorities be changed. The testimony of one llaurador, Matheu Cristofol of Burjassot, communicates the mood of the irrigators (especially those upstream) during drought conditions: "He has seen and sees today that the men of Puçol complain that they cannot have water and he sees many others, who are before Puçol in the turn, complain that the water of the canal does not suffice them. And the witness believes that were it not for the good order and administration which the canal has it would not be able to supply half of its irrigation district." [81]

From the point of view of the upstream man, then, the canal ordinances were the embodiment of equity and order; to the downstream man they were biased rules, made by and for the upstream irrigators and prejudiced against those downstream.

THE FAVARA CANAL

The physical and political morphologies of the Favara Canal were quite similar to those of Moncada. It was the longest canal (8.2 kilometers) on the right bank of the Guadalaviar, diverting water from the river at its dam in Mislata and flowing then to the Valencian suburb of Patraix and then to the Gully of Catarroja. Before the Gully it split into two branches, one crossing the Gully toward Albal, the other watering Massanassa and Alfafar.[82] As on the Moncada Canal the tail villages, Albal and Catarroja, were separated from those upstream by a gully that formed a natural division. Catarroja and Albal were analogues of Puig and Puçol and had similar conflicts with the upstream irrigators. Massanassa and Alfafar, on the northern side of the Gully, tended to side with Catarroja somewhat more than the *migans* of Moncada sided with Puig and Puçol.

The main Favara canal officially terminated at the divisor creating the branches of Albal and Massanassa. The hereters of these branches were responsible for cleaning them; in every other way, however, they participated fully in the affairs of the main canal, in the election of cequiers, inspectors and deputies, payment of cequiatge, "just like any upstream hereter," and election of one special inspector for each village for the surveillance of the annual cleaning work on the main canal.[83]

Even though they had representation, the downstream villages could always be outvoted. Without fronting on the main canal, they shared the responsibilities of those who did but were at a disadvantage in receiving their share of the water. In 1457 the *jusans* refused to pay cequiatge when they felt that their rights to water were being prejudiced by those upstream. When the canal ran full and when the branches and turn-outs upstream were open, no water reached the divisors of Catarroja, Albal, and Massanassa; when these places had the tanda the upstream branches were supposed to be shut. The guards provided for in the regulations were of no use, claimed the *jusans,* since the *sobirans* frightened them away with cudgels. To counter these "violences and oppressions" they sought the appointment of a porter by the Governor to enforce their rights[84] in the same way as Puig and Puçol had sought such assistance.

THE QUART–BENACHER–FAITANAR SYSTEM

The Quart dam diverted water from the Guadalaviar just below that of Moncada, but on the right bank of the river. The

Quart Canal then flowed through Manises and across a ravine (the Rambla of Manises) separating that village from the district of Quart over an aqueduct two hundred forty yards long. This section of the canal was known in the Middle Ages as the Canal of the Arches (Cequia dels Archs) because it was supported by twenty-eight arches.[85] In Quart the canal bifurcated by means of a divisor into the canal of Quart proper, which watered the district of Quart and part of Chirivella, and the canal of Benacher, which irrigated Aldaia and Alaquàs.[86] Faitanar was the eastern branch of Benacher and the largest of the three branches of the Quart system; it drained into the Favara Canal above Massanassa.[87]

The Quart system is of particular interest because its subdivision into three discrete physical units had clear administrative and jurisdictional ramifications. A longstanding dispute between Benacher and Faitanar over whether there ought to be a single cequier for both or separate jurisdictions was resolved in 1435 following a report by a panel of technical experts. During the first hearing of the case in August 1434,[88] the Governor appointed Mossen Jacme Gil, Jacme Perfeta, Arnau Faura, Pere Vives, and Berthomeu Casesnoves to study the matter and submit the results of their deliberations. Casesnoves was a surveyor and builder of irrigation works by profession;[89] Gil would become cequier of Rascanya in 1437;[90] Perfeta owned a mill on the Algirós branch of Mestalla;[91] Faura and Vives were both irrigators of Benacher and Faitanar[92] and were most likely included as representatives of the respective branches.

Details of the dispute are not recorded but can be surmised on the basis of the solution recommended by the experts: the best way to have done with disputes between the upstream (Benacher) and downstream (Faitanar) irrigators would be for the two canals to have only one cequier, who would exercise all the jurisdiction and power formerly exercised by two. They further declared that all the irrigators should pay cequiatge and other common expenses "equally, by cafizate," the first in order of irrigating, just like the middle and last, and the men of Faitanar just like those of Benacher. Finally, the commission declared that the two present cequiers would return all fines collected and that each would void all those pending from irrigators in the jurisdiction of the other.[93]

It is clear from the sentence that the downstream irrigators of Faitanar had refused to pay all or part of the cequiatge, on the familiar grounds, no doubt, that the *sobirans* had failed to give them their due share of the water. Second, there had ensued as an expression of endemic conflict between the two branches a jurisdictional free-for-all, with the cequier of each fining the hereters of the other with great abandon. The decision arrived at was probably the motive for drawing up new ordinances on May 29, 1435. The split jurisdiction had been a recent and temporary situation (ordinances of the combined canals were in force in 1421), a manifestation of the acuteness of upstream-downstream rivalry.[94]

INTER-RECH DISPUTES

Disputes between communities, like those between branches of a single canal, were based to a large degree on spatial factors. The map of the huerta reveals a certain arbitrariness in the composition of individual canal networks, particularly those of the right bank. Some branches could just as well have belonged to one system as another, and all the networks were tc an extent interlocking. There was lateral interlocking between parallel main canals and vertical joining through *escorredors* (drainage ditches). The closer two parallel systems were to each other, the greater the opportunity for confusion. When, for example, the Rovella commons sought to build a new extension between the river and the Favara Canal, the commons of the latter objected on the grounds that "the interval between one canal and another ought to be quite far." If not, drainage problems would increase (the groundwater level rises with each new canal), the banks of the channels would be weakened, and mutual destruction of the canals could result.[95]

For similar reasons there were statutory prohibitions against running water from one system into another: the cequier of Benacher and Faitanar, for example, was forbidden to run water into the Mislata system.[96] In some cases, however, irrigators in the rech of one system had limited rights to use the water of another. Several irrigators of Tormos in Burjassot were allowed to irrigate with Moncada water through a turn-out near the watering trough of that village. Moncada had the right to keep this *roll,* or canal turn-out, shut except "when three or four estates having irrigation rights from it are in need." But in this instance the Tormos men were keeping the turn-out open all the time, running Moncada's water into the Tormos Canal.[97]

Inasmuch as inter-rech disputes involved more than one community they were most often composed by appeal to the governor, who alone could enforce the sentence in both rechs. In the summer of 1435 the Governor imposed fines on officials of the Chirivella branch of the Mislata system for illegally diverting water from Faitanar.[98]

HYDRAULIC CAUSES OF CONFLICT AND TECHNICAL SOLUTIONS

Some conflicts arose from the very way in which water flows. When, for example, the normal surface slope is decreased as a result of the normal level being increased at some point downstream, a variable backwater results.[99] Such backwaters—*regolfs d'aygua,* in medieval Valencian—caused by various kinds of obstructions in the channels (checks, divisors, or mills, for example), had destructive effects upstream and produced conflicts between the harmed individuals and those responsible for the obstruction. Effects on the flow resulting from construction of mills, divisors, and other permanent structures had to be taken into consideration at the time of their emplacement. To this end, technical specialists called levelers were consulted, not only in the planning and building stages but also thereafter, when unforeseen hydraulic difficulties arose.[100]

A canal check is anything placed in a channel for the purpose of diverting water. To divert, the irrigator frequently needs to reduce the velocity of flow or to raise the level of the water or both. (The Valencian term *parada,* "stop," stresses the slowing function of a check, whereas the Murcian synonym *rafa,* "raising up," [101] stresses the second.) Checks were often improvised — any slat, rock, or even a pile of rubble would produce the desired effect. Occasionally checks were implanted permanently to reduce the velocity of the current. They were frequently placed downstream from mills, where they equalized the effects of a sharp drop upstream from the mill.[102] A check might also be placed at the head of a continuously flowing secondary channel to divert water into it.[103] In 1408 several men of Castellón, irrigating from a channel called The Cequiol, complained about a backwater resulting from a rock placed in the channel by Pere de Reus to divert water to a watering trough. The council ordered the rock inspected; if it projected upward so as to create a large backwater, it was to be lowered enough to eliminate the backwater.[104]

Backwaters could cause excessive silting, as occurred on the same canal two years later. Some irrigators complained that a check placed to divert water to the cathedral works caused The Cequiol to become stagnant; in fact, so much silt had gathered that the men who had contracted to clean the channel were now unwilling to do so unless the council would supplement the cequiatge funds.[105]

Mills also were common sources of conflict.[106] Mills need a constant water supply, which may not always be possible when turns are instated. Even in normal times millers and cultivators have different hydraulic requirements. The cultivator likes a uniform, steady, and level flow of water, but the miller must have a drop in the channel. He must build up head for his mill, but at the same time avoid backing up the water so strongly as to cause flooding and caving in the banks upstream. For these reasons, technical advice was needed at the time of construction. Anthoni de Veses, who purchased a mill on the Favara Canal and rebuilt it, was later accused by downstream irrigators of depriving them of water. Veses claimed that this could not be true inasmuch as he had rebuilt the mill with the help of "good masters and levelers of water." [107] But in spite of precautions mills were constantly causing damage to other water users. The town of Castellón took action in 1410 against Domingo Dolç, whose mill was causing backwaters great enough to break down the banks of the canal.[108]

If mills indeed had been properly "leveled" and their hydraulic problems taken into account at the time of their building, how could such damage ensue? In some instances, the original hydraulic balance was disrupted by the miller's changing the operation of the milling apparatus. In 1446 a dispute arose over the mill of En Ferrada on the Algirós branch of the Mestalla Canal, a fulling mill designed to run with two millstones (*rodes*). The present miller, Jacme Perfeta, was operating with only one stone, possibly because he wished to avoid the unpleasant vibration caused by two.[109] With only one wheel operating, the head needed to turn the mill was reduced by half, and a backwater resulted upstream. In this case the backwater was so great that the water overflowed the banks of the canal and spilled into the neighboring canal of Bonvehi.[110] A mill located on a drainage ditch could hinder proper drainage: the mill of En Caner ran on water from the moat of the city of Valencia, but because of the check placed in the moat to divert water to the mill a backwater was created which prevented the moat from draining properly. The council ordered the jurates and some master masons to investigate.[111]

Sometimes the miller was the aggrieved party. In another dispute on The Cequiol of Castellón, the owner of the mill of En Altamora, Domingo Mascho, claimed that he had the right to the unimpeded flow of water through his mill's canal without any obstruction upstream in the form of rocks, mud, rushes, or anything else. Some irrigators who had land near his mill and who took water from The Cequiol (which canal, he took pains to point out, had been "leveled and measured" in days of old) were trying to impede the flow of water to his mill by putting brush, mud, and rocks in The Cequiol to raise the water level.[112] The problem here was not the usual direct miller-farmer confrontation, in which the miller was accused of creating the backwater, but rather that the miller was suffering as a result of the cultivators' desire for more water.

Divisors perhaps even more than mills were natural centers

of conflict. Indeed, one divisor in the Favara system was called Na Piquabaralla (Lady Pick-a-Fight);[113] whether it was so named in anthropomorphization of a traditional potentiality for disruption, or whether such a pugnacious lady did in fact join in quarrels over that particular divisor, the association of divisors and strife made the name an apt symbol. Divisors, after all, divided more than just water. They divided groups of irrigators, as those of Benacher and Faitanar, and Algirós and Rambla.

Because water distribution was based on a principle of proportionality, the divisor — as the medium by which ratios representing area of irrigated land were translated into real quantities of water — was the key to the entire distributory system. For this reasons divisors were very carefully and precisely designed, leveled, measured, and built. They had to be constructed solidly enough to withstand any changes in the water level on either side of the "tongue" which might alter the traditionally established ratio. They were built of mortar (*argamassa*), stone (*pedra*), or flagstones (*loses*).[114] Another guarantee of stability was the recording of the measurements of divisors in a legal document. At a general meeting of the irrigators of the Favara Canal in 1362 it was decided that all the canal's divisors be "measured and leveled . . . in case they should ever be ruined or have to be rebuilt owing either to war or to the long passage of time." [115] The inventory, made by the leveler Andreu Julia, recorded only the width (in palms or fingers) of the ditches and an occasional indication of level, such as whether the floor of the canal on either side of the tongue was at a uniform elevation.

The flow on either side of a divisor was affected not only by the width of the channel, but by the level of the beds and the downstream course of the subchannels. The importance of a leveled divisor[116] can be appreciated from several examples of litigation in the huerta of Valencia. The Benacher Canal received its water, as we have seen, from the Canal of the Arches in Manises by means of a divisor which partitioned the water equally between the canals of Quart and Benacher.[117] In 1437 the irrigators of Benacher complained that Quart was taking more water than Benacher from the common channel; the officials of Quart replied that both sides of the divisor were equal and had originally been at the same level, and furthermore, that Quart's side had been well cleaned while that of Benacher was clogged, owing to neglect by the irrigators.

The original ratio of division of a divisor can be altered by deposits of silt on either side. The tongue of the divisor acts as an obstruction and creates a hydraulic jump. Where the flow becomes tranquil again (see Figure 3), silt is deposited; if it is not removed, water backs up above the obstruction and flows down the other side. In this particular case the court dispatched the levelers Berthomeu Casesnoves and Jacme Gallen to examine the disputed divisor and to determine the height of the water on both sides. The experts reported that they had seen the divisor, but with water flowing through it. When they had asked to see it without water, the officials had complained that emptying the canal at that time (early June) would cause great harm to the irrigators and the rice crop. Three weeks later the court, on the advice of Casesnoves and Gallen, ordered that the water be equally divided and that "the floors of the said divisors be raised one-half palm in equal level," and that above and below the division the floor and sides of the channels be strengthened with rock and mortar to ensure the continued equality of distribution.[118]

As a final example, a dispute between two branches of the Mestalla Canal in 1452 admirably illustrates the hydraulic problems created by the conjuncture of a mill and a divisor. The branches of Rambla and Algirós divide their water — in the words of the document — in a divisor constructed above the mills owned by Jacme Perfeta, and "each of the said canals, there where the tongue of the divisor is, has its channel and sides of stone, made at [the same] level, in which no change may be introduced, inasmuch as the ancient and accustomed course of the water ought to be preserved." [119]

Perfeta alleged that some irrigators of the Rambla Canal had been moved to introduce a change in the divisor; those of Algirós were determined to protect their rights to the ancient course of the water. Rambla's argument was as follows: Perfeta

Figure 3. A divisor of the Tormos Canal in Benimàmet (huerta of Valencia), dividing the flow into two equal parts. The swirling indicates a hydraulic jump; silt is deposited where the flow becomes tranquil again.

owned two mills on the Algirós branch and had schemed to divert more than the customary volume of water into Algirós' side, in prejudice to the whole canal of Rambla and particularly to the mill of En Bonvehi, owned by Johan Mercader.[120] Perfeta's two mills were a fulling mill (*hun moli pilater o draper*) called the mill of En Ferrada,[121] located very near the divisor, and a flour mill (*moli fariner*) located a short distance from the fulling mill.

The divisor, the argument of Rambla continues, was built "with a certain measure, with certain borders, and at a certain level." But Perfeta, to the great harm of the irrigators of Rambla, had caused more water to flow in the Algirós branch; consequently, much less water was now flowing into Rambla, which had more land to irrigate than did Algirós.[122] In former times the fulling mill had operated just as all fulling mills do; but since Perfeta had acquired the mill all of its channels had remained open, with the result that the water rushed into the Algirós branch "impetuously . . . and with great velocity." [123] Perfeta kept the fulling mill empty and nonoperative so that his flour mill, further downstream, would have a great abundance of water and would be able to move more millstones than it had originally been designed to move.

Perfeta's side of the story was then presented by his procurator, Berenguer Cardona, who had represented him in 1446. It should be pointed out that Cardona appears to have had more experience with irrigation affairs than his opponents in that his arguments are much more specific regarding hydraulic points. Indeed, it is possible to observe in lawyers a certain specialization in hydraulic matters; several men, Cardona included, repeatedly argued irrigation cases.[124]

Cardona argued, first, that this divisor had had the same dimensions for a hundred years or more and that no novelty had been introduced in it. Second, it was impertinent to sue Perfeta for *not* using his mill, since he alone stood to gain financially when it ran and to lose when it did not; it is the owner's prerogative to decide how to use his property. Third, in a suit successfully brought by Perfeta and the cequier of Algirós against

Mercader and the Rambla branch, he had gained the right to widen the channel of Algirós below the divisor in order to obtain more water, on the grounds that what happens below the divisor could not harm the other side.[125] The other side made much of the point that the Mestalla Canal and its divisors dated from the time of the Saracens[126] and that Cardona's client was upsetting this balance with his mill. Cardona believed this argument to be an absurdity: at the time the water had originally been divided there had certainly been no provision made "to give a certain advantage in the divisor of the Algirós Canal because of an uncertain impediment that might have been thought up" about some future mill.[127] "No man of this life," he mused, "can measure future needs."[128]

Cardona argued that all the mill's channels must remain open and gave a good description of hydraulic jump at a divisor and of a mill's need for a good head of water. It had been established by previous arbitration that the mill had to be run in such a way as not to cause a backwater by which the Algirós Canal would lose water. "The water had to jump at the divisor in such a way that once it had jumped, it could not flow back."[129] It may be, therefore, Cardona continued, that Algirós, by having "free jump" (*lo salt deliure*), drew more water than Rambla. If the irrigators of Rambla wished to alter their channel below the divisor in a like way, no one could complain. Cardona concluded that the banks of the two canals had been built "with such a level and measure" that neither could fail to receive its just portion unless one of the two should be poorly maintained and thus give rise to a backwater.[130]

In closing his argument, Cardona mentioned a pertinent structural point always taken into consideration when divisors are built:

> When the water of any main canal is divided equally between two canals, in the construction of the channel and of the divisors of those canals it is customary to note which of the two . . . runs most straight [in relation] to the course of the main canal. And if one of these canals should

be found to run more in line [with the main canal], so that it receives the water more freely and gets more than the other which does not receive it thus in a straight line, for this reason it is customary to give advantage of some small width to the [side of the] divisor of the canal which does not run so much in line, and the divisor of the canal which runs in line is made narrower, so that the latter does not receive more water than the former, but both of them receive the water with equality.[131]

To be sure, some of Cardona's argument was self-defeating: it is not true, for example, that what happens below a divisor cannot cause a change in the flow on the other side, as he himself argued with regard to poor maintenance. Cardona was arguing that the divisor was an entirely sufficient control, his opponents that it was not. Both arguments had basis in fact. Perfeta's failure to run the wheels of his fulling mill while running the water through the mill as a bleeder had the same effect as widening the Algirós Canal, increasing its capacity to Rambla's detriment. Rambla was insisting on the status quo immediately before Perfeta's purchase of the cloth mill, that is, before the original control on its side had been invalidated. Cardona was insisting on the older, letter-of-the-law arrangement: Perfeta had not actually altered the physical proportions of the divisor.

Thus with divisors, as with mills, solution to hydraulically based conflict was not always possible within the framework of the community. Consultation with outside experts was necessary, and often the matter became so complicated that only a formal court proceeding could straighten it out.

IV | The City's Role in Irrigation Development

Wittfogel's hydraulic theory links large-scale irrigation, centralized administration, and corvée labor. In medieval Valencia, as noted earlier, irrigation administration was cellular (one açut-cequia complex was the basic unit) and decentralized. Whether the irrigation community was municipally or autonomously controlled the degree of irrigator control over administration varied little, and it was considerable in both cases. Consensual authority underlay both administrative forms.

Nevertheless, a study of the financial requirements of these extensive irrigation works might be expected to show some dependence upon the possession of necessary wealth for the mounting of large-scale works by higher corporate authorities. Of course the medieval Valencians had not made the initial investment: the systems they operated were largely inherited intact from the Muslims, and most capital invested in irrigation in the thirteenth, fourteenth, and fifteenth centuries was applied only to maintenance and repair. Massive new plans were considered in the Middle Ages (the subject of Chapter V) but were not carried out. These plans compared in magnitude with the building of the Tibi Reservoir in Alicante in the sixteenth century and the extension of the Royal Canal of the Júcar in the eighteenth, but the medieval approach to irrigation financing was probably the major factor inhibiting the execution of the more grandiose plans. In those times expenses for both day-to-day maintenance and large-scale works were divided among the interested parties on a pro rata basis according to the amount of land held. In the Júcar diversion project discussed below the cost was to be shared among all the villages and lordly domains that stood to benefit from it: a democratic approach, but not one designed to get a canal built quickly and efficiently. All the energy was spent in endless *parlements* among the interested parties (some of whom were, in actuality, notably less than interested). Medieval Valencians, owing to the fragmentation of economic as well as political power, were able to devise, but not to execute, large-scale irrigation projects.

On the other hand, the system as it was inherited from the Arabs did not decline at all. In fact, it thrived and was extended. The extensions were continuous and relatively small in scale, and they resulted in a considerable enlargement of many Arab-irrigated huertas. The medieval Valencians may not have been able to build large-scale projects, but was it necessary for them to do so?

In this chapter the role of the city and, to a lesser extent, that of the Church and the king in the direction of regional water policy is examined. It is my view that while this direction was real, vital, and basic to the functioning of Valencian irrigation in its broadest sense, it was not predicated on capital input. The direction of the higher authorities was largely political. They did not generally make large investments in irrigation development, but even when they did they did not seek centralized controls. They functioned as an adjunct to, not a replacement of, traditional, cellular irrigation authority.

PATTERNS OF CAPITALIZATION

Towns invested in irrigation projects only when their citizens could not afford the cost. Even so, the capabilities of local initiative were often impressive. Sometime in the first quarter of the fifteenth century a group of landowners and others grouped and formed a company to build an irrigation canal in Benidorm.[1] Fourteen partners in all made an initial investment of 1450 florins (in eleven shares of 100 florins each, two double shares, and one half-share). Most held land in Benidorm, although the partners included Garcia Mezquita, the Valencian leveler, his son Anthoni, and the vicar of the neighboring village of Polop. Mezquita, who was also assigned a salary in his capacity of master of the works, was not involved on the same basis as the others; but they were enjoined to deal fairly with his share of the investment as if he, too, had been a landholder. Mezquita was to receive six sous in daily wage — an average salary for a skilled leveler — and his sons five (three for the elder, two for the younger). Another partner, Guillem Ramon Çaportella, was named chief mason at a salary of two sous, and a lump sum of three hundred thirty sous was assigned to Johan de Vilanova the Younger for his salary as account-keeper and treasurer. Fifteen florins per share were to be put in the "first sack of money" so that the work could begin. The company was governed by a board of elected deputies consisting of Mezquita, Çaportella, Vilanova, Jacme Gomiz, and the nobleman who held a double share, Johan de Luna.

Extraordinary expenses of single canals were usually met in the same way as ordinary ones — by apportionment among all the members of the community, in the manner of the cequiatge. In a dispute of 1414, for example, the governor's court upheld the general rule that the repair of diversion dams was to be accomplished at the expense of the irrigators and mill-owners, or all those using the water of the main canal.[2] In 1436 Luis Eximenis, owner of a mill on the Axera branch of the Rascanya Canal, agreed to contribute to the expense of changing the course of the canal (by moving Rascanya's diversion dam higher up the Guadalaviar in order to capture more water). Indeed, as a millowner he was obliged to contribute. Subsequent to his initial consent at a general meeting, the deputies of the community decided upon an alternate route, and Eximenis now refused to pay on the grounds that he had not been consulted. His argument was baseless by the very terms of the original agreement of 1436, which concluded with a declaration that all those with a legitimate interest in the canal had to pay a share of the expenses, whether present at the general meeting or not.[3]

Frequently circumstances required additional monies for maintenance expenses beyond the assessed rate of cequiatge. In Castellón in particular, where the cleaning of the canal was let out to contractors, the cequiatge was often unrealistically low. Once, because of a temporary diversion in The Cequiol to conduct water for building the cathedral, more silt than was normal accumulated in the main channel. As a result, the men who had contracted in advance to clean the canal were now unwilling to finish the job unless more money than the seven diners per fanecate paid by the hereters as cequiatge could be found. Thus several irrigators asked if the council "might wish to make some aid, so that the contractors would have more inclination to clean the said canal." The council, accordingly, "with concern that the town might benefit from the water of The Cequiol," granted eleven sous for the purpose requested for the current year, 1410, and for each succeeding year so long as work on the cathedral should continue.[4]

The major cleaning was always done in March or April, but the men were often hired long in advance. Money often had to be raised, therefore, in anticipation of cequiatge receipts, as was done by the syndic of the commons of Na Rovella in 1351. He had borrowed "quantities of money, not only from Jewish usurers, but from Christians as well, for the necessities of said canal, on the strength of which loan a collection was made among all the hereters of the canal." The two collectors were to hand over the money to Simon Çestrens, who was to pay all the debts contracted by the syndic.[5] Since the normal financial needs of irrigation were met by the cequiatge of each canal system, a city's direct financing of irrigation works was centered upon large projects in areas beyond the bounds of the traditional huerta or on hydraulic enterprises within the city walls themselves.

The entire effort to divert water from the Júcar River to the Guadalaviar (described in Chapter V) was directed by the city of Valencia. Had the project ever materialized, it would have been the city's responsibility to raise an estimated thirty-five to forty thousand pounds, although it was never the city's intent to put up the entire sum. The council agreed to seek a royal grant and then to negotiate an agreement between the city, the bishop, and the lords, villages, and individual landholders through whose property the proposed canal would pass, allocating the expenses among all according to the interest of each. There can be no doubt that the success of the project depended upon the willingness and ability of the important magnates to raise money, and its failure can be attributed to their unwillingness (or that of their vassals) to participate in such an expensive and daring project.

The same elements were present in a similar but more modest venture. The council of Valencia had been asked by the treasurer of the Order of Montesa in April 1404 to bring water from the Júcar through a siphon, benefiting the municipal districts of Alcira and Valencia alike. The Valencians were willing to cooperate but only if the project did not "tie up the city with great and fruitless expenses"; if the proposed canal could be brought up to the place surveyed, the city treasury would pay half the cost as long as the city had the right to half the water.[6] Several months later the jurates of the city wrote to Mossen Joffre of Tous reminding him that the city would pay half the expense and that the other half was to be borne by the hereters, of whom Mossen Joffre was "one, and the greatest." Specifically, according to the "accounting of jovates it falls your share to pay in the said work 66 pounds 15d.," Joffre was told.[7] The allocation — just like that of the cequiatge — was made according to the amount of land held. These plans did in fact come to fruition and the Cequia del Canó was inaugurated in December 1404.[8]

THE MARJALS OF VALENCIA

The principal hydraulic problem in the Marjals was that the Valencians — as Antonio Cavanilles pointed out — did not know how to build lasting channels in areas with high water tables. On the contrary, they built channels so susceptible to seepage that in several areas they paradoxically increased the swamp acreage while intending to bring the area under intensive irrigation cultivation.[9] The problem was practically insoluble; the channels inevitably broke down unless almost constant maintenance was provided. Since the area was demographically unstable, due to hydraulic difficulties no less than to the related problems of sanitation and disease, the city itself finally had to take over direction of the area in order for it not, literally, to go under.

The most important of the Marjal canals in the fourteenth and fifteenth centuries was the En Fluvia. Following the same course as that occupied since the nineteenth century by the Turia Canal (also called the Acequia del Oro), it took water from the Guadalaviar at a point which in the Middle Ages was called the Punta d'En Silvestre and ran from there into the Albufera.[10] The En Fluvia Canal dates at least from the second quarter of the fourteenth century, when it was recorded that one Guillermo de Fluvia "had made, and does not cease to make continually, new canals" in the Marjals, in an attempt to improve the land and bring it under cultivation.[11] Other canals, such as the Cequia del Cavaller and the Cequia de Castelló d'En Arrufat were even older.[12] These three, together with the Cequia del Vall,[13] the Cequia del Bisbe, and Cequia Morisca, were the Marjal canals most frequently mentioned in the documents. The course of each would be hard to trace, as they were all interlocking and connected by "transverse" branches and drainage ditches.[14]

This network of constantly deteriorating canals was the object of an extensive development project planned in the 1380's and executed in the 1390's. After long negotiation an agreement on the work to be done was made between the city and the Church, probably in 1386. The circumstances prompting their joint action were admirably summarized in the preamble to their agreement:

> In the name of the Holy Trinity.
>
> Since in the huerta or district of the city of Valencia a great area below the places of Russafa and Alfafar and others has become marshy and barren, especially because of the ruin of the canals, branch canals, and drainage ditches, which be-

> cause of lack of people who have been reduced in number and in strength by reason of wars, disease, and other past adversities, have not been cleaned or kept in repair as they should and as they were accustomed of old.
>
> And this is in great harm to the public weal, not only because of the crops which thus have been lost and are lost, but also due to the infection which ensues, the more so for the great extent of the said marshy region which is more than a league long and a half — or at least a third — wide.
>
> To redress this harm and to procure benefit it has been decided by the below-signed parties that the said region . . . be reduced to agriculture. This cannot be done without great work and expenses, according to what was seen and studied personally by some men of each group who were sent to the said region for that reason. . . .
>
> Therefore the things contained in the following chapters were agreed upon . . . by the reverend lord bishop and honorable chapter of the Church of Valencia, on their own behalf as well as on that of the Almshouse of En Conesa of which they were the administrators, on the one hand, and on the other, the honorable jurates and some leading citizens, in the name of and with full power for, the whole council and citizenry of the city of Valencia.[15]

The financial provisions agreed upon represent a fairly realistic apportioning of responsibility between Church, city, and residents. The bishop, the chapter, and the almshouse remitted in perpetuity all dues (*censals*) accruing from any land in the Marjals. Furthermore, tithes (*delmes*) and firstfruits (*primicies*) were waived for a period of ten years, beginning at Christmas, 1390. If any religious person refused to suspend collection of these dues, he would be held responsible for the cleaning and repair of all channels on which his property fronted just as if he were in actual residence there.

The jurates, for their part, promised that the city would undertake the initial overhauling ("the first repair") of the principal canals and the bridges and roads of the Marjals. They would also repair all the secondary channels and drainage ditches on which city land fronted, and they would keep all such channels perpetually in repair. They would be responsible for the maintenance of any bridge that had been erected by the city in the first place. None of the land in question was to be alienable or subject to financial burdens for any reason. No vineyards were to be allowed in the Marjals, nor rice plantations, but the fields were to be for wheat or any other crop except the above. Anyone settling in the Marjals would be obliged to plant poplars or willows on the borders of his field, and within the field itself fruit trees could be planted. The city would supply a cequier to administer the Marjals canals who, along with two inspectors, would have full jurisdiction.[16]

The actual work began on February 8, 1390, under direction of the Sotsobrer de Murs, Lois Menargues, and continued through August 19, 1393, often with twenty or thirty men working each day, with salaries as high as three sous and more for skilled laborers. When all the work had been completed, Menargues reported that he had expended by reason of the work of the Marjals, "not only in salaries of diggers and gleaners, in limestone, sand, bricks, rubble, baskets, but also for many other things [the sum of] 78,634 sous 9 diners." [17] Of the total amount, Menargues added, he had spent only 406 sous 3 diners more than what he had actually received from the city treasury.

Residents having property along the Marjals canals were obliged, as was customary in the whole huerta, to contribute cequiatge for the upkeep and general maintenance of the canals. But in many parts of the Marjals in the 1390's and later there were stretches of canal with no *frontalers,* in which case the city filled in with a yearly grant of one hundred pounds from the General Fund.[18] Still, in 1396 several residents complained that this annual authorized amount was not being made available easily and promptly enough to the cequier of the Marjals, with the result that the bridges were not being kept in repair and the canals were being broken down by the passage of cattle.[19]

Throughout the fifteenth century the city continued to pay frequent sums from the General Fund for further works in this perpetually vexing and financially demanding area. The Manuals de Consell are dotted with these expenditures, large and small but mostly large: 600 gold florins of Aragon in 1406 for a new canal to drain En Fluvia; 90 sous in 1441 to Berthomeu Casesnoves, the surveyor, for the work he had done in leveling the waters of the Marjals; 43,000 sous from the city treasury in 1445 for the New Canal joining the Castelló and En Fluvia canals and 60 pounds to Petro Veixo for surveying this canal; 165 sous 9 diners to the cequier for cleaning a new canal called La Traversa, in 1448.[20]

When new channels were excavated the city exercised the right to expropriate land in the path of construction, provided that remuneration was made to the owners; the land was carefully assessed and due restitution made. Thus, for the Cequia Traversa "first, there was taken from the estate of Berthomeu Guerau, one fanecate and a quarter, and eleven cubits, quit, at the rate of twenty pounds per cafizate; it was appraised at four pounds." [21] Indemnity was also paid, in addition to land actually taken, for land bordering this new canal which still belonged to the original owner but which he was not allowed to plant.[22] The total amount spent in these expropriations was 105 pounds 155 sous 49 diners.[23]

HYDRAULIC PROJECTS WITHIN THE CITY WALLS

Within the walls of the city, the town fathers had an active interest in hydraulic affairs which was expressed financially in a variety of endeavors. Most urban hydraulic problems either were in the realm of public health or directly involved the economic life of the city. The main canal of Na Rovella actually ran through the city — covered in some places, uncovered in others — and its water was used not only for industrial purposes but also for flushing the city sewers.[24] The main canals of Favara and Mislata ran near the city outside the walls, but their water did flow within the walls in some minor arteries. Dyers, tanners, cloth finishers, and other tradesmen had workshops on all three of these canals, not to mention the mills that used their water for power.[25]

The urban canals were a constant preoccupation because they were particularly likely to be contaminated by industrial waste and city sewage and consequently to give off offensive odors or occasion fear of disease. A proposal of 1419 expressed concern for the plight of the prisoners in the city jail because of the fetid water flowing in an adjacent *fillola.* This water came from the "canal which is near the meatmarket of the Moors," near the Rovella Canal. The matter was investigated by "masters expert in leveling," who recommended that stagnation in the prison's channel could be relieved by drawing water from Na Rovella at a cost of 1400 florins to be paid from the General Fund.[26] The council authorized construction of a watering trough within the city walls so that the townsmen would not have to take their animals outside in time of war. The trough was to be put "in the square in front of the mill of Na Rovella" (and to use the water of that canal, one assumes), and payment was authorized from tolls taken on the city's bridges and roads.[27]

On occasion the city encouraged the building of mills within its walls. In 1363 Johan Dolit and Domingo Borraç petitioned the council for license to build a flour mill of four or five millstones within the walls. The city officials granted the request, first because of shortage of mills and abundance of people in Valencia at the time; second because honor and fame accrued to a city with water mills within its walls; and last — and most important — because they anticipated a siege by the armies of Castile. The mill was to be built at the petitioners' own expense, but the city lent its help by granting free use of "all the stones of the Jewish cemetery of Valencia," to the end of making the diversion dam so strong that neither the river nor the enemy might be able to destroy it without great difficulty. Finally, all the land which had to be appropriated for the building of the mill and its canal were to be paid for by the city.[28]

Similar preoccupation with the grain supply in 1445 provoked the council to hire the leveler Petro Veixo to survey the Moncada Canal "to find out how much water could flow there to mill the flour on account of the drought." [29] The significance here is that the council, not the Moncada community, paid for this service. Except when specifically petitioned, the city rarely intervened in local water matters. It intervened only when forced by matters of overriding and general interest.

An accounting of the total municipal expenditures for water affairs would not be complete without including the frequent public processions of prayer, such as that to the church of St. Mary Magdalene "for health and rain" in 1448. These processions were organized and financed by the council.[30]

LABOR CORVÉES

The city of Valencia did not use corvée labor on irrigation projects, preferring hired hands even on mammoth works such as the Marjals reclamation.[31] In Castellón, however, where irrigation administration was centralized, the municipal authorities seemed more disposed to use the corvée available to it.[32] For purposes of defense as well as labor the population of Castellón was divided into groups of fifty (*cinquantenes*) and ten (*deenes*).[33] The *deenes* were used to solve several hydraulic problems. In April 1406 the diversion dam was destroyed by a flood and no water was entering the main canal of the town. Pere de Reus, the syndic and treasurer of the town, asked the council for money in order "to hire men with beasts to carry stone and wood for the rebuilding of the diversion dam." The council charged the jurates with attending to the repair of the dam, either by paying wages or by sending "three men as *deenes*." [34]

In July 1422 the jurates asked the council whether the mouth (*gola*) of the main canal, which had been "greatly ruined," ought to be cleaned out by hired men (*homens a lloguer*) or by the "*deenes* of the town." The council ruled that the work should be done with the corvée.[35] Later the same month Johan Vilaroia, who in the past had contracted to clean the main canal, was offered a hundred men of the *deenes* to help in cleaning the canal, the banks of which were in such poor condition that not enough water could flow through it.[36]

Castellón, therefore, was more willing than Valencia to resort to corvée labor in times of emergency instead of using town funds; but in the above cases the use of corvée labor rather than hired hands was an extraordinary measure and not basic to the normal functioning of the irrigation system.

V | The Search for New Sources of Water

The Guadalaviar and the Júcar rise less than twelve miles apart in the Sierra de Albarracin, the former in the province of Teruel, the latter in the northernmost part of the province of Cuenca. The course of the Júcar, winding through the mountains of Cuenca and Valencia, is more than one and three-quarters times that of the river of Valencia (532 to 297 kilometers) and its watershed is nearly four times greater (22,000 to 6,191 square kilometers). The greater watershed area and the higher rate of precipitation in its upper basin than those of the Guadalaviar contribute to a substantially higher volume of water.[1]

It is hardly more than forty kilometers from Antella on the Júcar to the southern boundary of the Valencian huerta at the Gully of Catarroja (see Map 2); so it is no mystery that the Júcar's waters should have captured the interest of the irrigators of Valencia. (In later times the Royal Canal of Alcira was extended right up to the Gully.) The propsect of augmenting the waters of the Guadalaviar with those of the Júcar appears first to have commanded the serious attention of the town fathers of Valencia during the drought of the mid-1370's. In October 1374 the jurates, together with some "eminent persons" and hydraulic experts, suggested to the council that it would be possible to take water from the Júcar near Tous (to the north of Antella, where the Escalona joins the Júcar) and make it flow through a canal to the Guadalaviar. With this water the Valencians could "irrigate many districts which are not now irrigated and could give increase to the water of this river and of that presently available to the city and the huerta, especially in times of

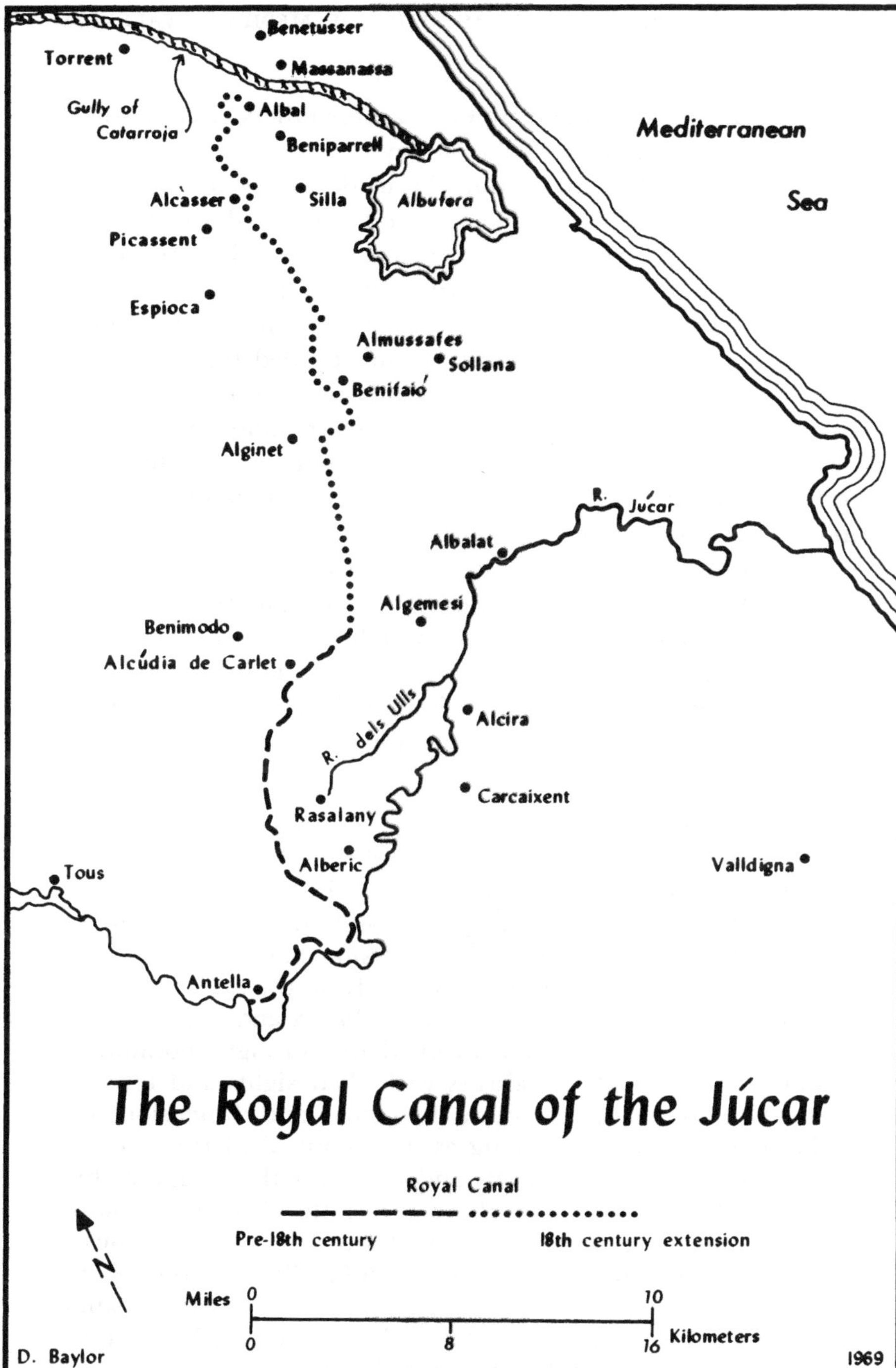

The Royal Canal of the Júcar

drought, as are the present times and some others already past." [2]

The council decided that a survey of the course of the projected diversionary canal should be made by levelers and other experts, at the expense of the city, and appointed a commission of four leading citizens to direct the project. The survey apparently was not made at the time, however, possibly because of rain in October. The drought continued into 1375, and in October of that year the council again decided that the undertaking of this "most arduous and costly task" should be studied by "persons expert in the art of geometry and the level" to determine "if and how the said work might be brought to perfection and with what cost." The council further ordered that, in spite of the fact that the city had good masters of this art, more masters ought to be sought: better that the survey be done by too many experts than too few. To this end messengers were dispatched to other areas to search out such masters.[3]

By February 1376 the survey had been made by a commission including eight levelers — four from Valencia, two from Barcelona, and two from Manresa. They reported to the council that:

> They had leveled and measured the whole project and the places for the diversion dam and the planned canal, beginning near and above the castle of Tous on the floor or plain of the said river [the Júcar], and continuing through the valley of Tous up to the diversion dam of the Royal Canal [of Alcira] which draws water from the said river and irrigates the districts of the cities of Játiva and Alcira.
>
> And from there they continued on through mountains and gullies and plains, always with their sights and levels,[4] and also taking into account the course and dimensions of the Royal Canal for as long as it extends. And they found according to the said levels and measures that digging the proposed canal near and above the castle of Tous it could come and bring water from the Júcar River, passing near the places of Antella, Rasalany, Alcúdia, Benimodo de Carlet, Alginet, Espioca, Picassent, Alcàsser, and Torrent, and flow into the Guadalaviar River above Manises, but below the diversion dam of the Moncada Canal.

> And all this work, according to the computation of the said masters and other experts, having given ample consideration, as indeed they ought, could cost from thirty-five to forty thousand pounds.[5]

The day following the report of the experts, the council discussed its contents and heard arguments *pro e contra*. It was decided that, first, a concession should be obtained from the king. Then, an agreement would be made with the "lords, villages and residents of the places through which the canal . . . was to pass." The expenses would have to be apportioned among all concerned and an agreement reached concerning the administration of the water. The jurates were ordered to designate a commission of twenty leading citizens to negotiate with the lords of the places in question.[6]

Nothing came of all this planning, apparently because the grant sought from King Peter IV (who ruled from 1336 to 1387) was not forthcoming. When the project is next heard of the jurates were dealing with Peter's successor, John I (ruled 1387–1395).[7] The new negotiation, in 1393, was a success, and in November the king granted the city the right to take water from the Júcar at Tous and conduct it to the Guadalaviar for irrigation purposes, so long as owners of the land in the route of the proposed canal were reimbursed for their losses in confiscated property.[8]

The next round of activity took place in the early years of the fifteenth century, again under drought conditions.[9] In May 1400 the jurates directed a letter to Hugh, bishop of Valencia, explaining that, at the urging of some men who had decided to bring a canal from the Júcar to the Guadalaviar at their expense, they now asked the Bishop to assign to the canal the firstfruits he received from crops in the districts to be served by the new canal.[10] Interest was still high in July, when the council noted that because of frequent droughts and lack of water a great portion of the crops was lost; grain had to be supplied from "other parts," since the mills of the city could not function. As a result the council resolved to continue in the attempt to divert Júcar water to the Guadalaviar and decided that a meeting of all those interested in the project, that is, not just the city officials but also lay and ecclesiastical lords who would have land on the proposed canal, be convened as soon as possible.[11]

Apparently no meeting was held that year, for nothing more is heard about the Júcar project until July 1401. On July 23 the jurates wrote to Bishop Hugh that, in order to avoid enmity and discord, a meeting (*parlement*) of all the interested parties would be held on August 5 at the monastery of Santa Maria del Carmen.[12] Three days later the Bishop received a new letter changing the date of the meeting to August 8 and the place to Torres Torres. The same letter was sent to a roster of magnates with interest in the new canal. Included were Berenguer March, master of the Order of Montesa; the Episcopal chapter of Valencia; Olfo of Proxida, lord of Alcàsser; the Mayoral of Quart; the Abbot of St. Bernard, a Cistercian monastery in the huerta, north of Valencia city; Eximen Perez, lord of Beniparrell; Mossen Otto, lord of Chiva; the Commander of Torrent; and Felip Boyl, lord of Manises.[13]

The results of the meeting were probably inconclusive, since some of the magnates did not bother to appear or send delegates. The jurates wrote to the Abbot of Valldigna on August 18 reminding him that in a letter addressed to him at his village of Almussafes he had been invited to the August 8 meeting at Torres Torres. Many of those who, like him, would have possessions in the irrigation district of the new canal, had appeared, but the abbot had not. They bade him appear himself or send a representative to the Cortes of the kingdom, to be held in Segorbe on September 1; there, together with the syndics of the city, they would discuss the project with the King. The same letter was sent to the Abbot of St. Bernard; to Ramoneta of Riusech, lady of Ribarroja;[14] to Na Çabata, lady of Benifaió; and to the lords of Carlet, Manises, Alaquàs, Antella, and Alginet.[15]

The project seems to have been abandoned again for the

time being, although several years later, in 1404 and 1406, the city was involved in several minor plans for diverting Júcar water into the districts of the city.[16] The major project — that terminating the Plain of Quart — was next discussed in 1415 when yet another report was commissioned by the Valencian council.[17]

One Thursday in September 1500, in front of the Apostles' Gate of the Cathedral of Valencia — customarily the time and place for irrigation business — the jurates of the city paid one hundred sous for "a map on which will be painted the water which will be conducted not only from Castellón but also from the region of the River Júcar." [18] Interest had not disappeared in the intervening years, and indeed, in 1529, another full-scale effort was made to get the project under way. In April the jurates and council engaged Andreu Puig, a mason (*obrer de vila*), to survey the route along which "he says that a canal of water was taken of old from the river Júcar which came to the Plain of Quart" (the meaning seems to be that Puig knew of the fourteenth-century survey and offered to repeat it).[19] For his efforts Puig was offered fifty ducats plus fifty cafizates of land irrigable from the new canal. The work was carried out, and Puig's report was made to the council on December 11.[20]

Puig's report, entitled "Memorial of the level whereby the water of the Júcar can come to Valencia without a diversion dam and without subterranean canalization," [21] begins by recalling the act of the jurates on April 16 when Puig was commissioned to investigate how this water might come from the Júcar "to the cequia [sic] of Quart." Puig found that the level of the river at Tous was 117 palms higher than the level at the head of the Royal Canal and stated that water could best be taken at the foot of the castle of Tous "whether with a large diversion dam or a small one [that is, the water did not have to be raised] and it is the best place in all the river." The course he recommended passed through Antella, Alberic, Rasalany, Alcúdia de Carlet, Algemesí, Alginet, Benifaió, Almussafes, Sollana, Espioca, Picassent, Alcàsser, Silla, Beniparrell, Albal, Catarroja, Benetússer,

Paiporta, Picanya, and Torrent — largely the route taken, that is to say, by the extension of the Royal Canal financed by the Duke of Hijar in the eighteenth century.[22]

The projected canal would terminate 146 palms above St. Onofre of Quart, where it would enter the Guadalaviar. The canal would have a capacity of 350 filas of water, of which 150 would suffice for all the above-mentioned places, leaving 200 for Valencia.[23] "The Plain of Quart," Puig's report continues, "could be completely irrigated in this manner: The water [would be] taken from Pedralba where the Plain of Quart was customarily irrigated of old[24] and the water with which the said plain is to be irrigated must be taken from the river of Valencia in this manner, that is that the four canals — Benacher, Favara, Mislata and Rovella — could be diverted to the Plain of Quart."[25] Since the capacity of these four canals, according to Puig's estimate, was only 114 filas and the new canal would have 200, there would be 86 filas left for the Plain of Quart or for other canals of the huerta.

An even more detailed analysis of the possibilities of irrigating previously dry lands with Júcar water was presented to the council in 1604–1605 by a commission of experts including Jerome Negret, who later served with the commission of Melchor Sisternes, which analyzed the irrigation potentialities of the entire Guadalaviar.[26]

The historian Gaspar Escolano wrote in 1611 that a new canal was being planned, a canal "which has been discussed since the conquest without ever beginning it." In his opinion it would be "a work so fruitful that even Sicily herself might envy us."[27] After tracing the history of the plans for the canal (the episodes of 1365 and 1401 related above), stating that Pope Benedict XIII of Luna had even conceded the benefits of a crusade to those who would help, but to no avail, Escolano continued that in his time the plan had again shown signs of life, "owing to the zeal of a son [of Valencia] named Fonte." It could be, the historian ventured, "that the birth of these so much desired waters might be owing to this Fonte, or fount."[28] The object of Escolano's pun was Antonio Pablo Font, whose efforts in the king's court were hailed in a pamphlet by a polemicist of the same surname, Joan Font.[29] Joan Font's reasoning was primarily economic; he argued that the huerta was then three leagues long by one and one-half across, or four and one-half leagues square in area, but that with the water provided by the new canal the area could be enlarged tenfold to forty leagues. The subsequent economic boom would affect Valencia in two ways: by vast increase of production and, more important, by driving down the price of grain in Valencia, which was the highest in Spain "and even in all Europe." [30] In spite of Font's plea the project again came to nought, altogether interest was never lost completely.[31]

RIVALRY WITH ELCHE AND ALICANTE

Residents of the capital were not the only citizens of the kingdom who coveted the waters of the Júcar. To the south, both Elche and Alicante entertained hopes of augmenting their own seriously limited water supplies (see Map 3).

Elche's first interest in the Júcar was recorded in a series of negotiations in 1420 with the mountain towns of Sax, Villena, and Chinchilla, which lay in the path of the proposed diversion. The council of Elche had sent two agents to Villena and Chinchilla to see if they would permit levelers to enter their territory and to ask if they would share the cost of the survey. The first negotiation was inconclusive, and so another messenger, Jacme Sanchez, of Lorca, was sent in May. On July 14 new letters were sent to Villena and Sax, and their replies were entered in the minutes of the council for July 30.[32] The officials of Villena had discussed the matter but were unable to determine "exactly where it [the water diversion canal] might be able to come." Sax promised its help, but in general terms only.

More letters were written in August,[33] but disappointing replies from Villena and Chinchilla in September seem to have put an end to the scheme. Villena replied that Elche could send all the levelers and men it pleased into the territory, and they would be given "favor and aid," but that "at the present we cannot contribute or give anything owing to the many doubts which

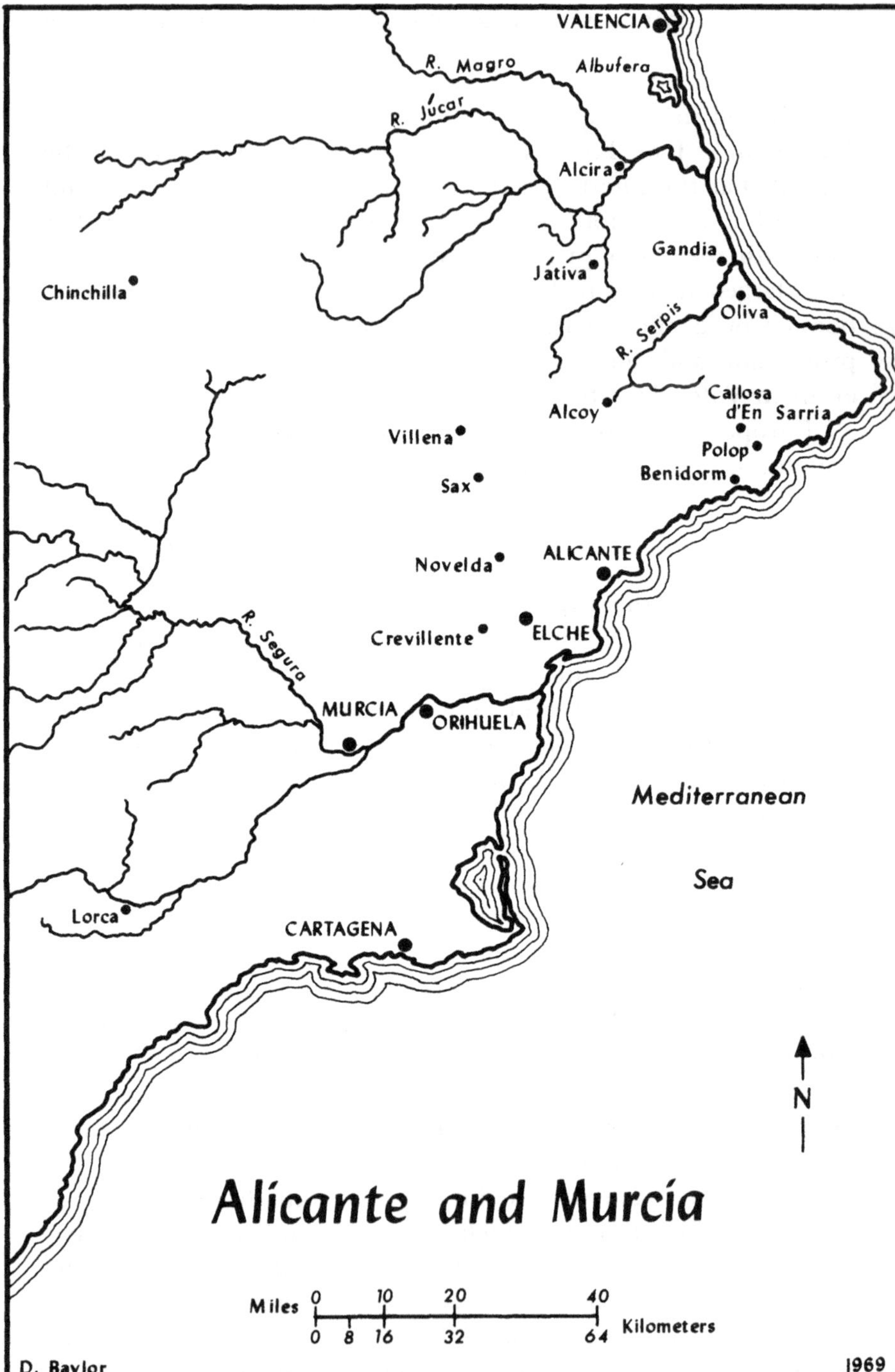
VALENCIA
R. Magro
Albufera
R. Júcar
Alcira
Gandia
Játiva
Chinchilla
Oliva
R. Serpis
Callosa
d'En Sarria
Alcoy
Villena
Polop
Benidorm
Sax
ALICANTE
Novelda
R. Segura
Crevillente
ELCHE
MURCIA
ORIHUELA
Mediterranean
Sea
Lorca
CARTAGENA
N
Alicante and Murcia
Miles 0 10 20 40
0 8 16 32 64 Kilometers
D. Baylor
1969

we have that the said water will not be able to flow through here, but if the water should so be able to flow . . . in due time we will pay, sou for pound, that which is reasonable to pay." [34] Chinchilla, on the other hand, wrote that it would be much pleased if Júcar water could in any way be made to flow through the town, but that at present the townspeople would be unable to act until they had addressed petitions to the King (Chinchilla was part of the kingdom of Castile) seeking license for such action.

Nothing more was heard of this Júcar project until 1528, when the council of Elche ordered Mossen Johan Vaca to undertake any procedures necessary for studying, leveling, or diverting "not only the water of the Júcar River and that of the Segura but the waters of Molinello and of the springs and swamps of Villena and of the Sierra de Crevillente" — and, indeed, of any other source of water which might be made to flow to Elche. A commission was sent to Villena at the same time.[35]

In the midst of a dire drought in 1668 the council called upon the architect Francisco Verde to solve the water problem.[36] Verde suggested that in view of the difficulties involved in bringing water from the Júcar a more modest interim measure ought to be undertaken. The town was in danger of depopulation because of severity of the drought, so Verde proposed an immediate solution: a canal twelve palms wide from the mill of Sax to the spring of Las Virtudes, crossing the lake in between. This canal would be useful when the Júcar Canal was eventually finished and in the meantime would collect the water of the swamps, springs, and drainage channels of Villena and would, in addition, drain the lake of surplus rainwater. The project was begun in 1669 by six groups of ten workers each.

Alicante's interest in the Júcar dated from the early seventeenth century. At the city's behest the possibilities of a diversion were studied in 1628 by Francisco Serrano and Juan Gil, who concluded that the entire project would cost 700,000 pounds, of which Alicante would have to put up only 150,000, the rest to be divided among the other beneficiaries. The council decided to proceed with the plans, but at this juncture the city of

Valencia brought suit against Alicante before the Real Audiencia on the grounds that, by the privilege of John I, Valencia had the right to use this water for the Plain of Quart. In 1640 the case was still in the courts.[37] The project was to have benefited both Alicante and Elche.

Alicante's patently serious intent to go ahead with the diversion plans became the immediate impulse to the Fonts' campaign to convince the Valencians of the importance of this project. Joan Font accused Alicante and Elche of attempting to take Júcar water, which was in fact the property of the city of Valencia, its cathedral chapter, and the lords of the plains of Quart, Liria, and Morvedre. The serious intent of the two southern towns was not doubted: the machinations of Alicante and Elche were well known in Valencia, "and their having syndics in Madrid, begging His Majesty to be able to divert the water, proves it." [38]

Two other potential water sources were of interest to the medieval Valencians. One was the Cabriel, a tributary of the Júcar,[39] from which the city had a privilege (from King Henry of Castile) authorizing diversion of water to the Guadalaviar.[40] In May 1372 a commission of jurates and experts was ordered to proceed "to the place where the Cabriel could be diverted" toward Valencia and to estimate the cost of such a diversion.[41] The other region of interest was the headwaters of the Guadalaviar. The city's interest in the lake of Tortajada in 1457 has already been noted.[42] To the south of Ademuz "near the place of Santa Cruz [de Moya] there was a lake of spring water, which with some expense could be diverted to this river of Guadalaviar," the council of Valencia noted in 1376.[43] Interest in Santa Cruz was revived during the drought years of 1413 and 1415. In the former the council ordered the jurates to proceed to Santa Cruz and, while investigating the water situation there, to deal with the village of Moya about the one hundred measures (*caffis*) of grain which they asked in return for the water.[44] In

May 1415 the jurates met in the monastery of the Valley of Christ, near Puçol, to discuss the matter further.[45]

Three elements that nearly all of these plans and deliberations had in common were, first, that the town councils assumed the overall direction of the projects; second, that drought spurred new ideas about water diversion and revived interest in old ones; and third, that almost none of the proposals ever came to fruition. Both geographical and political factors underlay failure of the large water projects. To administer water even greater cooperation is needed than for other economic and political plans; what happens in one place in a river system necessarily affects other places, often at great distances away. The physical discontinuity of the cultivated areas of the kingdom made regional coordination difficult[46] — coordination which in any case was also thwarted by the typically medieval fragmentation of jurisdictions and interests.

Nevertheless, the failure of grandiose projects should not lead to the underestimation of the extensive development achieved in the medieval huertas under the auspices of urban and autonomous corporate bodies. There is nothing to suggest that the guildlike structure of the autonomous communes or the fragmentation of jurisdictions among regional powers was not compatible with the practice of hydraulic agriculture. For example, the typically feudal concept that obligations of service attach to the soil when transferred to the irrigation community or town provided a socially built-in assurance of canal maintenance that obviated the need for a hydraulic corvée.[47] On balance, the society of the transplanted Catalan townsmen had little trouble adapting itself to the requirements of hydraulic agriculture.

VI | Regional Arrangements

In times of drought there was not enough water in the Guadalaviar to supply concurrently all the canals of the huerta and the canals of places upstream from the city of Valencia. Some agreement was necessary between all users on an equitable partition of the water through a system of turns or by voluntary or compulsory cession of water to downstream users by those upstream.

In August 1313 the jurates of Valencia took action to alleviate the plight of the lower canals of the huerta. Because of two years of drought the water in the river was in such short supply that the uppermost canals of the huerta — Moncada, Tormos, Quart, and Mislata — were receiving most of the water.[1] The upper canals had to yield to the lower ones seven *mulnars* (a measure sufficient to move one millstone) of water, of which three were to be taken from Moncada, two from Mislata, and one each from Quart and Tormos. This additional water was to be distributed in the following way: Mestalla (called "the canal of En Pere Mercer" [2] in the document that evolved) would have all seven mulnars from Monday morning to Wednesday evening, and then from Wednesday evening to Sunday evening the seven would be divided between the canals of Favara, Na Rovella, and Rascanya according to the mutual discretion of their officials. In addition, in order to succor the irrigators of Puig and Puçol, all the canal turn-outs and minor channels (rolls and filas) of the four upper canals were to be closed at night so that more water would remain for diversion to the tail of Moncada.[3]

These provisions, adopted in order to prevent "strife, hatred, and ill will" among the irrigators, from which might follow "harm, wounds, or deaths," were negotiated by all the important interests of the huerta. The jurates had secured the advice and consent (*consell e volentat*) of Raymond Gaston, bishop of Valencia; the chapter; the bailiff of the kingdom; and the members of the council, as well as of several lords and many of the irrigators themselves. Eleven inspectors (*veedors*) were appointed to execute the partitions of the water, "putting aside love, hatred, family feeling, corruption, and bribery." [4] In this case, therefore, the city took the initiative, but not without associating royal, noble, and ecclesiastical authority with its own.

As the result of another drought, James II sought to codify arrangements for division of the river's water in three separate privileges in 1321. The first, that of April 24,[5] was simply a refinement of the provisions adopted by the city in 1313 and was the result of a conflict between the landholders of Mestalla, Favara, Rascanya, and Na Rovella (here called Russafa because it flows through that suburb) and those of the Moncada Canal. The privilege provided that in time of need one of the four control gates of the almenara of the diversion dam of Moncada would have to be opened, thereby letting one-quarter of Moncada's water run to the lower canals, for a period of two days and two nights (Monday and Tuesday). In cases of even greater need two gates were to be opened.

James's second disposition, that of May 25, provided that in time of drought the Pueblos Castillos, that is, the villages upstream of the Moncada Canal (Pedralba, Vilamarxant, Benaguasil, and Ribarroja), would have to cede their water to the huerta "on certain days." [6] The third privilege, that of July 5, specified the form the turn was to take: the Pueblos Castillos would have all the water for four days continuously and the Valencian huerta would have the water for the next four days, at the completion of which the turn would begin again.[7] The Moncada Canal's relation to the city of Valencia resembled that of the Pueblos. Economically the Moncada system was an integral part of the huerta, but because of its initial designation as a royal canal it remained independent of the lower huerta and was recognized as a unique and separate jurisdictional entity. Most of the city's attention to it was concentrated on securing the release of some of its water on behalf of the lower canals.[8]

The King's provisions did not mean that other agreements could not be negotiated. In July 1401 the jurates of Valencia wrote to the officials of La Pobla de Vallbona (which came to be considered one of the Pueblos Castillos along with the four other villages mentioned in the privilege) and to the *alcayt* and elders of the largely Morisco village of Benaguasil that the city of Valencia — whose well-being was the concern of the whole kingdom, since Valencia was "mother" to it all — was suffering a water shortage. The city agreed with the cequier of Moncada that Moncada should have water for three days and Valencia for four. The arrangement would go into effect on the Saturday following but would have little value if the city could not have all the water of the river; the Pueblos Castillos were asked to take their water on the three days assigned to Moncada and to let this order last for three or four turns.[9] Similarly, in July 1415 the jurates secured from the Pueblos Castillos an alteration in the stipulated turn of four days and four nights by which the Pueblos granted the city one of its four days free (*graciosament*), so that the city now had five days and the Pueblos three.[10] This arrangement was completely ad hoc and was to be effective for only one turn. In both cases the jurates were clearly acting as *sobrecequier* of the huerta,[11] dealing with the Pueblos Castillos and the cequier of Moncada on behalf of the cequiers and irrigators of the other seven huerta canals.

THE CITY AS *SOBRECEQUIER* OF THE HUERTA

The city acted as *sobrecequier* in two ways. First, it represented the huerta in dealings with upstream irrigators — Moncada, the Pueblos Castillos, and others — and in matters of water control, as on tributaries of the Guadalaviar, when the "city's rights" (that is, the huerta's water supply) were threatened. Second, the city frequently acted within the huerta itself, to give "norm and form" to the cequiers in matters where the public good was concerned.

The decision to implement the turn provisions embodied

in the privilege of James II was taken by municipal officials. In July 1368 the jurates, noting that the river water was so low that only one or two of the uppermost canals of the huerta were supplied and that the mills of the city were unable to run, declared that implementation of the turn was necessary. The Pueblos Castillos were unwilling to comply, defending with arms their use of the water. In response the council held that, should the upstream villages continue in resistance and fail to release the water for four days, the city would preserve its "ancient possession" by force of arms. A crida was issued ordering all foot and horse soldiers of the city to stand in readiness to follow the banner of the city to defend the fueros.[12]

Even when the turn was not formally instituted the jurates were constantly writing to upstream officials to ask for water for the huerta. One of the earliest letters extant on the subject is to Pero Çapata, lord of Tous and *alcayt* of Benaguasil in 1335, stating that because of lack of water in the river the crops of the huerta were in great danger of being lost "if God and good friends did not come to the rescue." [13] Water was asked of the Pueblos for three of four days; but, although "a certain penalty" was mentioned, no appeal was made to the right of the city by James's privilege. Instead the hope was expressed that the lords and *alcayts* of the Pueblos would "observe the good duty of neighborliness." [14] The turn was to last as long as the drought prevailed, until the crops of the huerta could be restored. Similarly, in other drought years the city wrote one or more of the Pueblos asking for a certain percentage of their water or for an extra day.[15]

Valencia also had an interest in the Pueblos Castillos as a place to mill flour when water shortage prevented milling in the huerta; the city would send men upstream with grain and seek permission to mill it where hydraulic conditions were favorable.[16] Further, the huerta's wood supply came from the mountainous hinterland and was floated down the river. City officials had therefore to ensure its safe passage past the diversion dams of the Pueblos Castillos — if damage were caused to these dams by the city's wood merchants, the city was responsible.[17] If there was not enough water in the river below the Pueblos Castillos to float the wood to its destination, water was asked of the officials of those places.[18]

Even when the huerta was not directly concerned, the city, as mother of the realm, often intervened directly in the internal affairs of the Pueblos Castillos, usually on petition of Pueblo residents themselves. In 1437 the complaint was made by inhabitants of La Pobla de Vallbona to the Valencians that some Moors of Benaguasil, and others in higher land, had been growing rice, the drainage from which was creating pools of stagnant water (*ayguamolls*) in the fields of lower irrigators. The latter had kept their lands "well plowed and well fallowed" but because of the rice drainage were unable to sow. In order to prevent further strife, the jurates ordered the officials of Benaguasil to cease growing rice in areas that might give rise to the cited conditions. To this end a crida was ordered; if compliance with the city's will did not result, the rice would be torn out.[19] On other occasions it was the Moors of Benaguasil who complained to the city about irrigation practices of La Pobla, as in a dispute of 1454 "concerning the irrigating of vegetables on the canal of El Alguatzir." In this case the jurates directed Mossen Jacme Ferrer, *alcayt* and procurator of the city in Benaguasil, La Pobla, and Paterna, to go personally to that canal and inform the jurates of the situation.[20] The jurates were also involved in disputes between Benaguasil and Vilamarxant[21] and Benaguasil and Ribarroja.[22]

The interests of the jurates extended even farther upstream than Pedralba — to Bugarra, Chestalgar,[23] Chelva, and even as far up as Teruel.[24] The city intervened in the more remote areas particularly when diversion of water from the river threatened the huerta's supply. The city had men stationed upstream to guard the water[25] and sent frequent commissions to check on the water supply.[26] Sometimes municipal initiative was not enough, and the city would request the governor to send his representative upstream to demolish new irrigation works built contrary to customary practice.[27]

The city sought action against Chelva, not only for taking water directly from the Guadalaviar, but for impeding the water of gullies and spring which flowed into the Guadalaviar.[28] So attuned was the city to challenges of water rights that it intervened even where its own rights were not directly challenged. Valencia was suzerain of Morvedre and on occasion challenged the city of Segorbe for diverting water contrary to the "ancient use" of the inhabitants of Morvedre.[29]

Closer to home the city's interest was more general than in water supply and distribution, the details of which in any case were the responsibility of the cequiers of the individual canals. In areas marginal to the huerta proper — such as the Marjals — we have seen the city's exercise of control. But in the core area the jurates and council did act to give norm and form to the cequiers, as when they set the order of crop priorities in 1376[30] or when in 1373 they overrode the cequiers and granted general permission to wash cloth and wool in all canals.[31] The city also called general meetings of the cequiers to discuss "the affairs of the canals." [32]

ROYAL JUSTICE AND JURISDICTIONAL AMBIVALENCE

At one time or another all important local and regional powers became involved in the direction of irrigation. The function of royal justice should be considered, particularly the role played by the king's officials in Valencia, the bailiff (*batle*) and the governor.[33]

In certain instances the king intervened by giving a direct order to the cequiers; and hereters sometimes appealed directly to the king after appeals to the cequiers, the city, or royal officials had failed. In 1283 Peter III ordered the cequiers of Moncada and Favara not to let so much water run through their canals, in prejudice against the hereters (of the lower canals, one assumes), "because more than one *mulnar* of water can thus be lost in the swamps." [34] Later, in 1342, the magnates and nobles of the city and other villages of the realm complained to Peter IV of the power of the cequiers: the latter wronged them, they complained, giving the right to water to others, "and from their judgments there is none to whom recourse may be had." [35] The jurates of the city and the syndics of the royal towns (Burriana, for example) opposed the proposal of the nobles that each cequier be required to submit to an annual public inquest conducted by appointees of the lords.

On the question of cequiers' jurisdiction, the jurates always backed the cequiers. Upon petition from irrigators of the Moncada Canal in 1374 the King was about to promulgate two letters which, according to the jurates, were contrary to the fueros and for that reason had better "remained in his pen." If the water was to be well administered "and with equality, all will be sufficed with a good supply, but the direction has to be at the disposition of the cequier of Moncada, according as he wants to do it . . . wherefore, we beg you to revoke the said letters." [36]

The irrigators of the two lowest canals, Rascanya and Na Rovella, were in a position similar to that of the villages of Puig and Puçol at the tail of the Moncada Canal. Although the huerta canals were supposed to enjoy equal rights and to receive their traditional quotas of water according to the fueros, the practice nevertheless often worked to the disadvantage of those farthest downstream. James II's privilege obliging Moncada to cede one-fourth or one-half its water was granted in response to the petition of the four lower canals. On at least one occasion Rascanya and Na Rovella alone appealed to the king: claiming that the water of the river below the diversion dam of Moncada was "poorly divided by the cequiers of the lower canals," their hereters made such a "clamorous protest" (*clamosa insinuacio*) that King Martin, on December 3, 1390, ordered the Governor and Lieutenant Governor of the kingdom to appoint a commission of experts to investigate the claim. The commission was duly appointed and the experts surveyed the dams and canals and ascertained how many jovates of land were irrigated by each. Before the matter could be brought to conclusion, however, the Governor, his Lieutenant, and the King had all died; the suit was renewed under Alfonso V in 1432.[37]

The bailiff was the administrator of the Royal Patrimony. By a privilege of James II of May 1, 1321, he was forbidden to interfere in matters of irrigation, the jurisdiction over which was left solely to the cequiers of the canals of the city of Valencia. The bailiff, however, retained jurisdiction over anything touching the royal interest, such as water flowing to the Royal Palace or the water of the mills of the huerta, which were subject to royal emphyteusis.[38] Prior to 1321 the bailiff had apparently had some jurisdiction in irrigation matters. In 1310 the bishop and chapter complained that the bailiff had given water from the Moncada Canal to Russafa (that is, Na Rovella) to the detriment of Puçol. The bailiff replied by stating that "through long practice, by Moors as well as by Christians," it had been the custom to give water through the almenara of Moncada for the sewers (*valls*) and canals of the city of Valencia. This was assured by privilege; thus, upon request of the hereters of Russafa he had gone to the almenara of Moncada and let the water run to the city "in order to comply with the form of the privilege." [39] Even after 1321 the bailiff retained a measure of power over the Moncada Canal. Peter IV ordered in 1358 that in times of scarcity of water the jurates would divide the water among the canals of the huerta except for Moncada, whose water would be apportioned by the bailiff.[40] This measure was apparently taken to tighten control on Moncada, whose prime location ensured its getting more than enough water — so much, in fact, that even in times of drought a great part of its water flowed into the sea.[41]

In spite of strictures against the bailiff's interference with the seven lower canals, his concern with mills was sufficient to involve him with the cequiers of the huerta. In 1430 Pere Vinyes, cequier of Favara, complained to the Governor that he had received an order from the bailiff to repair the diversion dam "so that water of the river might enter the said canal in order that the mills which are built upon it can mill" under penalty of ten florins. The cequier averred, with due respect, that jurisdiction in such a matter does not pertain to the bailiff, nor can the bailiff interfere with the office of cequier, inasmuch as this is expressly forbidden by the fueros, notably that of James II of May 1, 1321.[42]

The governor was the chief royal officer in the kingdom.[43]

Throughout the Middle Ages he, not the bailiff, heard cases on appeal from the cequier's oral process.[44] As we have seen, the governor's court heard a wide variety of irrigation disputes, both those that were intercommunal and those dealing with the affairs of only one canal. On occasion the governor executed a royal order, as when James II ordered the procurator general (the title "governor" was not used until 1390) to destroy new diversion dams and canals that had been constructed in the Guadalaviar, returning conditions to their former state so that water could pass freely to Valencia and its huerta.[45]

The governor also heard cases involving millers: typically, the miller would ask the governor to prevent the cequier from interfering with operations of his mill.[46] This request often led to a discussion of who — governor or cequier — had jurisdiction in such cases. In 1446 the procurator of the Cathedral chapter, initiating a suit against Jacme Perfeta (owner of the fulling mill called En Ferrada, on the Mestalla Canal), stated that this dispute "ought to be argued before you [the Governor] and your court [because] you are judge ordinary of the matters which are disputed between the [two] parties, since indeed they have to do with mills, canals, waters, and their free flow." [47] Such cases were heard by the governor under the terms of a privilege of James II (September 11, 1300), according to which the governor heard appeals from all officials of the realm, cequiers included.[48]

The civil justiciar, as well as the bailiff, was prohibited from intervening in irrigation affairs.[49] James II ordered, on two separate occasions, that when violence or any other unjust action was committed in matters of canals and water only the cequiers and *veedors* should have jurisdiction.[50]

INTERCOMMUNITY PROBLEMS OF LA PLANA

The River Mijares rises in the Sierra de Gudar in the province of Teruel and flows through the province of Castellón and into the sea at Burriana.[51] The littoral plain it waters — La Plana — extends along the coast for eighteen kilometers at a width of five to eight kilometers. The major towns of the plain are Villarreal, the diversion dam of whose irrigation canal was the farthest upstream, on the right bank of the river; then Castellón and Almassora, on the left bank, whose common canal drew water from the Mijares at a point just upstream of the confluence of the river and the Rambla de la Viuda; and finally Burriana, whose dam was on the right side in the administrative district of Villarreal and the ducal village of Nules, which was irrigated both from the excess water of Villarreal and by a fixed turn from the Cequia Sobirana of Burriana.[52]

The huerta of La Plana was about one-third the size of the Valencian huerta and supported an urban population roughly two-thirds that of Valencia's. This urban population was divided among the towns of Castellón, the main regional center; its small satellite Almassora; and the towns of Villarreal and Burriana, whose combined population equaled that of Castellón.[53] The fragmentation of political and hydraulic power imposed upon the towns of La Plana the necessity for cooperation no less than an obvious potentiality for conflict. Castellón was not powerful enough to impose its will in water affairs on the other towns. Burriana, as the town farthest downstream, tended to oppose Castellón, Almassora, and Villarreal in river politics. (Burriana might have found an ally in Almassora, on the opposite bank of the river, but since Almassora shared its diversion dam with Castellón any structural predilection for siding with Burriana against Castellón was neutralized. Almassora originated her own share of disputes against Castellón, however.)

Rivalry among the La Plana towns was exacerbated by overlapping and confusing municipal and rech jurisdictions. What happened when these town-owned canals extended beyond municipal bounds, irrigating properties in another town's jurisdiction? To whom did the irrigator owe first loyalty, the town in which he lived, or that which controlled his water supply? For example, the cequia maior of Villarreal irrigated a small section of the district of Burriana called Els Vintens; similarly, the cequia maior of Castellón irrigated lands owned by residents of both Villarreal and Almassora. Under such conditions administration became greatly complicated. In the case of Castellón and Almassora the cequier of Castellón and his guards patrolled Almassora as well. Each year after the election of these officers the council dispatched a letter to the jurates of Almassora certifying the election.[54] But enforcement powers of the officials were hampered by the cequier's having to seek permission to impose penalties from the justiciar of Almassora.[55] This interposition of another authority between the cequier and the delinquent irrigator defeated one of the main principles of irrigation justice, namely that it be summary and speedy in order to promote the highest possible level of efficiency in water distribution.

Litigation among the four towns was constant throughout the Middle Ages, and it is clear that Burriana was the prime mover in initial steps toward a regional settlement. In order to put an end to conflict among the competing towns, James I guaranteed Burriana the right to take water from the canals of Castellón and Almassora for six consecutive days and nights in time of drought. Each town was to elect representatives to adjudicate disputes, and if they should fail to arrive at a decision two men from the neighboring town of Onda would arbitrate.[56]

The settlement that eventually governed hydraulic relations of the four towns from 1346 to the present was also the result of Burriana's initiative. On February 28, 1346, the syndics of Villarreal, Castellón, and Almassora on one side and the representatives of Burriana on the other agreed to submit their differences to the arbitration of Prince Peter, count of Ribagorza and a son of James II. On March 20, having heard the arguments, Peter proposed a two-stage division of the river similar to the concepts which governed division of the Guadalaviar. When water in the river was short it was to be divided into sixty filas above the diversion dam of Villarreal:[57] Villarreal received fourteen, Castellón fourteen and a half, Almassora twelve and a half, and Burriana nineteen.[58] But if the water was so short that less than one fila was reaching Almassora, then the second stage automatically went into effect. In these dire conditions all the water of the river was to be directed into each of the canals in turn, Villarreal receiving the water for twenty-eight hours continuously, Castellón for twenty-nine, Almassora for twenty-five, and Burriana for thirty-eight. That is, the water in times of severe drought was "doubled" (the ratio between hours and filas here is two to one).

The call for partition could be given by any of the four towns, by notifying the jurates of the others. In case the jurates could not be found, notification also could be given with a public writ in the presence of four or five persons of the town in question. Exactly one day after the hour of notification the towns were to send an elected *partidor* to the place of partition.[59]

The sentence was ordinarily put into effect by a letter to the jurates. Burriana instituted a partition in March 1415 by writing to the jurates of the other three towns, asking them to send their *partidors* to the accustomed place on the following Monday morning "to give each party his right" according to the terms of sentence.[60] It is clear, moreover, from the wording of the letter that Burriana still considered the sentence as an agreement not *among* the four towns but as between Burriana and the other three. As the party farthest downstream Burriana stood to gain most from implementation of the sentence.

The opposition of interest between Burriana and the other three is clear, too, from an incident in 1473 when it was found that the diversion dams of Villarreal and Castellón-Almassora were full of holes (*sportellats de molts portells*) and the control gates raised in such a fashion "that all the water was found in the canal of Burriana." It was no wonder that there should have been "some feeling that men of Burriana had done it." The council of Castellón considered the outrage and ordered that one man from Almassora be chosen to represent the three towns and to discuss the matter with the towns' lawyers in Valencia.[61]

Disputes also arose between the towns individually, as between Villarreal and Almassora over the right of residents of the former to take wood, brush, and stone from the banks of the river in the territory of the latter[62] and between Villarreal and Burriana concerning jurisdiction over residents of the former who irrigated fields from canals of the latter.[63] Relations between Castellón and Almassora and between Burriana and Nules were codified in long series of sentences — arbitration was the commonest way of settling intercommunity disputes in La Plana — and their problems resembled those of the head and tail villages of the longer canals of the Valencian huerta. The key point of dispute between Castellón and Almassora concerned the number and location of canal checks with which Almassora was allowed to divert water from the canal of Castellón.[64] That part of the canal passing through Almassora was administered by the irrigation officials of Castellón. The canal of Nules, though fed by the Cequia Sobirana of Burriana, was autonomously administered, and consequently the central factor in the relation between the two was a system of turns.[65]

The four towns preferred to keep control of their own affairs and to settle regional problems among themselves, especially by the device of mixed commissions. While ecclesiastical and noble personages were sometimes sought as arbitrators, their direct intervention was resented. The jurates of Burriana complained to those of Castellón in 1430 that the King had appointed a resident of Nules as cequier of Burriana, a move which, in the view of the jurates of Burriana, was harmful not only to their town but to all the towns of the plain taking water from the Mijares.[66] In more than one instance the council of Castellón acted to assert the authority in irrigation affairs of the jurates and municipal cequier over that of the bailiff. In April 1410 the cequier had been ordered by the town bailiff to revoke an order to the miller Domingo Dolç to close the intake gate of his mill. The council ruled that the bailiff had no jurisdiction and that the cequier was the judge ordinary of the canal.[67]

In spite of the commercial and political pre-eminence of Castellón, Burriana, the largest irrigation district and the one farthest downstream, was the most frequent protagonist in the water affairs of La Plana. Burriana instituted turn procedures and was the prime mover behind regional settlements. Indeed, in most of the watersheds of eastern Spain the power to settle water disputes was concentrated downstream — a response, in part, to the challenges from upstream users. Even in a small water district like the Vall de Segó authority in irrigation gradually passed from the regional political and economic center, Morvedre, to the village of Benifairó, the lowest irrigating district in the valley and the one that received the most water.[68]

The effective division of power in La Plana between Castellón and Burriana was in the long run a stabilizing factor. A similar situation prevailed in Valencia. With the exception of only a few acute crises — which forced readjuments in the power balance and initiated long periods of stability — the balance between the city, the Moncada Canal (in an intermediate position), and the upstream coalition was a factor promoting stability. The city's great political power acted as counterbalance to the leverage the Pueblos Castillos wielded by virtue of geography. The "tyranny" of the Pueblos Castillos, of which nineteenth-century Valencians complained,[69] became a real tyranny only in relatively recent times, when the submersion of local authority in a modern state significantly reduced the freedom of action enjoyed by the city in medieval times.

VII | The Huerta in Crisis

Clearly defined wet and dry seasons lend a characteristic rhythm to the Mediterranean year. The most striking change in the climatic cycle is an autumn break, which brings the dry season to an end. The percentage of rain days rises from the middle of August on, with cyclonic penetration into the northern Mediterranean in about mid-September.[1] The Valencian region is characterized by a single maximum rainfall[2] occurring usually in September but sometimes in October or November (see Appendix 5a). Valencia averages 56 rain days per year, compared to 72 for Barcelona, to the north, and only 42 for Cartagena, to the south.[3]

The transition to dry summer weather is more gradual than is the autumn break. The probability of rain continues close to the winter maximum until March and then lessens until the beginning of the dry season, usually in the latter half of June.[4] In an average year about seventy percent of total precipitation occurs between September and February, whereas July and August receive less than five percent of the total. There is extreme irregularity in monthly distribution of rainfall from year to year; and years with above- and below-average rainfall tend to cluster.[5] The dry season proper includes the second half of June and July, August, and the first part of September — months with uniformly high temperatures, fairly predictable from year to year (see Appendix 5b).

The opening weeks of the dry season were particularly difficult for the Valencian farmer because in June the fields in stubble from the winter wheat harvest had to be prepared for the summer sowing. Rain was so important at sowing time that the Valencians gave traditional names to the two or three rainstorms that typically fell during this period. Around the octave of Corpus Christi (nine and a half weeks after Easter) rain often fell with such force that it dissolved and carried downstream the red earth of the Guadalaviar's headwater regions, turning the water red. The river water at this time (early to mid-June) was called the Blood of Picanya (*la sang de Picanya*), after a huerta village which celebrated a festival of the Blood of Christ. The day of St. Peter, June 29, was associated in the popular mind with hailstorms, called St. Peter's Hail (*les pedres de Sant Pere*). Finally, driving rains (occurring in mid-July, around the feast of Saints Justa and Rufina, the potters) produced red water identified as the Blood of the Holy Plate-makers (*la sang de les Santes Escudelleres*).[6]

Seasonal distribution of drought (defined by Alice Foster as a period of two months or longer with a total rainfall of less than one inch) also varies widely in the Valencian region. Of sixty-nine droughts occurring between 1866 and 1926, thirty fell wholly within the period from June 1 through August 31; an additional seven began in the summer and lasted into September; and the rest were scattered throughout the remaining months.[7] The unequal distribution of rainfall, both monthly and yearly, is reflected in the water table and in the river regimes. The rivers of the kingdom of Valencia have a meager flow of 0.5 to 41 litres per square kilometer, with a late spring maximum and a secondary maximum in the autumn; moreover, the flows of the Mijares, Turia, and Júcar are complicated by karstic phenomena, making them all the more irregular. Floods, notorious in these rivers, tend to occur in February with minor ones in the late fall.[8]

HISTORICAL EVIDENCE OF DROUGHT

The fourteenth century and the first half of the fifteenth were marked in Europe by severe climatic disturbances in the form of catastrophic floods and droughts. The climate of the fourteenth century was "peculiarly continental." That is, western Europe, which usually feels the mitigating effects of the Atlantic (its "oceanic climate"), in that terrible century suffered the freezing winters and wet summers characteristic of central Europe.[9] In Valencia these disturbances took the form of frequent droughts, often lasting more than a year, and of fairly frequent floods and torrential rains, quite a few of which occurred in close conjunction with droughts.

Appendix 6 unites all the documentary evidence of droughts and floods in medieval Valencia known to me, up to the mid-fifteenth century. The drought data are based largely on municipal documents, which are only sporadic before the second quarter of the fourteenth century. Furthermore, most of these references occur in documents relating to specific problems of water use, such as irrigation, milling, and related hydraulic matters. Another type of reference, alluding to public prayers or rogatory processions for rain, does not occur in significant numbers until the 1440's and increases in frequency thereafter; these calls for public prayer soon became highly stylized (for example, identically worded letters to a stock list of religious houses beseeching the monks or nuns to pray for rain), and therefore I have preferred to end my tabulation of droughts in 1450. To have continued, recording a rapidly mounting percentage of procession references among all drought references, would have been to present material of different, and lesser, significance.[10] The word "drought" is used in this study not to describe objectively a climatic phenomenon, but to indicate shortage of water relative to the demand for it. Greatest reliance is placed, therefore, on citations alluding directly to demand.

The total number of drought and flood references contained in Appendix 6 is much too small to be statistically significant. But the distribution by months seems, *grosso modo,* to follow a pattern similar to the modern one described above. October is high in flood citations; May, July, and August are highest and September is lowest in drought.

As in modern times, droughts afflicted the medieval huerta in clusters of years, striking some decades severely and others not at all. Heavy rain tended to occur in apposition to drought. Thus, both drought and flooding is recorded in 1321; drought in 1351,

1352, 1355; and flooding in 1356 and 1358. The first two decades of the fifteenth century were extremely harsh, with drought recorded in 1400, 1401, 1403, 1406, and 1412–1415 and heavy rains in 1403, 1406, 1410, 1413, 1415, and 1417. The lack of recorded climatic disturbances in the 1380's and 1390's seems to indicate that these were fortunate times. On the other hand, this picture is primarily a function of the state of demand for water, not of the hydrological conditions themselves. Periods of less recorded demand may well have been arid; but no droughts would have been reported if there had been, for example, a decline in population and concurrent shrinking of the irrigated area.

The profiles of the two best-documented droughts in the fourteenth and fifteenth centuries illustrate the irregularity of the Valencian climate. The drought of 1374–1376 (Figure 4) is first cited in May 1374; it intensified during the summer and was mentioned again in October, typically a month of heavy rains, when the city fathers were moved to initiate a search for new water sources. References in October, February, and March indicate difficulties at the time of sowing and the period of greatest growth of winter cereals. Diversion projects attest concern for the seriousness of the situation. This dry spell lasted at least until July 1376, when arrangements for turns on the river were made.

The drought of 1412–1415 (Figure 5) also began in October and increased in severity through the spring until it reached a critical peak in July. Heavy rains in October apparently did not relieve the pressure, for another reference occurs in November. This particular series of dry years was evidently brought to an end by consistent, heavy rains in September and October of 1415.

PORTRAIT OF A DROUGHT (SUMMER 1413)

Extended drought thrust the huerta into a state of acute tension. As the pressure for water increased, so did the activities of the town fathers as they sought to ease the crisis with a variety of measures, legal, political, and technological. We have already examined some of the political and legal arrangements through which the city of Valencia attempted to assert the economic

Figure 4. Drought of May 1374–July 1376

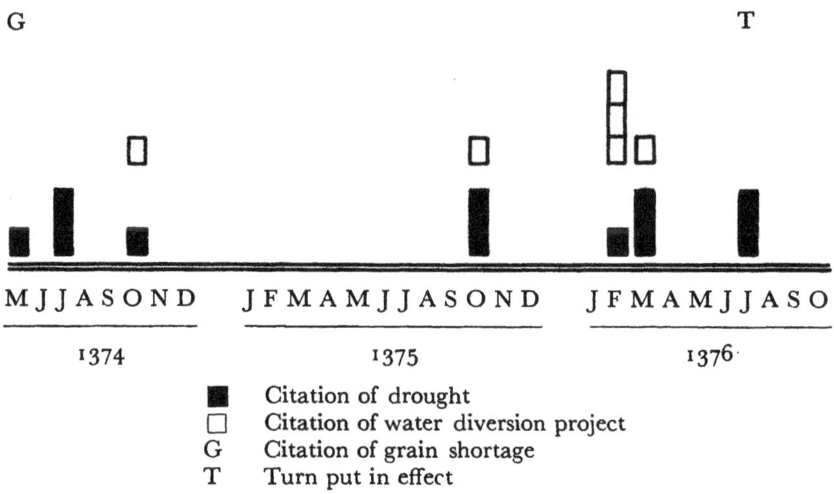

Figure 5. Drought of October 1412–October 1415

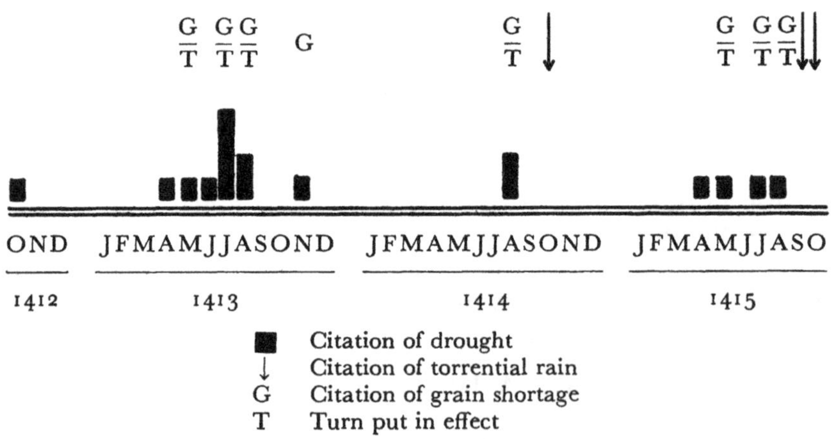

primacy of her huerta and its priority of right to the water of the river. So also did the city, usually spurred to action by drought, take the lead in exploring the entire region in an attempt to locate new sources of water for the huerta's benefit.

Here a specific drought is examined in detail in order to illustrate the interplay of the city's interests, as well as the quickening pace of developments as the drought progressed. After four dry years at the beginning of the fifteenth century,[11] a largely drought-free period followed until the autumn of 1412, when water in the river was said to be low.[12] This drought lasted throughout 1413 and recurred in the summers of 1414 and 1415. The summer of 1413 was apparently the most severe of the three (see Figure 5); document citations began in mid-April and continued with increasing frequency throughout the summer and into the fall. The developments will be followed here day by day, insofar as possible; the documents speak for themselves.

April 19. The jurates of the city wrote to the cequier of Moncada that "great necessity constrains us, not only because of scarcity and lack of water but also because of diminution of flour for the provision of the city, the more so because, owing to the imminent visit of the king, we must make plans for having a great abundance of flour." The jurates accordingly asked the cequier to let them have Moncada's water on the following day and throughout the night, assuring the cequier that no prejudice to the rights and privileges of the Moncada Canal was intended.[13] It should be stressed that the lack of flour did not signify crop failure: comparatively little bread grain was grown in the huerta, and grain for the city's needs was largely imported. The problem was that when water was short the mills of the huerta were often unable to function. Municipal documents indicate that the city was very dependent on water mills (hand and animal mills are never mentioned), but this may reflect a bias of the sources. Millers and bakers — the former especially were men of considerable power — would naturally tend to make their voices heard in the council and would be among those pressing most strongly for enforcement of river turns.

May 2. The syndic of the canals of Benacher, Faitanar, and Quart requested the Lieutenant Governor to send a letter to all officials and nobles between Alaquàs and Manises ordering them to submit to the jurisdiction of the two canal guards Marti Darmelles and Miquel Ametler.[14]

May 5. The jurates of Valencia dispatched a letter to Garcia dé Montagut, the *alcayt* of Ribarroja, to Mossen Pere de Moncada, lord of Vilamarxant, and to the officials of La Pobla, stating that bearers had been sent with the city's wheat to be milled in the mills of the above-mentioned places so that the city would have an abundance of flour.[15]

May 25. The subsyndic of the city of Valencia reported that a new canal, taking water illegally from the Guadalaviar, was being constructed in the Valley of Chelva. He asked the governor to send officials to destroy these works and order the men of Chelva not to place any impediments in the gullies and springs flowing into the river.[16]

May 30. The lieutenant governor issued a crida to the effect that work had begun on a new canal in the Valley of Chelva, taking water that usually ran into the Guadalaviar, which river "flows to the canals which irrigate the entire huerta" of Valencia. He ordered the Lord of Chelva not to undertake any more work on this canal for any reason, under penalty of ten thousand florins.[17]

June 3. The subsyndic reported to the council of Valencia that he had carried out the orders of the jurates in seeking an injunction from the governor against the Lord of Chelva, who had been attempting to build a new canal, to the mortal danger of the city — the more so "in this time of great drought and penury of water." [18]

June 6. The Valencia jurates wrote to the officials of Ribarroja, Benaguasil, and Vilamarxant that they were sending their representative Pere Colomines, "well-informed of our intent"; they begged the officials to "give him faith and belief, especially concerning that which he tells them pertaining to the said matter of the water, and to obey him by works, until the matter is finished." [19] What the "said matter of the water" was is nowhere specified, but it must have concerned the rotation arrangements between the Pueblos Castillos and the huerta.

June 10. The council decreed that the jurates and the city's lawyers should review the report of the leveler Garcia Mezquita about bringing water from the Júcar at the diversion dam of Antella.[20]

June 14. The jurates of Valencia, extending their concern beyond the huerta proper (the area between Puig and Puçol and the Morvedre irrigation district), wrote to officials of the town of Segorbe about certain innovations in the diversion of the water of their river that had given rise to "depopulation and destruction of the places of Altura, Les Alcubles, and even the monastery of the Valley of Christ," whose residents were citizens of Valencia "in certain cases." The jurates had sent their agent, Pere Colomines, to gather information on the scene personally.[21]

June 16. Segorbe replied to Valencia that in fact Altura and the monastery had been able to irrigate, whereas several small districts of Segorbe (Castajo, La Loma, and Lagunas) had not irrigated at all and their grain and grapevines had been lost.[22]

June 23. The council of Valencia ordered the jurates to make a study of the water that could be diverted from Santa Cruz de Moya to the Guadalaviar and of diversion dams necessitated in the river by such an eventuality.[23] On the same day the jurates wrote to the prior of the monastery of St. Jerome, near Puçol, stating that they needed the services of "that brother of yours who knows how to level and conduct water, wherefore we beg you to send him to us so that he might be here with us tomorrow, before lunch." [24]

July 4. The subsyndic of Valencia asked the Governor or his Lieutenant to go up the river within the kingdom of Valencia and, in those places "where new diversion dams and new, or widened, canals" have been made, to destroy such works and see that the water flowed to the huerta.[25]

July 13. The jurates of Valencia wrote to the jurates of Paterna that they were sending three bakers to mill grain at the mill of Paterna, since the city was short of flour.[26]

July 14. The jurates of Valencia wrote to the officials of Benaguasil and Vilamarxant that they were sending the subsyndic to talk of "pressing affairs." [27] Another letter, to the Lady of Manises, spoke again of the great need for flour especially among the city's bakers, for making bread and bread dough "on which a great part of the people of the city live." The jurates asked her to grant the use of a millstone to all the bakers of Valencia who might come to her mill.[28] The flour shortage was so severe that the jurates also wrote to the officials of Alcira. They had heard that the intakes of the Júcar's diversion dams were empty so that Alcira's mills were also stilled. The Valencians asked the Alcira officials to find a way to get their mills working again, in which case Valencian grain could be sent there by beast to be milled. As the Júcar's capacity was greater than that of the Guadalaviar, the Valencians hoped that the level, though low, could be raised artificially and that Alcira's mills could supply flour to "the great multitude of people[29] who, by divine grace, inhabit the city." [30]

July 24. The jurates wrote to Jacme Serra, a fellow jurate who was out of the city: "We are worried lest you do not come, for the pressure here which we suffer owing to lack of flour due to shortage of water is great. And as you are not here, they will make us ransom [that is, pay dearly for] [the grain] whether milled or not; for this reason we pray you try to be here sometime this week, ordering and decreeing, so that . . . you might have communion and participation in our anxiety and pressure and so that we might be able to defend ourselves from riots or give some remedy and alleviation to the people." [31]

July 26. The council of Valencia again considered diverting water from Santa Cruz de Moya and empowered the jurates to negotiate and terminate the matter once and for all in the best interests of the city. Money from the General Fund was allotted to this end.[32]

July 28. The Valencia jurates wrote to the bailiff of the Barony of Oliva, reminding him of their present anxiety about the flour shortage and stating that they had heard that there was abundance of water in the barony "and many mills milling." They asked the barony to supply them with as much flour weekly as it could.[33] At the same time another letter was written to the Lady of Manises, again asking her to cooperate with the bakers of the city.[34]

July 31. The council considered appointment of a commission to investigate bringing water of the Júcar to the Guadalaviar through a siphon at Alcira.[35]

The first phase of the summer's drought crisis was marked, first, by anxiety over the increasingly short supply of flour due to the inactivity of the huerta's mills, and second, by a variety of plans for augmenting the water supply. In August the anxiety of the jurates proved well founded, as tension erupted into violence and the city had to impose a solution by force.

August 16. On this Wednesday the jurates asked Johan Scriva, the lieutenant governor, to accompany them upstream to La Pobla, Ribarroja, and Vilamarxant so that the city could have all the water of the river freely on the days authorized by royal privilege.[36]

August 17. On Thursday morning Scriva, Bernat Gili, subsyndic of the city, and two jurates, Jacme de Celma and Johan Valleriola, were present in La Pobla; Scriva ordered La Pobla's officials not to take water from the river, but to leave the almenares open during Valencia's turn under penalty of a thousand florins, payable to the king's coffers. That same evening the two porters of the governor's court, Alfonso de Teraçona and Johan Coltell, reported that they had gone to Benaguasil and had ordered the lieutenant alami and the elders not to hinder the passage of water to Valencia. They then proceeded to the almenara called Jofari, where they found the two cequiers of the Benaguasil Canal and a Moor armed with a lance and dagger. The porters ordered the almenara opened under penalty of 1,000 florins and, when the cequiers refused to comply, opened it themselves. They then ordered the cequiers not to divert the water into their own canal, under penalty now of 2,000 florins; but the cequiers and others closed the almenara again forcibly. After two more turns, with the fine standing at 4,000 florins, the cequiers still noncompliant, and the planks of the control gate floating in the river, the porters departed for the almenara called La Fenosa in La Pobla, where the same scene was repeated with the same cequiers. The porters then came to the almenara of the canal of Vilamarxant, which they opened in the presence of its lord,

Pere de Moncada, who notwithstanding the Governor's order, commanded a Moor (who was cequier) to close the almenara. The orders volleyed back and forth until the fine again stood at 4,000 florins. Moncada said he would not comply but would instead defend his possessions. The porters proceeded to Ribarroja and opened its five almenares; the *alcayt* there, too, refused compliance.[37]

August 19. On Saturday the Valencia council met to consider events of the past two days. Action had originally been taken so that the residents of Valencia would not have to continue paying for their milling with "undue ransoms and extortions" and so that crops of the huerta would not perish. The prevailing conditions were the cause of "great murmuring, conflict, and rebelliousness in the city and huerta."[38] The council then heard the reports of the jurates and porters and, in view of the "great incivility and violence" of the officials of Vilamarxant, Benaguasil, and La Pobla, decided upon the following measures to secure the city's right to the water: the criminal justiciar and the jurates would proceed personally to the places in question with armed men — both mounted and on foot — to "remove the impediment" to free flow of water. Salaries of 5 sous daily were authorized for 100 horsemen and 3 sous for 1,000 footsoldiers, including archers and lancers; beasts were provided for carrying picks, other tools, and artillery necessary for destroying any obstacle found in the river. The salary authorization was for three days. The town crier was to call throughout the city and huerta for volunteers to assemble in the Plaça de les Corts to receive wages in advance from the justiciars and jurates. By Saturday evening a public table had been set up in the Plaça, in front of the house of Bernat Boix, where salaries were paid to the volunteers. At the same time it came to the attention of the town fathers that Pere de Moncada and Anthoni de Castellet, the lords of Vilamarxant, were in the city; accordingly they were seized by the two justiciars and imprisoned "high in the Chamber of the Arms of the City," along with several of their vassals. In addition, some men of Benaguasil and many of their beasts were thrown into the common prison.[39] At this juncture, letters patent from King

Ferdinand were presented by the representatives of Count Frederick of Luna (lord of La Pobla and Benaguasil) and of Moncada and Castellet, lending royal authority to their positions of defiance to the city.[40] The council listened to the letters and concluded that the petitioners had willfully misled the king, claiming rights which they did not have.[41] The day ended with an agreement on the part of the Count of Luna and the two lords of Vilamarxant to submit the entire question to a general council of Valencian officials on the morrow.

August 20. On Sunday morning more men were signed up to fight. The general council decided that the salaries of archers and horsemen had been set too low and raised that of the archers to 4 sous daily and that of the horsemen to 5 sous 6 diners. The next order of business was to hear written statements from the lords of Vilamarxant and the Count of Luna, who claimed in effect that even the water stipulated for Valencia in the four-day turn was *aygua de gracia.* The lords begged the city not to send armed men, assured the Valencians that no violence would be done, and suggested that two cequiers be sent instead to oversee implementation of the prescribed turn. The justiciars, jurates, council, and syndic of Valencia all concurred in rejecting the allegations contained in the dispositions, affirming that the city indeed had right to the water by virtue of fueros and privileges. All three lords then swore, through their procurators, to respect the decision of the general council and placed their vassals in the hands of the city as surety. The justiciars and one jurate went to where the nobles were imprisoned and received their oath.[42] That same Sunday the jurates and council dispatched a letter to the King. First, general arguments were repeated: the crops were dying and, more important, there was not enough water to mill the wheat. Then the events of the past few days were described. It is interesting that the city did not deem it necessary to cite the privilege of James II; the letter was merely to inform the King, not to defend the city's actions or justify its position. After congratulations to the King for his progress in the siege of Balaguer,[43] and the usual compliments, the letter ended.[44]

August 21. On Monday Pere Colomines was sent upstream.

He went first to Benaguasil, where he opened the almenares "in such a way that all the water flowed in the river." [45]

August 22. The following day, Tuesday, Colomines proceeded to Vilamarxant and Ribarroja, opened the almenares, and ordered the officials not to touch them. Meanwhile in the city, Pere de Moncada was brought down from his confinement to make a statement to the council. He unburdened himself of "many good and apt words of excuse" and declared that as a son of the city he was willing and ready to obey her; he wished only that the sinister things which were said of him not be believed and other bad reports not be attributed to him. Having spoken his piece, the nobleman was returned to his cell. The council then resolved to free both him and Castellet, as well as the Moors who had been imprisoned with their beasts in the common jail. It was agreed that the other recalcitrant officials mentioned in the porters' report should be arrested by the justiciars and placed in the common jail.[46]

August 24. Now that the disobedience had been quelled and the malefactors punished, the city could institute the turn with the Pueblos Castillos in the fashion prescribed in the privilege of James II. On Thursday, the jurates ordered Pere Colomines to proceed to Ribarroja, Vilamarxant, Benaguasil, and La Pobla and "to take the water on the four days which belong to the city." Colomines accordingly went first to La Pobla and commanded the justiciars and jurates to comply. He proceeded to Benaguasil and Vilamarxant, and, on Friday morning, to Ribarroja, encountering no resistance. Colomines did not return to make his report until Monday, August 28, which would have been the first of Valencia's four days in the turn.[47]

Drought conditions prevailed throughout 1415, and lack of water for milling continued to be the biggest problem.[48] Tensions were much eased, however, with the normalization of relations between the huerta and the Pueblos Castillos. In the summer of 1415 turn arrangements were negotiated peacefully.[49]

The drought crisis of 1413 was resolved by the city's imposition of a solution. The town fathers, acting without outside support (and, in fact, in opposition to the King's will) successfully defended the proposition that the huerta of Valencia had priority in use of the water of the Guadalaviar. The city's ability to put an army quickly into the field and to force impotent nobles to back down demonstrates the balance of power in the fifteenth-century kingdom. The city was in full ascendancy, and its power over the nobles of the Pueblos Castillos was complete. The king was a remote figure; his governor issued at the city's pleasure edicts apparently at odds with the king's own privileges. The affair of 1413 presaged in many aspects the war of the Germanías of the next century (1519–1521). Both episodes were characterized by the opposition of the urban middle class — artisans and yeoman cultivators — to the landed aristocracy and their Muslim tenants. Both, indirectly at least, were movements of resistance to the crown. The outcome of the Germanías, however, was just the opposite of that of the earlier struggle. The victory of the nobles over the urban brotherhoods altered the balance of power, diminished the autonomy of the city of Valencia and the kingdom alike, and brought the king's viceroy to a position of dominance.

Part Two | Cultural Continuity in Irrigation

VIII | Arabs, Romans, or Christians? The Historiography of Spanish Irrigation in the Nineteenth Century

L'histoire des irrigations est partout populaire.
—Maurice Aymard, 1864

Spanish scholarship of this century and the last has elaborated two rival views of culture contact between Al-Andalus and Christian Spain. One tradition descends from Francisco Javier Simonet, the nineteenth-century historian of the Mozarabs of Granada, to Claudio Sánchez-Albornoz, who views the Islamic invasion as a catastrophe that deflected Spain from her true historic path. This orthodox judgment denies the influence of Islamic culture upon the Christians and stresses instead the eternal Iberian blood coursing in the veins of most Andalusí Muslims. The second tradition includes the works of Richard Ford, the English hispanist and polygraph, Francisco Fernández y González, the Spanish arabist — both predecessors of the contemporary theory of Américo Castro — and a host of local histories that have interpreted Spanish culture largely as a product of the Islamic heritage. The local historians, in whom a strong *costumbrista* element predominates, have in fact been the strongest and most faithful of the pro-Muslim band. Irrigation historians of the nineteenth and twentieth centuries have drawn inspiration from both schools, as well as from the ever burgeoning literature of archeology, which advances both Roman and pre-Roman claims.

FROM POPULAR TO SEMILEARNED: ORTIZ DE MENDOZA

In the Middle Ages the irrigation system was generally recognized as an Islamic inheritance (see Chapter XIII). The provisions of the fueros that the irrigated areas were to be restricted to pre-Reconquest bounds and that the customs of distribution were to be unchanged from those of the Moors were accepted literally and almost universally. While perceptive irrigators did on occasion point out that customs had in fact changed through time, and that a process of modification could therefore be discerned, the nearly unanimous judgment of medieval society was that the principles of distribution and most of the physical plant antedated the Reconquest. In the sixteenth century, however, widespread repugnance of things Moorish and a desire to repudiate the infidel past led to a bifurcation between popular tradition, which to this day ascribes a Moorish paternity to all that is venerable, and a semilearned scholarship the aim of which was to assert the purity of Spanish Christian culture and deny all Islamic influence. What the medieval irrigator had held true the sixteenth-century scholar rejected.

Baltasar Ortiz de Mendoza, writing in 1589, devoted the first chapter of his *Claridad de la acequia de la villa de Elche*[1] to the question of who, Moors or Christians, were the originators of the division of the waters of that town, and "of what might be the origin of a *hilo* [measurement unit] of water's coming in a turn of thirty-six days." Ortiz, in seeking to refute the generally held proposition that Muslims had devised the irrigation system of Elche, argued that both the institutions of distribution and the canals themselves antedated the Muslims and could be shown to be of Christian foundation:

> Many have been of the opinion that the order set down in the books of names of the direct lords of the water of this

> town was a Moorish invention, which cannot be proven, but, rather, the contrary [is true], owing to many and diverse reasons. Three proofs and arguments will suffice those who believe that truth ought to prevail, which, when heard, I believe should accord to the Christians both victory and reason.
>
> Since in a trial both sides must be heard, it is proper to inquire how these books came to be, and whether this be as I have heard alleged, that some of the divisors have Arabic names, such as Alingasa, Atufá, and Palombar.[2]
>
> To those [who so allege] I reply that Spain was first Christian before it was Moorish, and already [before Spain was Moorish] there were *acequias* through which the water ran. Thus when [the Moors] took possession they found at Alingasa a stand of pear trees, and thus they called it "divisor of the pear-grove." And at Atufá they found an apple-grove, and they named it "divisor of the apple-grove." And in Alausa they found an almond tree and therefore called it "divisor of the almond tree." And at the divisor of El Palombar there was a dovecot wherein many doves were raised, and for this reason it was given this title.
>
> Leaving aside the rest [of the names of the divisors] which had their names given them of old since they are twenty-five in number, simply because the Moors changed four names the argument that the Moors must be the authors who signed the above-mentioned register of books is not convincing.
>
> [The second proof is that the books are ordered according to the Dominical letters, which is clearly a Christian custom.]
>
> The last reason by which is excluded the possibility that this method [of distributing water] might be Moorish is that this order is holy and just in that all have equality without causing strife, nor do some have more rights than others. And [since] it is certain that in their [Islamic] sect there is nothing good, clearly it is demonstrated that the said register of books is the invention of Christians and not of Moors, as some have understood it.[3]

Despite its anti-Muslim bias, this passage is notable for its centering of scholarly discussion on Arabic toponyms and on aspects of the institutions of distribution. In another section,

Ortiz carries the etymological argument further by insisting that "*acequia* is a Valencian word; its etymology is 'where it is guided' [donde ella es guiada, that is, a se guia(?)]." [4]

NINETEENTH-CENTURY VIEWS: ARABOPHILES, PROTOHYDRAULISTS, AND THEIR CRITICS

All of the major nineteenth-century writers on Spanish irrigation were convinced of Islamic origins. The basic supposition accepted by them all is strikingly similar, though more naïvely stated, to that proposed today by the proponents of hydraulic theories of civilization. It is that the requisite of any large and complex irrigation system, comprising river barrages and large main canals, and hydraulic devices such as siphons and aqueducts, is a stable centralized government with enough wealth and control over labor to construct such a system. Simply stated, the assumption was that a great work must have been built by a great king.

The roots of this attitude toward public works are modern, or in any case not medieval. It is typical of the central authority's attitude toward such projects that in medieval Valencia a "Royal Canal" was not one built by the king but one from which he received revenue. Nevertheless, at least by the eighteenth century royal participation seemed implicit in any large irrigation project. Benito Feijóo, discussing the providence of rivers in an essay entitled "The Honor and Bounty of Agriculture," asserted:

> It is true that this providence is most laborious and costly. It requires, for the most part, intelligence superior to that which farmers have and more money than that which individuals have. Farmers can only give information concerning the sites which stand in need of the benefits of irrigation and concerning the local rivers. The possible use of the water of these is a matter for experts in Geometry and Hydrostatics. And, finally, the cost must be met either by the Prince or the Public, respectively, of the territory which is to receive the benefit. Application and zeal for the common good can conquer all.[5]

Such a "protohydraulic" theory was first and most influentially expounded by Francisco Javier Borrull in the course of a stirring defense of the Tribunal of Waters (entitled *Discurso sobre la distribución de las aguas del Turia y deber conservarse el Tribunal de los Acequieros de Valencia*) before the Cortes of Cádiz in 1813.[6] Borrull, unable to find any record of the early history of the Valencian system in the works of Islamic historians (who, to be sure, he read in translation or as discussed in the works of Spanish arabists), set out to collate all the available data and thus "form a sure concept" of Valencia's irrigation history. His method was simply to locate a moment of peace when a huge financial treasure could be applied to the benefit of the public.[7] This moment could not have been in pre-Roman times because the principal settlers, the Carthaginians, were commercial folk who were continuously at war.[8] Nor could the distribution be attributed to the Romans, who had no fondness for agriculture and against whom the overtaxed Hispanians arose in rebellion. If in these unhappy times Columella lamented the lowly state of agriculture in Italy, how could that of Spain have fared better? Public funds were spent not on irrigation, but for roads bridges, monuments, and aqueducts for urban water supply.[9] The Goths, of course, being a pastoral, warlike people with a slave economy, were unlikely founders of an irrigation system.[10]

Borrull looked to the time of the Muslims for the system's beginnings. There is no need even to guess, for James I tells us himself, in his fueros, that Valencian irrigation custom dates from the time of the Saracens. It would have been easy to ascertain which princes were responsible for the system, if only the Arabic manuscripts (those of the Escorial) had not been burned.[11] The dates of the foundation of the system had therefore to be ascertained by deduction. It could not have occurred before the tenth century because too much energy had been diverted to warlike endeavors such as the conquest of France; nor could it have taken place after the tenth century, for beginning with Almanzor most of the energies of the Andalusis were absorbed in wars with the Christians. Borrull concluded that the only possible period for the system's creation was during the reign of

'Abd al-Raḥman III (912–961), who put an end to civil strife, and that of his son, al-Ḥakam II (961–976). In explaining whether the father or the son played the most important role, Borrull deduced that the father must have ordered the initial studies and begun the works that were finished in the fifteen-year reign of the son.[12]

Underlying Borrull's argument throughout is the supposition that public works projects at any time and in any society always are carried out in the same way, that is, in the manner of early nineteenth-century Spain. Such projects are initiated by royal order, studied by a commission, planned by an engineer, and finally executed with public funds. He knew, moreover, that no such works were possible if the central authority has been involved in war:

> It follows also that war, far from yielding funds to be employed in [agricultural improvements], consumes all the funds of the state and obliges the princes to seek loans to sustain it. Experience proves this; and should we question a witness as authoritative as our own monarch, he would answer us in the context of the Royal Order of May 19, 1819, that is, "that the six-year war which the monarchy waged has left the state without resources to finance the costly works of the *acequias*." Far less could the Saracen monarchy have had them in the first and third epochs [of Muslim rule] in which it maintained, with notable valor, war not only with different Christian princes but also with its own vassals, not for six years, but for the lengthy period of three centuries.[13]

After forming his theory, Borrull found support in Josef Antonio Conde's *Historia,* in which can be found several indications of agricultural improvements, without benefit, however, of indication of sources.[14] The coincidence of nearly a century of peace with the rule of two great monarchs (the test of greatness was the meriting of elegies) was conclusive for Borrull. Upon this reasoning, and not upon any documentary evidence, stood a theory that gained immediate and general acceptance in the nineteenth century and official recognition even in the twentieth.

(In 1960 a celebration commemorating the "Millennium of the Waters" was held in Valencia, indicating public recognition of the establishment of the irrigation system and specifically of the Tribunal of Waters in the reign of ʿAbd al-Rahman III.[15] In the years intervening between the publication of Borrull's discourse and the millennial celebration, however, no new light had been shed upon the history of irrigation in Al-Andalus.)

Soon after Borrull's *Discurso* was printed a copy reached the hands of a French provincial official, François-Jacques Jaubert de Passa.[16] Jaubert entered into correspondence with Borrull [17] and in the years 1816–1819 made an extensive tour of eastern Spain, gathering information for a study of irrigation law commissioned by the Société Royale d'Agriculture of Paris. The results of his studies were published in 1823[18] and a Spanish translation appeared in 1844.[19] Jaubert introduced his study with some general remarks on the history of agriculture in Spain, citing in passing some of the more distinguished Iberian-born agricultural writers: Columella and several Arabs, notably ibn al-ʿAwwām.[20] The sum of his investigations convinced him "that only one people (the Arabs) have known the soil of Spain perfectly, and the proof of this will become more perceptible as I approach the center of their empire and the places which have been subject to their rule for the longest period of time." [21] The study begins with a review of the irrigation districts of Catalonia, which had been subject only minimally to Islamic rule, and proceeds down the coast to the highly islamized regions of the kingdom of Valencia.

Nevertheless, there is no greater precision in Jaubert's discussion than there was in Borrull's. He attributed to the Moors, for example, the building of the Canal of the Arches in Manises[22] and likewise the redaction of some of the ordinances of the Quart Canal. Although he cited no documental evidence, his reasoning concerning the conservativeness of agrarian law is worthy of note:

> The ordinances of Cuarte [Quart] Canal are still in part owing to the Moors, and their antiquity, though most worthy

> of attention, can be explained very easily. In effect, rural laws ought to be more permanent and obligatory than civil and criminal law; for this stability is always the product of time and experience. They are established without resistance only owing to the good offered and as a consequence of the unanimous consent of all those whom they protect. Civil laws, on the other hand, are dependent upon cultural progress. Useful to the generation which receives them, they are often contrary to the interests of the succeeding generation; and history shows us that a nation needs not even a hundred years to change and even corrupt its customs. For this reason, the kingdom of Valencia, poor in civil laws, would be poorer still were it not for the rural laws which it has.[23]

The customary and consensual nature of agrarian law led Jaubert to conclude that those ordinances of Quart already present in the corpus before the redaction of 1350 most likely antedated the Conquest. As for the origins of the Tribunal of Waters, Jaubert was content to summarize Borrull's view assuming its existence at the time of the Reconquest.[24]

The method followed by commentators throughout the rest of the century was to accept Borrull's basic proposition as obvious and then to support or embellish it with etymological evidence and, in a few cases, comparisons with contemporary irrigation in Islamic countries.

Of all nineteenth-century writers on Spain, Richard Ford was the most deeply convinced of the Islamic substratum of the Spanish culture that he observed and described in all its aspects. "Spain," he wrote, "first civilized by the Phoenicians, and long possessed by the Moors, has indelibly retained the original impressions. Test her, therefore, and her natives by an Oriental standard." [25] Nowhere did he apply his standard more eloquently than in a brilliant gloss on Borrull in the section of his *Handbook for Travellers in Spain* devoted to Valencia.

> The whole system of artificial watering is Oriental and Moorish, as the still existing technical names and machinery prove: thus the common and most picturesque *noria* (Ara-

bicè, *anaoura*) is the Cairo *sáckiyeh,* or large water-wheel, which, armed with jars, descends into the well, and as it rises discharges the contents into a reservoir. The Egyptian *"shadoof,"* the pole and bucket, or galley-pump, such as is seen in our market gardens near Hammersmith, is also very frequent.

In the *Huerta* of Valencia, a main-trunk artery or principal *canal, "mucannalin,"* supplies all the smaller veins, *acequias, "ciquia,"* of the circulation: this is managed in a reticulated net-work of minute ramifications, and dams, *azudas — sudd.* The idea is simple, but the execution is most difficult; and often the greatest triumph of the hydraulist is where his works are least apparent. The chief object was to secure a fair distribution, so that none should be left dry, none overflooded. Thus when the engineer ceased, the legislator began, and both were Moorish. The supply was divided into days of the week and hours of the day. The owner of each plot knew his appointed period and was ready to receive his share. Since water here, as in the East, is the life blood of the soil, and equivalent to fertility and wealth, the apportionment becomes a constant source of solicitude and contention. Similar instances are recorded in the Old Testament, and rivalry has been well derived from *Rivus,* the bickering about water-brooks; so wells in Genesis xxvi were named *Esek,* contention, and *Sitnah,* hatred. Accordingly, in this irritable climate, where the knife is always ready, precautions have long been taken to keep the peace. The regulating tribunal, *de los acequieros,* or *del riego,* said to have been instituted by the Moor, Alhaken Almonstansir Billar [sic], was wisely retained by Jaime I. It is truly primitive and Oriental: seven syndics or judges are chosen by each other, out of the yeomen and irrigators, the *labradores y acequieros* of the *Huerta;* they sit at 12 o'clock every Thursday, in the open air, on benches at *La puerta de los apostoles,* at "the gate" of the cathedral; all complaints respecting irrigation are brought before these Solomons and decided in a summary way. There must be no law's delay, for water here gives daily bread, and if the suit went into our Court of Chancery, land and cultivators would be ruined; time accordingly is saved by prohibiting the use of pen, ink, and paper; there are no bills and answers, no special pleadings. In this — oh rare court of common sense! — no attorneys are allowed to practise, no barristers are allowed to obfuscate the plain matter-of-fact. The patriarchal judges understand the subject practically, and decide without appeal; the discussion is carried on *viva voce* in public and in the "Lemosin," or the dialect of the people: consult for curious details the *Memorias* of Fro. Javier Borrull.[26]

Ford distrusted Spanish historians who were ignorant of Arabic,[27] though he did not know much himself; most of his comparative data is based, in this quotation, on E. W. Lane's description of Egyptian irrigation.[28] Ford's effort, apart from its literary value, was a pioneering attempt to establish the Islamic character of Spanish irrigation by comparison with other Islamic systems.

In the mid-1860's, several Englishmen motivated primarily by interest in Indian irrigation problems journeyed to Spain in order to report on techniques and institutions which might be of use to the British government in the subcontinent. Foremost was Sir Clements Markham, geographer, historian, translator, and an expert in the history and culture of another great irrigation society, Inca Peru.[29] Markham's delightful *Report on the Irrigation of Eastern Spain* (1867) was based upon his own observations — guided in Valencia by the historian and *costumbrista* Vicente Boix,[30] who accompanied him to some of the irrigation works and helped him understand "the scarcely intelligible Lemosin of the villagers," and the bibliophile, Pedro Salvá, owner of "the finest library in the city." [31]

Markham's second chapter, devoted to the "Moorish origin of works of irrigation in Eastern Spain," is little more than an elaboration of some of the themes developed by Borrull: the periods of strife and internal dissension after the period 760–960 "were not times for originating public works, though the irrigation system that had been instituted in the days of the caliphs was perpetuated and confirmed under the succeeding dynasties, until, when the Christian conquerors appeared in the thirteenth century, it recommended itself for adoption, backed by the experienced benefits of five hundred years." He did, however, distinguish between eastern Spain (Valencia and Murcia) and the kingdom of Granada, where the public authority continued to push forward with mining and irrigation works well after decadence had overcome the rest of Al-Andalus. Yet, Granada lacked, for Markham, the prime interest of Valencia and Murcia, namely the "interesting series of local rules and customs which are met with in connection with the more ancient systems established in Murcia and Valencia." [32] Then, too, the chief object of the Granada water supply system was not the irrigation of crops but the provisioning of the fountains and baths of the capital.

Markham was more original in his description of the reservoir systems of Elche and Alicante, which interested him greatly, first because of their similarity to the tank systems of the Madras Presidency in India and, second, because sure proofs of their Arabic origin could — he thought — be adduced. In Elche the arabism of hydraulic toponyms was the key to his argument: "The names of these *partidores* give more distinct proofs of the Arabian origin of the irrigation system of Elche than could be obtained either from traditions or even from written chronicles. I obtained these names from the mouths of millers and cultivators who live on the banks of the channels and use their water; and I am indebted for the following explanations of their meaning to that eminent Arabic scholar, the Rev. George P. Badger." [33] Markham's interest in Islamic Spain dated from 1855, while he was still working on the manuscript of his book *Cuzco and Lima* and before he had embarked on the cinchona expedition. He thus began the study of Arabic with a Lebanese Maronite named Joseph Churi and was said to have "soon mastered the verbs." [34] But by the time he reached Spain some ten years later his Arabic was not up to the task, and he had to rely upon his friend the Reverend Mr. Badger[35] for the deciphering of the unusual and Oriental-sounding place names. Badger's etymologies suffer from excessive enthusiasm, lack of knowledge of Andalusí Arabic, and ignorance of Catalan: he derived Forat (Catalan "hole") from the Arabic *farat* — plenty, superfluity —

and Aval (Catalan "down," that is, lower or downstream) from *awwal,* first.[36] The pitfalls of trying to stretch a word to fit a thesis are clear from the following brave attempt: "*Candaliz.* There is an Arabic word, one of the rare instances of a noun with five radical letters, which is something like this — namely, *Khandalis;* but, as it means a number of lean camels, it is hardly likely that it can be the name of this channel. Candaliz may be a corruption of Kanz-al-izz, 'the treasure of dignity or happiness.' "[37] This uncritical etymologizing was a departure from the usual approach and methodologically was even an advance.[38] In addition, Markham was skeptical about the Borrull thesis, at least as it related to the Tribunal, whose origin, he reported, "is referred, *but on doubtful authority,* to the reign of the Caliph Alhaken II, A.D. 961–76."[39]

The most critical of the mid-century commentators was Maurice Aymard.[40] He believed that the emphasis laid on Arabic origins was over-rated, that in any case it was based on a misconception of the nature of Islamic irrigation, and that the crucial developments in Valencian irrigation institutions were subsequent to the Reconquest. Like Markham, Aymard found significance in the difference between Valencian and Granadan institutions. In Granada the system was still "to an exceptional degree the same as in the time of the Moors. . . . [But] one is astonished by the incoherence of the ideas which govern the repartitions. The practice is not in line with the principles. The bonds of association are ill-joined; duties are vague; policing hardly exists; penalties are poorly defined. One feels, in a word, that the synthesizing genius of the Latin race has not passed through there."[41] If Granada represents a more genuinely "Moorish" stage of development, then the Valencian stage must be the result of post-Islamic refinements:

> It is apparently believed by everybody that the Spanish conquerors of the Moors only applied the institutions of their predecessors in matters of irrigation; that the honor of their creation belongs to the Moors; that Spain actually did nothing but inherit a splendid heritage, by which she benefits. . . . But we must reflect on the continual im-

provements that have been in progress all over the country; on the crude ways yet to be seen in Granada. In every century regulations have been repeatedly revised; and everywhere, in spite of the diversity of fundamental principles, a methodical administration has been obtained, regular, strong, and concentrated. Does this honor belong to the Moors or to the Spaniards? The example of Granada leaves no doubt in our mind on the subject. We find there undisturbed the institutions practiced by the Moors themselves. The work of Loaysa,[42] written but a little while after the conquest, largely by dictation from the old tillers of the soil, may be considered as a photograph of the irrigation practiced by the Moors.

It becomes possible, therefore, to make a sort of comparison between the irrigation institutions of the Moors and of the Spaniards. On the side of the Moors, there is nearly a complete absence of first principles of regulation — not any centralization, hardly any policing, divisions of water made without fixed plan; surely this was a time when anarchy prevailed. If this anarchy has ceased, it is because the disorder itself, ungoverned by a powerful hand, has become orderly after a long lapse of time.

But this is not the sign by which we recognize good institutions. Such institutions should be good from the first, without the aid of time, without support of power; they should be susceptible of assuring regular working from the start. The systems of Valencia, Alicante, and Murcia, reformed and gradually made anew by the Spaniards, would do this. The system of the Moors, preserved by Granada, would not.

But we must be just to all. Without doubt the Moors irrigated before the Spaniards, but to these latter incontestably belongs the greater part of the honor of those institutions of administration that are admired today.[43]

Though we can praise Aymard for his critical acuity and refusal to accept what most everyone else held to be true, nevertheless one must ask whether the comparison offered is truly meaningful. While the situation recorded by Loaysa may well be a faithful description of the Islamic system of Granada, the environmental differences between Valencia and Granada were great and led to different institutional arrangements. Irrigation institutions in Islam were far from homogeneous. Then too, Aymard, like his predecessors, was unaware of substantial medieval documentation of the Valencian systems, photographic in the same measure as that of Granada. Moreover, the anarchic situation prevailing in Granada might have been the result of degeneration from its pristine state; and in any case, many of the mechanisms of social control in Islam would not be explicit in water regulations. Without a study of Islamic institutions in themselves the comparison does not hold up. The selection of Granada as "Moorish" seems rather arbitrary, when in fact the basis of the selection is a value judgment, similar to that of the *Claridad,* that order and Islam are somehow antithetical.[44]

Two works written at the end of the 1880's, one by Pedro Diaz Cassou and one by Melchor Bellver and Vicente del Cacho, proposed interesting criteria for the evaluation of the Islamic past of Spanish irrigation. Diaz Cassou, perhaps the most original local historian in nineteenth-century Spain, touched briefly in his works *Memoria sobre los riegos de Segura* (1877) and *La huerta de Murcia* (1887) on the Islamic past of the irrigation system, pointing out that the secondary canals with Beni names (Beniaxán, Benialel, and so on) indicated that originally these channels had carried water to tribal lands. With the passage of time collective ownership of the land became individual, but the channels and water continued to be held communally in the domain of the tribe or family.[45]

In 1889 Diaz Cassou expanded his observations in a sequel to the earlier studies, *Ordenanzas y costumbres de la huerta de Murcia.* Beginning his discussion with an echo of Borrull, he observed that the largeness and complexity of the Murcian system, with a main channel and subdivisions on either side of the river, "supposes the intervention of the public authority, which did not have sufficient strength except during the reigns of Abd-er-Rahman III and Al-Hakem II, in whose time canals were dug in Murcia, according to some Arab historian among those translated by Conde." Leaving hypothesis aside, Diaz Cassou then pressed ahead on his own, discussing the taxonomy of the Murcian canals. The area of the huerta cultivated by the Arabs must have been smaller than that presently cultivated. The derivations outside the central core were due to extensions made either by late-arriving tribes (evidently Berber) or by the Christians. The main canals have generic names (such as Aljufia, from the Arabic *al-jaufī,* northern), while their extensions bear proper names and the patronymics of the tribes responsible for the prolongation of the channels. In the upper part of the huerta and in the center of the valley, the channel names are taken from topographical or geographical characteristics (Algualeja, orchards; Nacar or Nacra, pond, reservoir). On higher grounds, where irrigation was more difficult and where latecoming Muslims were obliged to settle, the canals bear tribal names (mostly Berber, at that), such as Benetucer (Beni Tussef), Benicoto and Benicomay (Beni Komaya), Bendame (Beni Aḥmad), Zenetes (Zenata). Diaz Cassou warned, however, against reconstructing the history of the huerta on the basis of canal names alone. Many minor toponyms have been disfigured (Al-Ferraira became Herrera, for example), and others were transmuted into pseudoarabisms, as in the case of the divisor originally named Dardallah after the man who built it that became corrupted on the basis of a false Arabic etymology to Ardallah (land of God).

On the basis of these considerations, Diaz Cassou proposed three criteria for establishing the history of the Murcian canal network: 1) Canals with the highest right (those which are not subjected to a turn and in which water always flows, called *toma abierta*) belong to the nuclear huerta of the epoch of the caliphate. 2) Canals of lesser right that draw water from the river or main canal at certain times only (*toma cerrada*) had their origin in subsequent concessions of late Islamic or Christian times. 3) In the canals whose first sections are not subject to a turn but whose lower sections are, the latter sections are modern extensions of the primitive channels.[46]

Diaz Cassou's criteria, based on toponyms and the structure of water rights, bear careful consideration. To my knowledge no one has since attempted to prove or disprove his theory, for such an attempt would involve careful evaluation of the toponyms of the secondary channels in addition to studies of the documentation of early Christian settlement. The names of the derivations of the Vernisa Canal (Gandia) suggest that the structure described by Diaz Cassou may not have been limited to Murcia. On the Pellerias branch of Vernisa four of the six secondary feeders have Beni names, while their tertiary subdivisions have a predominance of romance topographical names (among the subdivisions of Beniopa are Montaña, Palmes, Olivera, and Quintana).[47]

In the same year that Diaz Cassou published his theory of the development of the Murcian irrigation system, Bellver and Cacho published their treatise, which had won a prize offered by Lo Rat Penat (an erudite society) of Valencia in 1888, on the influence of the Arabs upon the agriculture, industry, and commerce of Castellón de la Plana.[48] The primary significance of their work for irrigation history is that it drew broad comparisons between Roman and Islamic agricultural styles and placed irrigation within the broader framework of economic and agrarian organization. Their basic premise was that Islamic rule brought about the replacement of the Roman extensive fallowing regime by an intensive regime of continuous cultivation based upon irrigation.[49] While admitting that no concrete proof could be brought to bear upon the problem of Islamic origins, Bellver and Cacho indicated a wide range of secondary evidence in support of their theory. These points — some plainly more convincing than others — were, in summary:

1) The selective replacement of Latin terminology with arabisms to indicate irrigation connotations of water-carriers (the arabism *acequia* replaced the Latin *canalis,* use of which was subsequently restricted to the sense of "gutter") or the introduction of arabisms where no suitable Latin expression existed (*azud,* "diversion dam")
2) the wording of the fueros of James I
3) the unanimous opinion of experts, Spanish and foreign alike (Borrull, Conde, Jaubert de Passa)

4) toponyms such as *huerta de los moros*
5) the widespread adoption of Arab-introduced hydraulic devices such as *norias*
6) the preponderance of irrigation agriculture in towns of Islamic foundation
7) the introduction by the Arabs of exotic plant species, most with Arabic names and most requiring irrigation: lupine (*altramuz*), flat pea (*almorta*), carob (*algarrobo*), orange (*naranja;* Catalan, *taronga*), date palms, alfalfa, cotton, safron, and so on.[50]

These data, when added to such other agricultural modifications of great importance as replacement of the ox by the horse as the common plow animal,[51] add up to an impressive and convincing picture of change in spite of the lack of documentary evidence.[52]

Bellver and Cacho's discussion of agriculture concludes with a suggestive and important modification of the hydraulic theory outlined by Borrull. Accepting the generally held proposition that the Golden Age of Andalusí agriculture took place between 940 and 1001 under Umayyad rule,[53] they questioned whether the agricultural renaissance of Islamic Castellón was due to the direct action of the rulers or to the individual efforts of the cultivators themselves: "We dare to risk the assertion that this renaissance was not the result of a plan conceived and executed by the central administration but the result of the slow, but intelligent and tenaciously sustained, striving of some active cultivators who were desirous of deriving from the soil the greatest possible return." [54] While accepting the intervention of the Muslim princes in the construction of "the magnificent canals and dams which fertilize the privileged *huertas* of Valencia, Murcia and Granada," they nevertheless doubted the willingness of any prince to expend interest and capital on the works of "obscure rivers" such as the Mijares and the Palancia. History reveals no intervention by the central authority in the irrigation works of the Castellón area, whose most impressive work, the great siphon which passes underneath the Rambla de la Viuda, is known to have had a purely private origin. As important a departure from the Borrull hypothesis as this is, Bellver and Cacho still seem to have been influenced — even as Borrull himself had been — more by the example of the present than by concrete data from the past:

> If it was the Caliphs who directly and immediately originated the agricultural improvements introduced in the regional district of Castellón during the Mohammedan period, we can do no more than present them as a model for modern governors who have hesitated to take certain measures favorable to cultivation.
>
> If, on the other hand, their cooperation was limited to conceding opportune permission for the use of wastelands and public waters, whose disposition was in their hands, and the cultivators, either individually or communally . . . were those who conceived and carried out those improvements, we can likewise offer them as an example most worthy of imitation by present-day farmers and proprietors of Castellón, accustomed to expect everything from the tutelary action of the government.[55]

Bellver and Cacho's effort received a harsh review from another local historian, Roque Chabás. Writing in his own historical journal, *El archivo,* Chabás said that the work reminded him "of those discourses of Feijóo where he discussed electrical phenomena, in which the erudition is extraordinary, his ingenuity the sharpest, no matter how much the result had to be worthless. What can one who lives in a time when electricity is unknown tell us about lightning?" [56] What individual resources did the Arabs — "that apathetic people" — have that were lacking to the wealthy Roman owners of latifundia? Chabás could not see how the continuity between Islamic and Christian agriculture could have been effected, since the conquering Catalans and Aragonese "naturally followed the custom of their own land, adapting themselves to the exigencies of the Valencian climate and soil." The Muslims, moreover, lived apart, with little influence on their neighbors. "It is to be noted that Moors never hired Christians, and it is always the master who imposes the form of work and not the servant."[57] In sum, Chabás lamented that the study of the genesis of these customs had been restricted by the "absurd theme proposed," namely, their arabism.

Indeed, with Diaz Cassou, Bellver, and Cacho the Borrull hypothesis had run its course and was beginning to try severely the patience of serious scholars. Without documentary evidence concerning the role of the Muslims, the hypothesis could not meet the requirements of modern scholarship and was relegated to the guidebook, the magazine, and other organs of popular and semilearned history. In the present century the two traditions have bifurcated, the Islam-rejecting tradition having gained the approbation of academic history on the basis of testimony from the two fields best qualified to evaluate whatever evidence existed: Islamic studies and archeology.

The most influential Spanish arabist to turn his attention to irrigation was a member of the generation of 1898, Julián Ribera Tarragó (1858–1934). The critical point raised by the arabists was that Hispano-Arabic sources were almost completely silent on the subject of irrigation: the problem thus became one of evaluating an area of culture for which no documents existed. Ribera, therefore, was duly and harshly skeptical when word reached him of the topic stipulated by the Diputación Provincial of Castellón for the prize later won by Bellver and Cacho. In an article entitled *"Mons parturiens"* Ribera warned in the strongest possible terms that even the most immense labors expended on the proposed topic could not be expected to bring forth more than a mouse. The first criticism he offered concerned regional history in general and regional history of Al-Andalus in particular: to carry out the assignment the writer would have to know what Castellón had been like both before and after the Islamic invasion. "The mere stating of this proposition," he commented, "seems to me enough to discourage the bravest." The administrative, social, and economic questions which must be asked of Islamic settlement in Castellón could not be answered on the basis of the sources available. Moreover, Ribera scoffed at the tendency of provincial towns to invent illustrious pasts, citing in particular Sueca and Gandia as towns which attacked "anyone who dares to question their hypothetical antiquities." Apropos of the scanty citations in geographical works such as that of Idrisi, Ribera turned to the subject of irrigation with a specific warning: "Let us hope that no one attributes to [Idrisi] that we owe our canals and irrigation system to ʿAbd al-Raḥman III and Al-Ḥakam II, because such a thing never occurred to him. If Conde so affirms, he alone knows why; up to now, no qualified person has believed this to be a historical fact, though many have believed it possible." With all the indignation of a scholar whose field had too long been dominated by legend, bias, and facile hypothesis, Ribera knew only too well the kind of evidence which passed for scholarly in the writings of popularizers of Islamic Spain: "Four common places, with their names transliterated in the French style, taken at third or fourth hand, ostentatiously embellished with some nebulous rhetoric, are not going to fool anyone." [58]

A decade later, while occupied with the phenomena of culture contact and acculturation, Ribera addressed himself to the institutions of the huerta of Zaragoza. In his *Orígenes del justicia de Aragón*[59] he developed a "theory of imitation" to account for the diffusion of elements from Islamic to Christian culture in medieval Spain.[60] Asserting first that imitation occurs in direct relation to the ease of intercultural communication, he concluded: "One has only to recall with whom Aragón communicated [that is, Al-Andalus] in order to affirm *a priori* that it could not fail to make imitations from Islam. Such would have been unavoidable; its geographic position and the well-known characteristics of its history make it clear." [61] It was altogether consonant with this theory to find evidence of cultural borrowing and continuity in Aragonese customs of irrigation:

> If we look from the banks of the Ebro toward the huerta we find the complicated mechanism of irrigation which had to continue [after the Reconquest] with the same organization and regulations; for neither were the conquerors in a position to promulgate rules for matters with which they were scarcely acquainted, nor, even when they were in a

> position to do so, were these matters susceptible to radical and instantaneous change.
>
> [There was no motive for any immediate change: the Moors remained owners of their properties and lands; they only lost and were uprooted from the houses they had within the walls of the city at the conclusion of one year after the capitulations.]
>
> That they continued along the same paths as before is proven by the fact that for a long time thereafter (and up to the present in some towns of Aragón) the directors of these functions are called *zabacequias;* and even today the Arabic words *adula* and *ador* — meaning irrigation turns — are common words in the huerta of Zaragoza, as well as the words *acequia, azud, alberca, almenara,* etc.; and even the tax or due which is paid conserves the Arabic name of *alfarda,* which continues being a technical term in foral legislation.[62]

By the time he addressed himself to similar irrigation institutions in the Kingdom of Valencia, he had abandoned the original rigor of his theory of imitation, feeling perhaps that he had overemphasized Islamic influence to the prejudice of pre-Islamic and independent inventions.[63] In 1908 he concluded that the system of the huerta of Valencia *"no es cosa de moros."* [64]

Ribera's arguments rejecting Islamic influence in the Valencian case are uneven. He wisely recognized the value of comparative studies, but his choice of Marrakesh as subject was inopportune. Physically it is quite dissimilar from Valencia; it has no river like the Turia, has no well-defined huerta, and is inland, not coastal. Technologically, the difference between irrigation in Valencia and Marrakesh is clear: the Moroccan system is based on conduction of water through subterranean channels (*qanats*), that of Valencia through surface canals. Legally, the communal institutions of Valencia contrast with private, individual appropriation in Marrakesh.[65] Finally, Valencia was not settled by Berbers or Magribi Arabs but by Syrians. Ribera, like Aymard, evidently thought that all Islamic irrigation systems were the same. Had he chosen Damascus for his comparison he would have reached quite different conclusions.

More telling, to continue Ribera's argument, is the silence of Arab historians. Why would they, who expressed "such exaggerated and pompous admiration" for the insignificant, private channels of Marrakesh, remain silent regarding the great public works of Valencia? "No Arab historian of Valencia — and there are many who were erudite and well-informed, whose works are still read — says a word about the construction of a single canal. This silence is very significant." Moreover, when water problems appear in legal literature, eastern jurisprudents are usually cited. Then again, the Tribunal of Waters has no equal in Islamic justice, of which individual, rather than collegiate, authority is more typical.[66] Ribera concluded that the Arabs when conquering Valencia did as the Christians were later to do: they found an established irrigation system and "had the prudence to respect local traditions." The clear effect that Arab settlement did have was to promote an intensification of cultivation as a response to an increase of population.[67]

In 1925 Ribera turned once again to the theme of irrigation and combined, although not in a completely satisfactory way, the two threads of origin and continuity.

> That which has been said — that the irrigation of the huerta continued after the conquest just as before in the time of the Moors — does not authorize the belief that the Moors constructed the canals, nor, much less, that the distribution of their water was administered as democratically or popularly as later became the norm or custom. The individualist and autocratic criterion of public works in the Arab empires and the unitary character of their administrative offices do not permit us to admit that the Muslims of Valencia constructed the canals of the irrigation system, nor less that in the regimen of water, questions would be decided by a collegiate tribunal.[68]

Thus Ribera (who here participates in a common misconception concerning the nature of the Tribunal) came almost full circle.

As the sternest critic of the Borrull hypothesis, his view, like that of Borrull's, was ultimately determined by his conception of Islamic society rather than on knowledge of Islamic irrigation.

Twentieth-century arabists have added little to Ribera's views, although some scant new material from Arabic sources has been brought to light. More significant, perhaps, for the course that views on irrigation in this century have taken is the work of the engineer Rafael Valls David, who won a prize in 1897 for his monograph *Pallantia*.[69]

Valls David discusses the significance of and elaborates upon the discovery and mapping by Francisco de Paula Jaldero of an ancient network of canals irrigating the Plain of Quart from the Turia, upstream of Valencia. Historiographically, the importance of this discovery was to change the emphasis in irrigation studies from Islamic to Roman and pre-Roman times. Valls David pointed out that if one asks a simple cultivator about the origins of anything old, be it an amphitheatre or an aqueduct, he is sure to reply: "Those ruins are the work of Moors. . . . If we repeat the same question to a half-educated person we can be sure that his answer, upon seeing the ruins blackened by the inclemencies of the weather, will be that those ruins were the works of the Romans. . . .[70] That is to say, for the one person the oldest thing which he believes to exist is Arabic civilization or domination, and for the other Roman domination." [71]

Valls David was a cultural diffusionist; he pointed out the elusiveness of firm genealogies of such phenomena as aqueducts because few peoples are truly original — it is dangerous, therefore, to base hypotheses concerning irrigation origins on deductions from what we know of ancient societies. For in truth everyone built aqueducts, and those of Megara and Athens antedated those of Rome.[72] The works in question had been ascribed to the third century B.C. If this was so, thought Valls, they were most probably of pre-Roman construction, for only at the end of that century did the legions arrive in full force.[73] Valls David's work was an augury of the twentieth-century approach: to leave the question of Islamic content up in the air and concentrate on archeological evidence.

"HYDRAULIC" HISTORIOGRAPHY

Although Karl Wittfogel's approach to irrigation has been most influential among anthropologists, it is in essence a historical view and is based on specific historiographical suppositions. Moreover, his use of the Spanish case as an example of the conquest and replacement of a hydraulic society (that of Al-Andalus) by a nonhydraulic one (that of the Christian kingdoms) offers an opportunity to test some of the historical implications of his theory.

In chapter vi of *Oriental Despotism* Wittfogel characterized Spain as a society that "crosses the institutional divide." The Arabs, as they expanded in the eighth century from the Near Eastern centers of hydraulic society, took with them across North Africa orientally despotic forms of government which they introduced in turn into the Iberian peninsula. Spain, which under the Romans had been only a marginally hydraulic area (having small-scale irrigation with only a few of the institutional trappings of agromanagerial bureaucracy), now became the home of a people familiar with hydraulic agriculture who, in their new habitat, continued using technologies they had successfully employed in their original homelands. As a result, Islamic Spain became more than marginally Oriental. "It became a genuine hydraulic society, ruled despotically by appointed officials and taxed by agromanagerial methods of acquisition." [74] Characteristic forms of hydraulic culture, such as the high development of astronomy and mathematics[75] and despotic military and civil administrative practices, pervaded Andalusí society.

Turning to the great transition, Wittfogel portrays the Reconquest (from the thirteenth century on) as the transformation of a great, urbanized, hydraulic civilization into a late feudal society. Militarily and administratively, the base of the Reconquest was never "orientalized." [76] An economy based on intensive irrigation agriculture was replaced by one of cattle-breeding.

Wittfogel's discussion of the Reconquest is superficial and deceptive. On the Islamic side, it fails to take account of the political transformation subsequent to the disintegration of the Caliphate, which severely modified the monolithic, despotic organization that the Caliphate certainly once had. The decentralized, evanescent Party Kingdoms (Taifas) were politically quite different from the Caliphate and were to a significant degree homologous to the Christian "feudal" kingdoms of the north. The key transformation of the Reconquest was not from the Islamic absolutism of the Caliphate to the Christian absolutism of the Catholic kings, but rather from the cellular structure of the Islamic Taifa to that of the Christian Taifa.

Wittfogel gives an eccentric interpretation of Castilian "feudalism" (some Spanish institutional historians have denied the existence of such an entity) and of relations between the frontier existence and the rise of Castilian despotism.[77] The importance of the shift in center of power from the irrigated huertas of the south to the plains of the north is clear. But what significance does this fact hold for the historian of irrigation? Just as the social implications of the hydraulic theory were tested by anthropologists, the historical implications, too, stand in need of commentary and questioning. From the point of view of the irrigation historian Wittfogel's approach is unscientific. It takes intensive irrigation as indicative of a range of social, economic, and administrative responses without delineating the nature of the actual institutions of water distribution or their place in the overall context of social life. As E. R. Leach pointed out in a trenchant critique of Wittfogel's historical method, evidence for most of the great hydraulic civilizations tends to be archeological; and, while archeology may reveal "a good deal about the nature of the irrigation works in question . . . it tells us very little about the kind of society with which they were associated."[78] "Hydraulic" characteristics ought not to be imputed to an administrative system simply on the basis of physical evidence; nor should it be assumed that all irrigation systems in despotic states have developed in the same way. The physical aspects of a system can be deceiving, and Wittfogel (like Borrull) seems to say that an impressive irrigation system must be the work of an impressive government, an agromanagerial despotism commanding an unpaid labor force. Leach points up absurd misinterpretations that can result from this kind of reasoning. The fifty-five-mile long Kalāwewa canal system of Ceylon "*looks* like a colossal and highly organised piece of bureaucratic planning, the work of one of Wittfogel's idealised Oriental Despots. But if so, the planning must have been done by a kind of Durkheimian group mind! The system took about 1400 years to build." [79] The large systems of eastern Spain had similarly long histories, their development continuing under a variety of sociopolitical systems. Even when documents are lacking, the evidence of place names suggests such a slow, informal act of creation. Indeed, as Leach points out, the fact that a given hydraulic work must necessarily be the result of a great number of man-hours "does not *in itself* imply any massive control over labour resources by the 'bureaucratic rulers.' " [80]

The work of Borrull and his followers, too, was permeated by the unhistorical supposition that only a certain kind of society could have produced the Valencian irrigation system. The possibility that the system might be the product of a gradual accretion, the composite child of several very different societies, was never explored in any systematic way, with the result that similar social structures were attributed to Romans and Muslims by the partisans of each on the basis of the type of irrigation each was assumed to have practiced. Thus what is known of the social, political, and economic structure of Roman and Islamic Spain is made to explain how the fully developed irrigation system came about.

Institutions of water distribution reflect greatly upon the society which originates or practices them. When understood, they can add to our knowledge of society and culture. But they cannot be deduced from indirect evidence, based wholly on archeological remains, nor upon theories of public works and administrative power.

IX | The Classical Inheritance

The Arab conquests of the eighth century initiated a great era of agricultural revival resulting in an intensification of irrigation practice throughout the Islamic world. Technologically, the civilization of Islam was a synthesizing one. The Arabs may have invented little themselves, but they preserved, refined, developed, and intensified the technological practices of the ancient world. The spread of Greek scientific ideas to the West through the medium of the Arabic language is well known, and a parallel process can be described in the preservation and extension of ancient Iranian, Babylonian, and Nabatean agricultural techniques.[1] The westward spread of irrigation technology across the Mediterranean followed a distinct pattern, one which might also serve to illustrate the diffusion of eastern concepts of water distribution and measurement.

TECHNOLOGICAL CONTINUITY: FOUR CASES OF DIFFUSION

In order to evaluate the role of the Arabs, and later the Spaniards, in the diffusion of irrigation technology, techniques must be studied individually. In the present study, the diffusion of three devices and one crop associated with irrigation are analyzed: 1) the dam as a method of raising water for diversion into irrigation canals; 2) the *noria*, or Persian wheel, a device for lifting water; 3) the *qanat*, or horizontal well, a method of concentrating groundwater and bringing it to the surface by gravity flow; and 4) the cultivation of oranges as an example of the spread of a domesticated plant that requires artificial irrigation. In each case the pattern of diffusion is the same, or strikingly similar, and can be summarized in the following chronological scheme:

1) the locus of invention is the ancient Near East, particularly Persia; then
2) a stage of limited diffusion through the Mediterranean in classical times, most typically under Roman aegis; next,
3) an intensification of use and perfection of the technique by the Muslims, eighth to tenth centuries;
4) a particularly dramatic intensity of use in Al-Andalus, which in turn becomes
5) a new center of diffusion back to North Africa and, later, to the New World.

This pattern could be extended to other, less well documented aspects of agricultural and irrigation practice.[2]

Diversion Dams. The history of the invention and early spread of diversion dams is problematical. Most probably it is a case of independent invention and homologous development in various localities. Henri Goblot nevertheless sees a certain Iranian influence in this hydraulic device, as in others, more easily traced, that date to the Achaemenian era.[3]

The crucial distinction in the history of dams is between storage dams, a late invention, and diversion dams, which have a long history in antiquity. The *azuds* of eastern and southern Spain are part of this tradition, and the word (as well as the Arabic *sudd,* from which it is derived) has always had the connotation of diversion, never of storage. The word *sudd* or *sadd* (from *sadda,* to close) is most typically associated with South Arabia.[4] The most famous diversion dam in history was the Marib dam (*sudd Mārib*) in the Yemen. Its date of construction is not clear, although Goblot believes it could be as early as the fifth century B.C. (in which case Iranian influence in its building would be more probable than for a later date of foundation).[5] The destruction of the dam in the sixth century A.D. was immortalized in the Koran. The Marib dam was not designed for storage, but purely for diverison of the water of the Wadi Dhāna into two canal distribution systems.[6] Like all diversion dams its purpose was, first, to slow down water — a matter of importance in diversion from intermittent streams that flow with great velocity during flash floods — and, second, to raise it to a level suitable for diversion. The technique is similar, whether used in run-off systems as in ancient South Arabia, Roman North Africa, or Umayyad Syria, or in irrigation from perennial streams as in Spain. It was a technique well known to the Arabs and particularly to the Yemenites, who settled in eastern Al-Andalus.[7]

The fifteenth-century author al-Ḥimyarī (probably a native of Ceuta in spite of his South Arabian name), gives an excellent description of the raising function of diversion dams near Lorca:

> Lorca is located on a river which flows eastward . . . and whose water is used for irrigation in the same way as in Egypt. This river, in the region of Lorca, has two different beds,[8] one higher than the other. When there is need to irrigate the high ground, the level of the river is raised by means of dams (*bi'l-sadād*) until it reaches the higher channel. Then it can be used for irrigation. At several places on the river there are *norias* for the irrigation of gardens. It also feeds some important canals (*jadāwal*), each of which permits the irrigation of ten parasangs of land, or more.[9]

The masonry dam was a usual feature of Spanish irrigation systems in colonial Latin America. Spanish dam-building activity was particularly characteristic of Mexico, New Mexico, and California, where Spanish systems of water control were generally more efficient than the pristine native ones.[10]

Norias. The origin of the *noria,* a wheel fitted with buckets for lifting water, is most likely to have been in Persia, as the generic term "Persian wheel" indicates.[11] The Persian wheel was known in the Mediterranean in late antiquity; Vitruvius (circa 30 B.C.) clearly described a current-driven *noria.*[12] This distribution of water-lifting devices, as well as *qanats* and other such hydraulic apparatus, is directly contingent upon the economic return promised relative to their cost. The wheel moved by stream current was large and expensive and was confined to the zone between the twentieth and fortieth parallels. It was almost wholly associated with irrigation, the high return which irrigated crops bring in arid lands more than sufficing to offset the initial investment.[13]

The wheel moved by animals or men (called *sāqiya* in Egypt, *sāniya* in North Africa, and, hence, *aceña* and *sinia* in Spain[14]) requires gearing and therefore was probably invented later than the current wheel, which does not. Nevertheless it enjoys a much wider distribution, having diffused not only along the Mediterranean coast from Greece to Spain but also to the north of the irrigated zones (as in France) where it is applied to nonagricultural uses.[15] The inexpensive and simple *shadūf,* or balanced bucket with counterweight (Latin *ciconia,* as in the *Etymologies* of Isidore of Seville; Castilian *cigueñal*) has an extremely wide distribution, even in northern lands (Scandinavia, Poland, Canada) where *norias* are not found.[16]

The diffusion of the *noria* is typically associated with the Arabs as agents of diffusion. In North Africa the *noria* is found in the coastal zones of Morocco and, in less dense concentration, along the northern edge of the Sahara from the Tafilalet to Tripolitania, but not farther south.[17] The distribution of the current wheel in the Iberian peninsula conforms to the area of stabilized Islamic settlement (below the line of the Duero and Ebro rivers; see Map 4), although in Portugal there is some diffusion north of the Duero. In Spain the large current wheels are particularly characteristic of Andalusia and Murcia; in Andalusia they were used to take water directly from large streams (the Guadalquivir and its tributaries, for example), whereas in Murcia they were generally located on large irrigation canals. The famous wheel at La Ñora, driven by the current of the Aljufia Canal, is the prime example of the Murcian style.[18]

The animal-powered *noria,* still a common device all over the peninsula and perhaps the most frequently remarked visible symbol of the Islamic imprint upon irrigation, was first introduced with the Muslim conquerors in the eighth century. (The

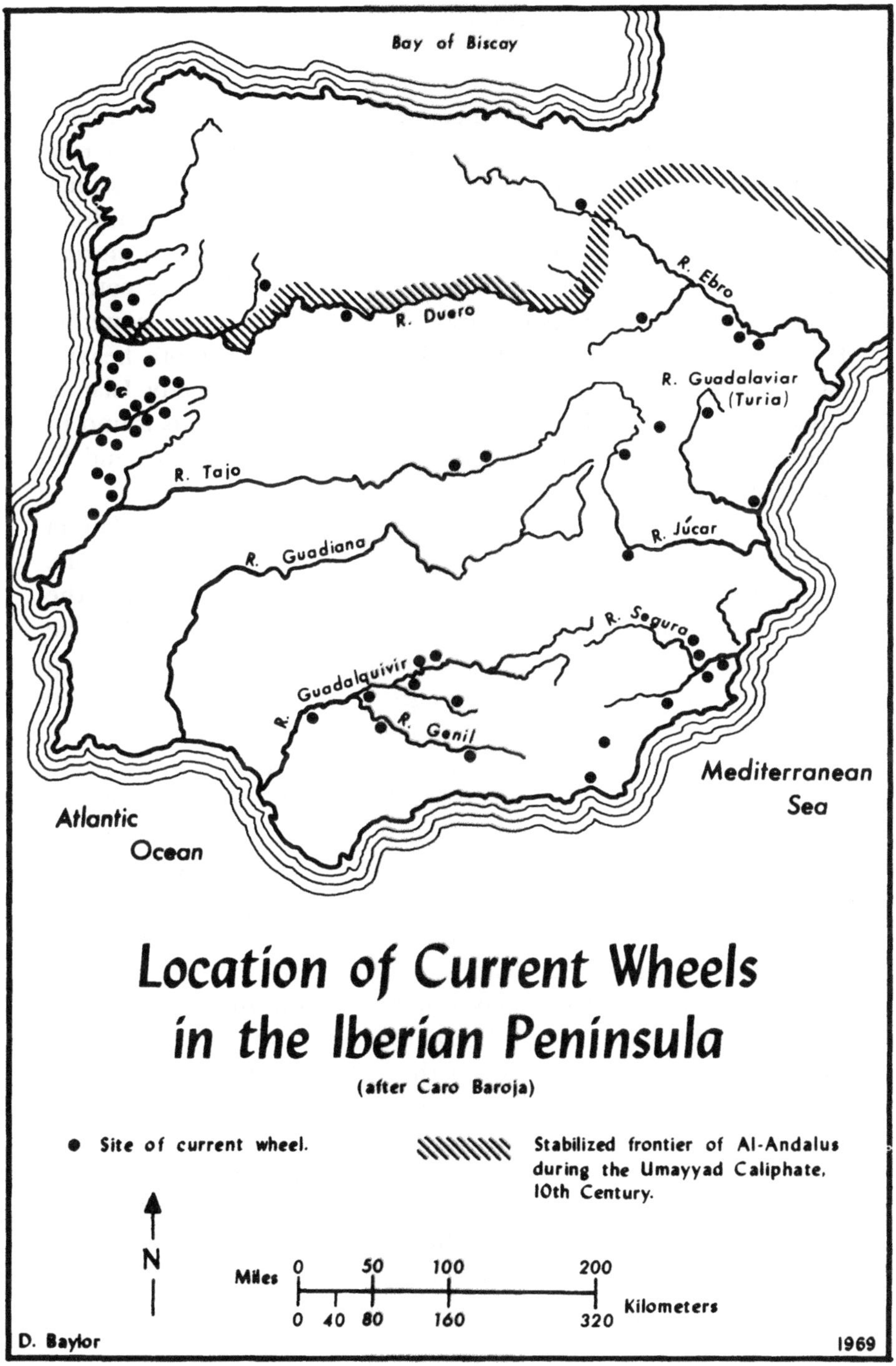

Location of Current Wheels in the Iberian Peninsula

(after Caro Baroja)

Romans had used current wheels in Spain in imperial times.) Its impact on the countryside was immense. As Caro Baroja points out, it was "a machine which, owing to its cheapness, has permitted more than any other the development of the smallhold or familial exploitation [of the land]." [19] These small *norias,* although found widely in areas which escaped Islamic control at an early date (such as the area around Gerona), are nonetheless most intensely concentrated in the southern, more heavily islamized areas. In Portugal, for example, the typically Islamic wheel with a chain of buckets (Portuguese *alcatruzes*) is found in especially high concentration in the southwest (Algarve) and in the central region, north of Lisbon (see Map 5).[20]

There are several references in the Repartimiento of Murcia to the association of wheels and irrigation canals, for example, "la açenna que tira agua de la açequia." [21] In 1271, after the council of Orihuela had sought the King's aid in regard to "the *norias* and the *sinias* which cost them much to make and keep in repair," Alfonso X of Castile agreed to cede certain dues "to all those who have their estates and who irrigate with *añoras* or with *acenyas,* or who build them from now on." [22] The Repartimiento of Valencia also refers to a wheel: a grant to the Dominicans in 1239 provides for "one wheel only in that canal which goes to the mill of B. de Turol, and they might raise it so that it could draw out a good irrigation of water, by day and night, continuously and always; and [they must contribute] something to the maintenance and cleaning of the canal." [23] Nevertheless, hydraulic wheels did not play much of a role in the Valencian huerta.[24] They seem to have been employed only extraordinarily, as in the plans of the officials of the city to build a *sinia* to irrigate the garden of the Lonja, in 1529 (Appendix 7).

Islamic and, later, Christian Spain became a new center for the diffusion of the *noria.* The first hydraulic wheel in Fez, Morocco, built in the thirteenth century, was the work of an Andalusian, Muḥammad ibn al-Ḥajj of Seville, possibly on the model of the great current wheel in the River Tajo at Toledo.[25] The Spaniards and Portuguese, in turn, introduced the *noria* into the

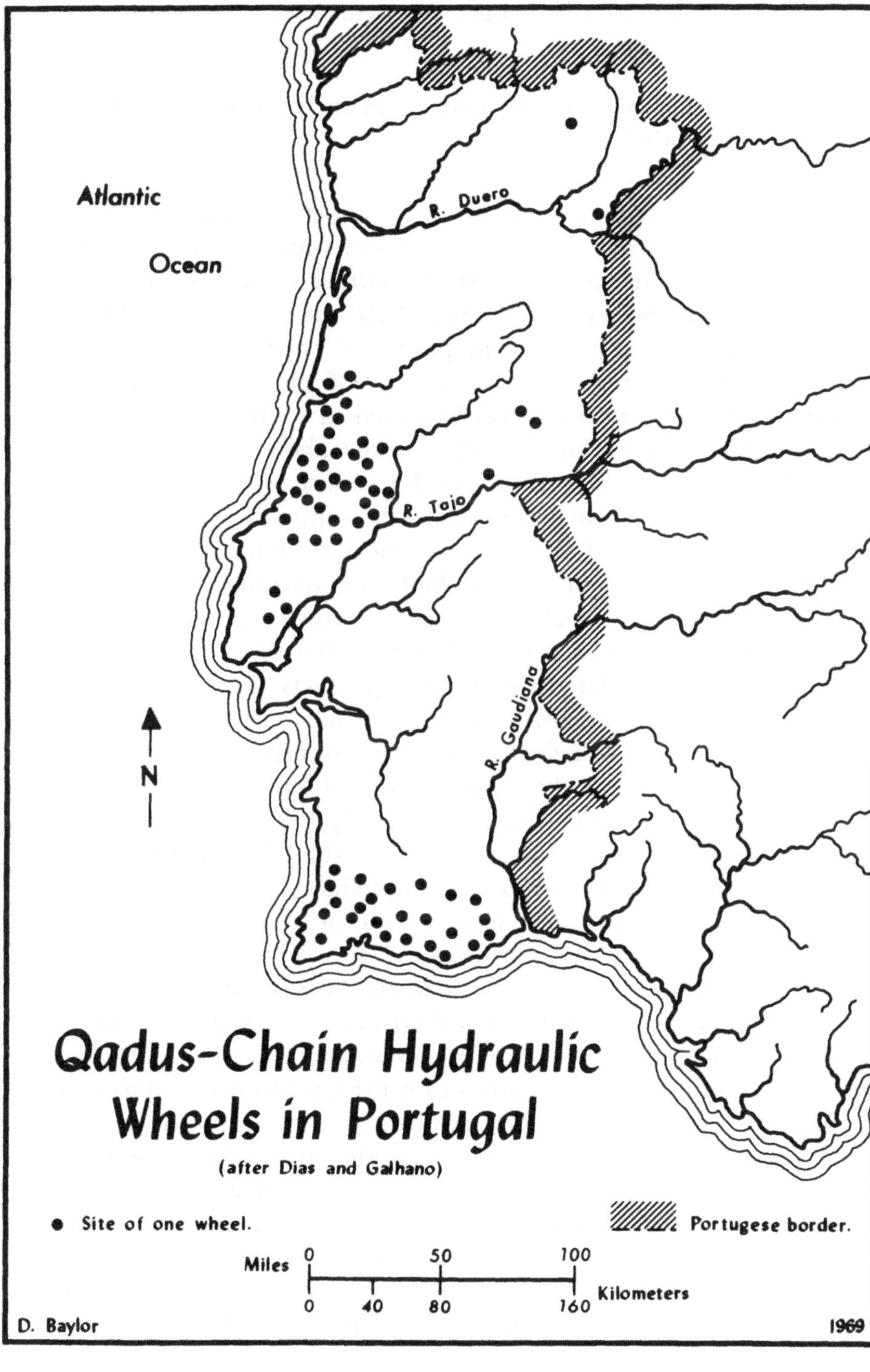
Atlantic
Ocean
R. Duero
R. Tajo
R. Gaudiana
N
Qadus-Chain Hydraulic
Wheels in Portugal
(after Dias and Galhano)
Site of one wheel.
Portugese border.
Miles
0
50
100
0
40
80
160
Kilometers
D. Baylor
1969

New World, where it was used extensively in Mexico and less widely in South America.[26]

Qanats. A *qanat* is an underground conduit for tapping groundwater made by sinking a series of wells and connecting them underground; hence, they are sometimes called "chain-wells." Only the first well (the "Mother Well") is a true well, the rest serving as sources of air and points of entry for maintenance workers (see Figure 6).[27] The history of the *qanat* closely parallels that of the *noria.* It was first used in western Persia, northern Mesopotamia, and eastern Turkey about twenty-five hundred years ago.[28] The technique for its use was widely diffused under the Achaemenian and Assyrian empires, and both archeological remains and literary references — Herodotus described them — are abundant. *Qanats* had the advantage of using less slope than surface canals, of preventing loss by evaporation, and of avoiding the construction of lifting devices.

The *qanat* antedated the use of aqueducts and was associated with their development. The Romans put *qanat* technology to two uses. First, as a device for tapping groundwater and conducting it from its source they often preferred *qanats* to wells, which would have necessitated large and costly lifting devices; such a system was constructed near Tunis on the plain of La Soukra. Second, the subterranean parts of aqueducts were often supplied with the breathing wells characteristic of *qanats.* Roman aqueducts in Tunisia and Lyons (France) survive as examples.[29]

The Arabs brought the *qanat* to the West on a large scale. In North Africa, where they are called *foggaras* (in Algeria, that is — they are *hattaras* in Morocco), *qanats* are typical of the Saharan oasis environment. The *qanats* of southern Morocco (Marrakesh and the Tafilalet) and southern Algeria represent the greatest development of this technology outside the Persian core area.[30] The two Spanish *qanat* systems that have been studied, those of Madrid and Crevillente, were also built by Arabs.[31] The Madrid system, which still functions, has been investigated by Jaime Oliver Asín, who associates the name of the capital with the Latin *matrix,* source of water, developing his etymological hy-

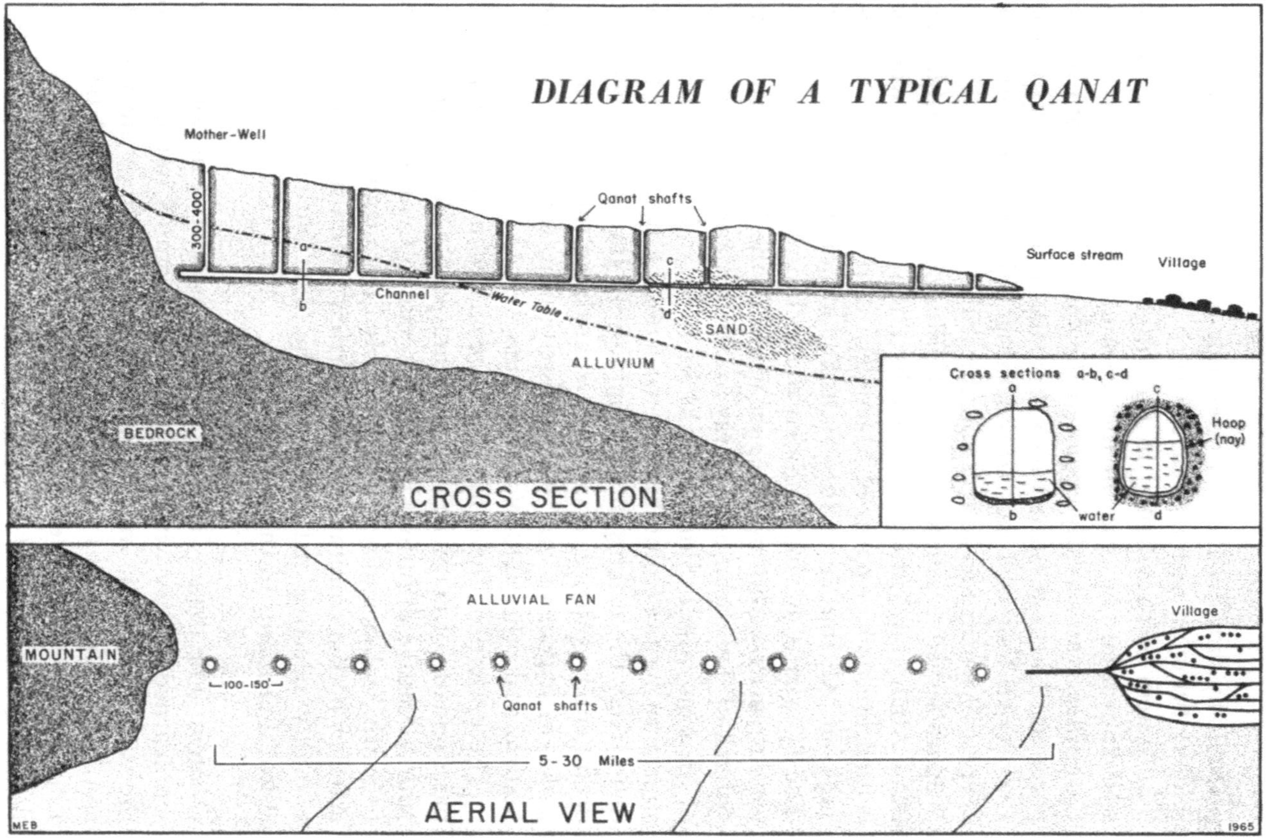
DIAGRAM OF A TYPICAL QANAT
Mother-Well
300-400'
Qanat shafts
a
b
c
d
Channel
Water Table
SAND
ALLUVIUM
BEDROCK
Surface stream
Village
Cross sections a-b, c-d
a
b
c
d
Hoop (nay)
water
CROSS SECTION
MOUNTAIN
ALLUVIAL FAN
100-150'
Qanat shafts
5-30 Miles
Village
AERIAL VIEW
MEB
1965

pothesis by showing the city's dependence on a system of *qanats* (*viajes*) bringing water from the springs of the Guadarrama.[32] The system of Crevillente, in the kingdom of Valencia, is less well known and lacks an up-to-date study.[33] It was described by Cavanilles, who actually explored the conduit to a depth of 200 paces until he was turned back by bats. This *qanat* was nearly 1,500 yards long and had 19 breathing wells (*respiraderos*), the deepest of which was 40 feet and the shallowest 11.[34]

Spain served as a new nucleus of diffusion. It is probable that ʿAbd Allāh ibn Yūnus al-Muhandis ("the engineer"), who executed the *qanat* system of Marrakesh during the reign (1107–1143) of ʿAli ibn Yūsuf ibn Tāshufin, was of Andalusí origin.[35] The Spaniards, having obtained the technique from the Arabs, were themselves authors of *qanat* works in the New World. The systems of Tehuacan and Parras in Mexico are almost certainly of Spanish origin. Similar works in Peru and Chile, however, appear to be pre-Columbian. It will not be possible to resolve definitively the origins of these putatively Inca systems without documental evidence, or without comparing the institutions of *qanat* management and water distribution with those of the Madrid and other Old-World systems.[36]

The pattern of the diffusion of *qanat* technology (Map 6) is paradigmatic of many irrigation techniques: a Persian core area with successive diffusions outward — east to Central Asia, south to South Arabia, and west across the Mediterranean. The Atlantic acted as both a conduit of and a barrier to diffusion, producing a highly selective patterning in the distribution of Spanish cultural elements in the New World.

Oranges. The orange, like the *noria,* is associated in popular or semilearned history with the Arabs, to whom its importation to the West from India in the tenth century was almost universally ascribed by historians of the nineteenth century and before.[37] Like so many other elements of Islamic horticulture, it was present in Roman times, in Italy as early as the first century A.D.[38] Citriculture was revived by the Arabs, and under their rule it spread from Persia to Spain. (It should be pointed out that,

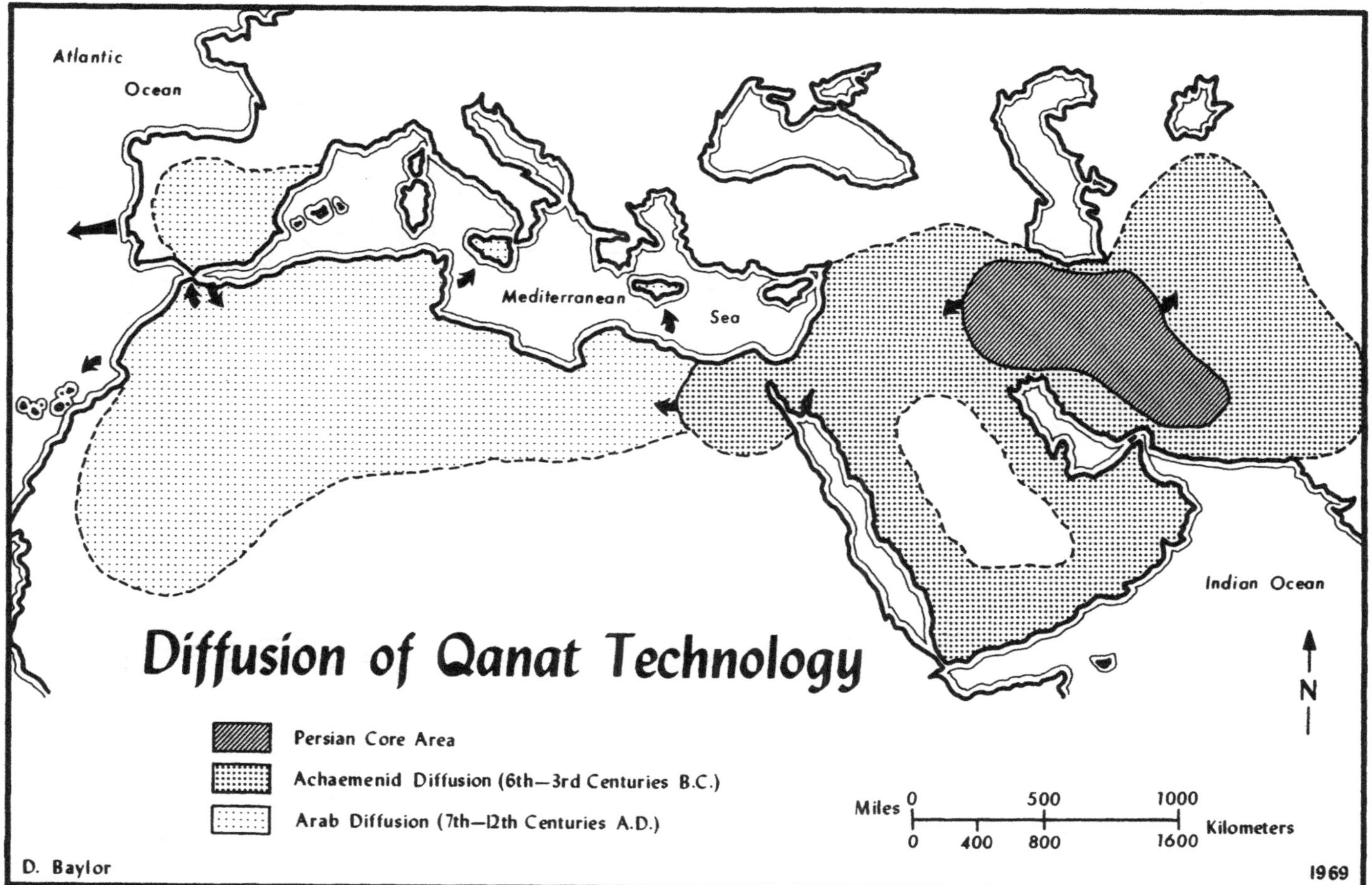

Atlantic
Ocean
Mediterranean
Sea
Indian Ocean
Diffusion of Qanat Technology
Persian Core Area
Achaemenid Diffusion (6th—3rd Centuries B.C.)
Arab Diffusion (7th—12th Centuries A.D.)
Miles 0 500 1000
Kilometers 0 400 800 1600
N
D. Baylor
1969

according to ibn Khaldūn the fourteenth-century Arab historian, oranges were considered inedible and were planted only for appearance.[39]) The Patio de los Naranjos of the Córdoba mosque remains a symbol of Islamic devotion to the orange. Al-Shakundī, a thirteenth-century Andalusí, wrote in his famous *Risāla* that "most of the houses of Seville, not to say all, were abundantly planted with fruit trees, such as the orange, the lemon, the lime, and the citron." [40] Oranges continued to be grown in Christian Valencia as an ornamental luxury crop.[41] From Spain the Conquistadors carried orange seeds to the New World, where by 1600 there were important groves in Peru and Chile. Ponce de León probably was responsible for the first Florida oranges and Spanish missionaries for those of California.[42] The Valencia orange has now become the "universal orange" in the United States.[43]

The agents of diffusion were, first, the irrigators themselves. The process of colonization in the Mediterranean world, as Fernand Braudel characterized it, meant "to plant the same trees, the same plants, to have before one's eyes the same landscapes and to place the same foods on the table . . . to live beneath the same sky and to watch the familiar succession of the seasons" as in the land of one's origin.[44] The Arabs were particularly conscious of the deliberate recreation of eastern landscapes in the West. Under the rule of the displaced Syrian Umayyad dynasty, the Arabs of Valencia and Córdoba strove to duplicate the environment of Damascus and its lush, irrigated Ghūṭa. They did so with the same techniques and institutions as those they had perfected in the old country. The same was true of the Yemenites who settled the huertas of Alicante, Elche, and Novelda. Simple cultivators could transmit customary arrangements regarding water distribution and knowledge of the more simple hydraulic techniques. A relatively untrained man could build a simple hydraulic wheel. Sophisticated devices such as large current wheels must have been built by professionals such as Muḥammad ibn al-Ḥajj. *Qanats* must not only be built by professionals (ibn Yūnus is an example), but must be carefully planned inasmuch as it takes years to build one. This would indicate the need for some formal direction by an administrative or political agency.

INSTITUTIONAL CONTINUITY

The "hydraulic" approach to irrigation is from the top; it concentrates on the centralized administrative direction of hydraulic works. Yet, if examined from another point of view, irrigation systems in the ancient Near East and in the classical and post-classical Mediterranean worlds reveal institutional patterns of a local nature which exist wherever irrigation agriculture is practiced and which are ecologically, more than culturally, derived.

Similar arrangements dictated by the necessity for fair distribution of water and the avoidance of conflict have been elaborated in quite different cultures. Mediterranean peoples have consistently adopted certain principles reflecting the need for cooperation. These concepts are found not only in the Code of Hammurabi (Middle Assyrian Laws), but are also characteristic of the medieval Valencian system. Three such principles follow.

1) The concept of proportional distribution: the cultivator receives water in proportion to the amount of land he works.

2) The concept of the individual's responsibility to the whole community of irrigators: each irrigator has specific responsibilities for the upkeep of canals (he is usually responsible for maintaining those sections of public canals on which his property borders). The individual irrigator accepts certain responsibilities in water use, such as willingness to submit to a turn, and he is liable for damages caused to neighbors (especially those downstream) owing to his negligence or malice.

3) The idea that apportionment of water and policing of irrigation arrangements within a given system are the responsibility of the irrigators collectively. Local irrigation authorities are usually members of the community of irrigators and are rarely appointed from outside. Ordinary justice in irrigation is of a local nature, with recourse to higher organs of justice only in the case of appeal. Justice in irrigation must be local because first, owing

to the exigencies of cultivation it must be speedy, and second, only local men know the local custom (that is, the traditional arrangements for distribution based on the local topography and settlement pattern). Day-to-day control of irrigation matters is most typically invested in a council of local elders, irrigators themselves, with intervention by the central authority only in extraordinary circumstances. Irrigation systems, therefore, tend to be politically autonomous units which elaborate their own institutions of self-government.

Most of these elements are present in the Code of Hammurabi (even though only four terse paragraphs of it are specifically devoted to irrigation) and in some pertinent letters of the same ruler. Two provisions of the Code are devoted to the responsibilities of the irrigator for keeping his bank of the canal in repair; if he fails to do so, and a breach is opened which floods the fields of others, the delinquent must make restitution for the lost crops.[45] The remaining paragraphs deal with negligence in the act of irrigating, by which water floods the fields of neighbors. These concerns are evident in medieval Valencian ordinances defining the duties of riparian cultivators (*frontalers*) and in prohibitions against "over-irrigating" (*sorregar*) the crops, fields, or roads adjoining one's own field.

In regard to local justice, Hammurabi ordered an official to investigate a matter of irrigation and to see that "the elders of the city [and] the tenants of the waterlands hold a court." [46] Implicit in this order is the recognition that irrigation can be controlled only through cooperation of the irrigators according to custom as interpreted by a body of elders. Even though the Babylonian despots may at times have directed large-scale building and overhauls of canals and other hydraulic works with the use of corvée labor,[47] Hammurabi himself nevertheless sought to ensure the autonomy of everyday local institutions of control.

PATTERNS OF ROMAN IRRIGATION

Irrigation was widely known in classical Europe, but it was not essential to agricultural practice. "In the classical world,"

R. J. Forbes comments, "irrigation was not a complement but rather a supplement of farming." [48] In Greece[49] and Italy economy was the keynote of agricultural use of water. Cross-plowing and other techniques designed to conserve water in the soil from the winter rains were generally appreciated, and most plants were watered sparingly. Exotic fruit trees were planted in places where more than average moisture accumulated, at the base of mountains, for example, and their introduction from the East did not lead — at least, not immediately — to any intensification of the practice of irrigation. The prime locus of irrigation in classical times was the vegetable garden.[50]

Classical irrigation was highly developed only where climatic conditions were severe. In the provinces of Numidia and Mauretania, which owed their well-known prosperity in Roman times to intensive irrigation, diversion dams were placed at the mouths of wadis to catch run-off and hold it for a short period so that it could be distributed through a network of branch canals.[51] Roman irrigation in Iberia was probably quite similar to that of the North African provinces, which Iberia generally resembled in topography and climate although it was much better endowed with perennial water sources. But irrigation on the peninsula was mentioned only infrequently by classical writers. A few scattered notes in Strabo and Pliny[52] confirm the archeological findings that irrigation was confined largely to the coastal areas of Phoenician and Greek settlement.

The Valencian region is comparatively rich in evidence of Roman hydraulic engineering, the visible remains of which have long been noted by local scholars. Interest in their origin and use has been consistent since Antonio José Cavanilles studied the aqueduct of Chelva at the end of the eighteenth century.[53] Studies by Francisco Jaldero, Rafael Valls David, and Domingo Fletcher Valls describe an extensive canal network irrigating the Plain of Quart between the present sites of Vilamarxant and Pedralba and then following the Turia downstream to the site of the present city of Valencia.[54] The dating and function of these conduits raise questions not yet answered satisfactorily. Valls David thought the works described by Jaldero and himself were pre-Roman; but once the aqueduct traces entered the urbanized area served by the modern canals, he could not distinguish the ruins from the operative canals — which, in part at least, had been built from the remains of the older system. Nicolau Primitiu studied a section of this network within the city, in conjunction with the Roman sewage system and the canal of Rovella. The aqueduct described by Valls, he found, must have supplied water to the city in Roman times. But Rovella was found to be of definite Islamic construction, "simply because the said canal is found at a higher level than the sewers and, therefore, quite a bit higher than the Roman floor." (see Figure 7).[55] Unless archeological levels are clearly delineated, it is difficult to judge the age of surviving hydraulic works. It is hard, for one thing, to tell the difference between Roman and Arab construction simply by looking at it. Many hydraulic works in North Africa thought once to be Roman remains have now been proven Islamic.[56] The only sure way of proof is by chemical analysis of the mortar, and this has not yet been done in Valencia.

It would be difficult to tell by physical appearance alone whether the Quart network was used for irrigation of crops, for drainage, primarily for household uses of the owners of latifundia, or for milling.[57] If irrigation was indeed the prime function, one must still ask how intensive the cultivation was and how the water was distributed. Miquel Tarradell believes that the remains of rural dwellings are too dense for a latifundia agrarian structure and that, therefore, small- and medium-sized rural property must have predominated. The network of canals was capable of sustaining a relatively numerous population, on land (the Plain of Quart) that is today practically deserted.[58]

The physical means of distributing water must have been similar to those described by La Blanchère in North Africa — notably the use of masonry barrages — adapted to stream irrigation. A recent archeological find proves a striking continuity in distribution devices from Roman times to the present: this was the discovery in 1948 of five masonry sluice gates or turn-out partitions (*talladoras, tablachos*) from the third or fourth centuries A.D. that closely resemble their modern counterparts.[59]

Archaeological Levels of the City of Valencia

(after Nicolau Primitiu)

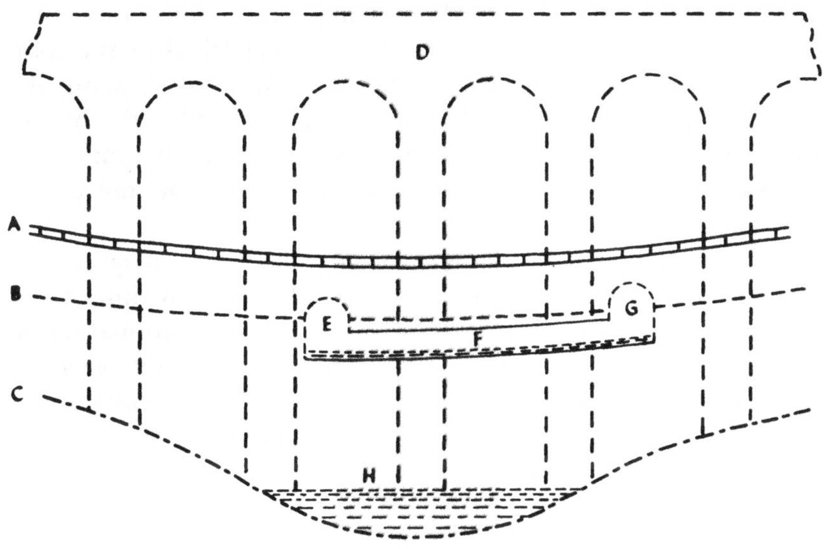

LEGEND:

A—Present-day street level.
B—Level of Islamic period.
C—Level of Roman period.
D—Roman aqueduct described by Valls David.
E—The Vall Vell ('Old Sewer').
F—Return-ditch from the Rovella Canal to the Vall.
G—Rovella Canal.
H—Present level of the R. Guadalaviar (Turia).

D. Baylor 1969

Finally, the history of alluviation at the mouth of the Turia makes it extremely improbable that there was much intensive irrigation in Valencia before the Middle Ages. J. M. Houston points out that "the coastline has been extended seaward over two and a half miles since the first century A.D. Evidence from the fourth century A.D. shows that the coastline has advanced only half a mile since then. This demonstrates that alluviation was much more rapid when the river was unchecked by the complicated pattern of irrigation canals over the coastal plain, developed chiefly during the Moorish occupation." [60] If the huerta could be excavated now and Roman vestigial canals found, the likelihood is that they would be remains of drainage, not irrigation, canals.

These problems — economic, archeological, and geographical — could not, even were our knowledge of them ten times more perfect, yield any information at all about the institutional mechanisms of distribution employed by the Romans. Yet we must attempt to form some opinion about what these institutions might have been: without knowledge of them we should be able to say very little indeed about the place of the Romans in the history of irrigation institutions.

HOW THE ROMANS DISTRIBUTED WATER

The Roman Law of Water was a compilation based upon the customs of many peoples living and irrigating within the Empire. Its basic doctrine, that of "riparian right," asserted the right of a man living on the bank of a stream to divert water for his own use, so long as he did not cause damage to those living downstream from him. Many concepts of the Roman water law were in evidence in medieval Valencia, whether as a result of continuity of practice or through reintroduction during the Roman legal revival fathered by James I. Evidence concerning specific customs for water distribution can only be deduced from the general propositions of the *Corpus*. Nevertheless specific concepts underlying the Valencian system of turns are present in the provisions of the *Digest*: 1) recognition of the effects on irrigation of the variability of temperature and water supply; 2) the concept of priority in usage; 3) the notion that rights must often be allotted for restricted periods only in order to achieve both equity and efficiency in the distribution of a limited supply of water.

That turns were basic to the Roman concept of water distribution is proven by references to the right to use water daily, every other day, every third day, and so forth.[61] The law recognized that rights could be granted to varying quantities of water expressed in specific and measurable time units, providing thereby a means for the control of limited resources.[62] The basic principle of allocation was that each irrigator was entitled to water in proportion to the size of his holding;[63] priority according to use, however, could take precedence over that of size.[64]

A few surviving literary references and inscriptions confirm that the principles of law represented actual practice. Frontinus described an orderly regimen of distribution to landowners drawing water from a brook near Rome "by turns on fixed days and in determined quantities." [65] Pliny, in the *Natural History,* describes irrigation of crops in North Africa: "There is a city-state of Africa called Tacape, in the middle of the desert on the route to the Syrtes and Great Leptis, which has the exceptionally marvelous blessing of a well-watered soil. There is a spring that distributes water over a space of about three miles in every direction, giving a generous supply, but nevertheless it is distributed among the population *only at special fixed periods of the day.*" [66] The most perfect description extant of a Roman irrigation system is preserved in a lengthy inscription found at Lamasba, in southern Numidia (now Algeria, near Batna).[67] The inscription documents in detail the distribution of water in turns among a group of fifty-two irrigating proprietors during the reign of the emperor Elagabalus (A.D. 200–222). The order of turns regulated water flowing from a spring called Claudiana and was established by a decree of the local senate and settlers ("ex decreto ordinis et colonorum"). That is to say, the irrigators themselves established the regulations by which they used the water. The water flowed down from its source into a main canal, "ad matriceriganda." The parcels were arranged in rows, probably along branch canals, some of which were above and some below the level of the main canal (the water is said to flow up to some and down to others).

The description of the turn begins: "Scala I. Ex VII Kal. Octob. primo mane quo Claudiana descendit ad matriceriganda h. I." That is, the water of the Claudiana flowed down to the main canal at sunrise, September 24. It is striking that the turn should be tied to specific dates rather than that it be variable, the cycle beginning again at whatever time everyone has had water. The significance of the September starting date — the equinox — is doubtless that it was a convenient date to assign as the customary beginning for the sowing of winter wheat.[68] There is further significance in beginning the turn at sunrise. The Romans usually computed the day from midnight to the following midnight, but in this case the custom was altered and the day reckoned from sunrise to sunrise, a change dictated by the practice of irrigators (as those of Valencia, who irrigated from "sun to sun").

Each proprietor is then listed by name, along with his rights to a certain amount of water:

MATTIVS FORTIS	K CCCVIII [69]	EX · H · I · D · VII KAL · OCTOBR IN H VS · D · EIVSDEM · P P · S · H · IIIIS
FLAVIVS ADIVTOR VET	K CCCL	EX · H · VS · D · VII · KAL · OCTOBR IN · H · XS · D EIVSDEM P P · S H V

The main problem of interpretation is the meaning of "K." Most probably it is a measure of land — *capite, capitationis,* as a synonym of *iugo, iugationis.*[70] This makes sense because water was granted in proportion to the amount of land to be watered, and in this inscription the ratio of K numbers to hours is constant.

Mattius Fortis had a field of 308 units, and he received water from dawn, ("ex hora prima") on September 24 ("diei VII Kal. Oct.") up to the fifth hour and a half of the same day ("in horam

V semis, diei eiusdem"): that is, for his share, "pro parte sua," four and one-half hours. Flavius Adiutor, the veteran, who was next in order, had 350 units. This was 42 more that Mattius Fortis' units; so he received five hours of water, from when Mattius' turn ended at the fifth hour and a half up to the tenth hour and a half. The turn continued up to the Kalends of December. During that time only four names are repeated, but probably with reference to different holdings, inasmuch as in each case the figures for the value of K differ. Each parcel apparently received only one watering during the autumn season (although, clearly, most would have received their water well after the actual sowing).

The many problems of interpretation aside, this fascinating inscription makes clear that communities of irrigators were formally constituted (Heitland refers to the "water-leet" of *aqua Claudiana*) and that they regulated by the imposition of a system of turns what probably was a modest supply of water in an intensively cultivated area.

Thus most of the basic elements of Valencia irrigation institutions were present in Roman law and practice. But the most significant evidence is from North Africa, and any claim concerning institutional continuity between Roman and medieval Spain would be conjecture.

WHAT MUSA SAW

In reply to an inquiry of Caliph al-Walīd, Musa ibn Nusayr, one of the conquerors of Visigothic Spain, was said to have answered: "There is not a single canal in this country." [71] Musa had strategic rather than economic considerations in mind, and, it is true, he did not campaign in the Valencian region. Nevertheless, this doubtlessly apocryphal anecdote serves to underscore the general impression that the Iberia which fell to the Muslim troops was not intensively irrigated. How different were the recorded impressions of the Christian conquerors centuries later who referred pointedly to the difficulty of campaigning in irrigated country!

Andalusí Muslims did, however, have some appreciation of Roman hydraulic remains. In the *Rawd al-mi'ṭār* there is a passage curiously reminiscent of late medieval Christian attitudes toward the Islamic past:

> A canal flows from the river of Murcia near the bridge called Ḳanṭarat Ashkābo, and it follows a conduit dug by the ancients (*al-'ūwal*) across the rock of the mountain, which they drilled for about a mile. This is the canal which irrigates the lands to the south of Murcia. The ancients likewise have dug another canalization across the mountain facing the preceding one, over a distance of about two miles, for the passage of water. This is the one which irrigates the lands to the north of Murcia. These two subterranean canals (*jadwalain*) are equipped with air wells (*manāfis*), which communicate with the crest of the mountain, and with sluice-gates (*manāhid*) cut in the direction of the river. . . .[72]

This description of a *qanat* system that has since disappeared is illustrative of the continuity perceived by the Arabs in Roman irrigation — the base on which they built and expanded in most lands of the Mediterranean they conquered.[73]

Al-Andalus, under Arab rule, experienced the full impact of the diffusion of irrigation technology from the East. But students of Islamic Spain (following Ribera's lead) have been reluctant to devote their energies to an aspect of social life almost ignored by Arabic writers. The dearth of Arabic source material has limited most writers to a repetition of the few allusions to irrigation in Al-Andalus known to all, most of which are found in the works of the famous geographers, such as Idrisi, and in poetic descriptions of the Spanish countryside.[74] Medieval Arab historians all but ignored irrigation, and even the writers of agricultural treatises restricted their comments on water to generalities concerning its properties and horticultural uses.[75] The kinds of documents which interest social and institutional historians — notarial records, contracts, bills of sale, records of litigation and arbitration, which supply the details of the daily practice of irrigation — have perished almost totally. Scarcely half a dozen have been found,[76] and these alone, while helpful, cannot fill such an immense void.

Taken as a whole, Arabic references to water — scant though they are — reinforce the image of a land where irrigation agriculture was taken for granted as part of the life of an urban society, where all cities of economic, cultural, and administrative importance were surrounded by irrigated "gardens." In the characterization of Leopoldo Torres Balbás, the cities of Al-Andalus were extroverted, opening outward to the lush gardens of flowers, trees, and vegetables which surrounded them.[77]

Geographers, in describing irrigated areas, were interested in the physical attributes of the system rather than institutional peculiarities. Al-ʿUdhrī's description of the Murcian system described above is typical:

> Its land is irrigated from a river like the Nile of Egypt which flows eastward and has its origin in the spring called Mulnaḥasha. . . . On the river of Tudmir [the Segura] are the *norias* (*nawaʿīr*) which irrigate its crops. The irrigation canal (*sāqiya*) extracted from it starts at Alcantarilla[78] and arrives at the estates of the people of the city of Murcia at the edge of the village of Tawṣ, which is one of the villages of Orihuela. Then the people of Orihuela begin to draw a canal from this river through its district until it comes to an end in a place called Catrál. The length of this canal . . . is twenty-eight miles.[79]

But when the geographers reached Valencia they invariably lost interest in canals (al-ʿUdhrī was fascinated by the town walls). A fleeting reference to a hydraulic wheel on the "Great Canal" near the Bāb al-Hanash is the most detailed reference to the Valencian network in the literature.[80]

For want of further detail, we can only reread the elegy of ibn al-Abbar, with its nostalgia for the verdant huerta,[81] or the plaintive lyric of the eleventh-century poet al-Waqāshī, who lamented the destruction of the noble and lush huerta by the ravages of war.[82]

X | Irrigation Administration in Al-Andalus: The Ṣāḥib al-Sāqiya

Of the officials who presided over the distribution of water in Al-Andalus, and of the administrative apparatus, procedure, and regulations that guided them in their duties, next to nothing is known.

Ibn Ḥaiyān recounted the story of the eleventh-century freedmen Mubārak and Muẓaffar, who were "in charge of the administration of irrigation [or of the canal, that is, *wakālat al-sāqiya*] in Valencia" and who then rose to the dignity of emir, in Valencia and Játiva, respectively.[1] We know also, from a sixteenth-century translation of an Arabic document, that waters of the Genil in Granada were apportioned in 1219 (616 of the Hejira), during the Almohade period, by a certain Aḥmad b. ʿAbd Allah, "the well-known apportioner of said river."[2]

Records of two water disputes between Andalusí towns fail to shed light on administrative processes, although they provide some detail regarding settlement of regional water conflict. Litigation in 1187 (583 H.) between two Andalusian towns, Bartillana and Lubros, was resolved by agreement between their sheiks and other irrigators upon the advice of experts and men who knew the local custom (a pattern ordinarily followed in Christian Spain except that the sheik's role would be taken by a lawyer or syndic representing the town and the cequier or other administrative officials would usually figure among the experts). Only after the disputants reached agreement was the document taken before the qadi of Guadix for official ratification.[3] A similar settlement was achieved in 1223 (619 H.) between the towns of Torres Torres and Carcer, in the Valencian region. After it had dragged on for twenty years, this dispute was finally argued before the chief qadi of Morvedre, Abū al-Ḥasan b. Fath, and the faqih, Muḥammad b. Ḥazben.[4] Again no administrators are mentioned in this document. Here the qadi represents a level of authority higher than the local level, to which irrigators had recourse when local means failed. In the same way, cases in Christian Spain which could not be solved by the cequier and the local administrative apparatus reached the court of the governor of the kingdom.

Even though a picture of everyday administration of irrigation in Al-Andalus cannot be outlined on the basis of these meager references, still a reconstruction of Islamic practice can be attempted on the basis of terminology applied by the Christians to their irrigation officials (*çabacequies, alamí,* and *alcalde de las aguas,* for example).

THE ṢĀḤIB AL-SĀQIYA

There are no actual references in extant Hispano-Arabic sources to an official called *ṣāḥib al-sāqiya.*[5] A derived arabism (*çabacequia*) first appeared in thirteenth-century Aragon,[6] and related forms were current in Valencia (*çabacequies, çabacequier*) and Murcia (*sobrecequiero, juez sobrecequiero*) soon after the two towns were captured from the Muslims.

As was true of other irrigation terms as well, the functions of these Christian officials and the semantic fluidity of their titles varied considerably from region to region. The *çabacequies-sobrecequiero* was the chief irrigation official in some locales (Burriana, Orihuela), whereas in others, such as in the Valencian huerta, the *çabacequies* was subordinate to the cequier. In some places, moreover, the irrigation officer served the autonomous community of irrigators (huerta of Valencia), while in others (Castellón, Orihuela) he was a municipal functionary. There seem to have been, thus, two models for the medieval Christian organization of irrigation administration: a cellular one, based on irrigation communities with their own elected officials, and a centralized one, in which irrigation administration was a branch of a higher jurisdiction, typically that of the town.

The municipal model is the one which more closely corresponds to my conception of the Andalusí ṣāḥib al-sāqiya. In the Islamic judicial system the representative of ordinary justice was the qadi, but specialized areas of law were entrusted to special magistrates, subordinate to the *qaḍā'*. Under the Umayyads in particular the administrative justice of the *qaḍā'* was broken down and specialized jurisdictions were created. These officials were favored in western Islam more so than in the East; particularly in Al-Andalus special organs of justice flourished. Al-Harashī, a jurisprudent of the Malikite rite (that prevailing in Al-Andalus) stated that the prince was empowered to institute special jurisdictions in determined categories of judicial process: he cites, for example, a qadi for matrimonial affairs, a qadi of the police, and, significantly, a "qadi of water." [7] This *qāḍi al-miyah* of Al-Andalus survived into the post-Islamic epoch as the *alcalde de las aguas* of Lorca, a municipal official whose duties and jurisdiction were similar to those of the *sobrecequiero* of Orihuela.[8]

In the caliphate of the Andalusí Umayyads, various of these magistracies subordinate to the *qaḍā'*, all with the title of *ṣāḥib*, came to play an important role in municipal life. Indeed, their function was to fill an administrative void in Islamic law, which lacked a specific body of municipal law and did not provide for any distinct municipal jurisdiction. The marketplace became the province of the *ṣāḥib al-sūq* (master of the market), whose surveillance also included guilds and matters of public health and morality. Criminal procedure and the town police were the charge of the *ṣāḥib al-shurṭa* (master of the police), also known in some cities of western Islam as *ṣāḥib al-madīna* or master of the city. The essential characteristic of these magistrates was that — unlike the qadi — they could initiate action. Characteristically they were out on the beat, in the marketplace and on the streets of the city, seeking out misdemeanors and exacting punishment. The qadi was not competent in criminal matters, and therefore these autonomous magistracies rounded out the *qaḍā'* jurisdiction and responded to the practical necessities of urban life.[9]

These three *ṣāḥib* offices were adopted, at one time or another, by Christian towns neighboring Al-Andalus. The *ṣāḥib al-shurṭa* first appeared in Christian Spain in the late tenth century (*zahbascorta,* in a document of 998), the *ṣāḥib al-sūq* in 1020 (*zavazouke*), and the *ṣāḥib al-madīna* in the twelfth century (*çalmedina, çavalmedina*).[10] The *çavacequia* first appeared a century later but survived considerably longer than the others. The tradition of the *zavazouke,* however, was carried on by another Islamic transplant, the mustasaf (in Valencian and Catalan usage; *almotacen* in Castilian[11]), paralleling the supplantation of the Umayyad *ṣāḥib al-sūq* by the Abbasid *muḥtasib.* The *muḥtasib's* functions were identical to those of his predecessor but were executed under the guise of specific religious precepts.

Because the modus operandi of the Christian cequier was similar to that of the mustasaf, and because the mustasaf inherited the administrative procedure of the *muḥtasib* with only slight modification,[12] there are grounds for assuming that the Andalusí ṣāḥib al-sāqiya acted in ways similar to those of the muḥtasib. The cequier, the mustasaf, and the *muḥtasib* all initiated actions; they all heard small disputes (small, that is, in terms of the fine required by the misdemeanor involved) based upon customary law as embodied in municipal, communal, or guild regulations or custom. They could both sentence and fine summarily, and their legal business was conducted by oral process (perforce: in Islamic Law only the qadi has jurisdiction in cases where any evidence in writing is admitted).[13]

The affinities between the two Christian officials in jurisdiction as well as in procedure points to their common background as municipal magistracies in Al-Andalus. In medieval Valencia the mustasaf of the city and the cequiers of the canals which passed through it (Rovella and Favara) came into frequent jurisdictional conflict. The mustasaf was inspector of the mills, as of all economic activity, but mills were users of canal water and therefore also under the surveillance of the cequiers. Both cequier and mustasaf had competence over the condition of the roads, the mustasaf being interested in their state generally with an eye toward the comfort of pedestrians and travelers, the cequier in seeing that canal water did not flood them. Both officials were interested in matters of public health, such as might arise from stagnant water, sewers, and industrial waste produced by tanners, dyers, and other urban craftsmen who used canal water. Both were regularly on patrol, searching out misdemeanors, the mustasaf on behalf of the town weal, the cequier carrying out his charge from the community of irrigators.

Even though no direct evidence can be brought to bear upon the Andalusí ṣāḥib al-sāqiya, therefore, comparison with other officials of similar name, procedure, and jurisdiction makes possible an educated guess concerning the nature of the office. Like the other ṣāḥib magistracies which passed into Spanish usage, the ṣāḥib al-sāqiya of Al-Andalus must have been an urban officer with limited police powers and the ability to judge summarily everyday infractions of water law or other regulations contingent upon water use. He would have been charged with ensuring equity in the distribution of water, just as the *muḥtasib* was the guarantor of the equity of the marketplace. He would not have been elected by a community of irrigators but would have been the appointee of the governor or prince. In addition, he would have been charged with carrying out basic Koranic precepts concerning the public nature of water and the duty to share it equally and keep it clean and pure. His charge to do so would have been in general terms—just as the *muḥtasib* was enjoined "to encourage what is good and to discourage what is evil"—which were not necessarily embodied in written regulations. On the other hand, and beyond his discretionary powers, he would have had to enforce specific customs, arrangements, and legal precedents regarding turns, the repair and cleaning of canals and diversion dams, and the stealing of water and other common irrigation infractions.

Like the *muḥtasib,* the ṣāḥib al-sāqiya would have come into frequent jurisdictional conflict with officials representing other levels of administration or social groups, especially tribal authorities. His authority therefore would have diminished (or have been increasingly disputed) as he moved farther from the main urban canal or canals. His primary responsibilities would have been care of the diversion dam, cleaning and policing the main canal, and establishing the turns on the main canal. The secondary canals, such as the "Beni" canals of Murcia, would have been more the responsibility of tribal leaders (this would not preclude the *ṣāḥib's* role as coordinator).

THE AMĪN AL-MĀ'

The ṣāḥib al-sāqiya was not the only irrigation official known to Al-Andalus. A lesser functionary, the *amīn al-mā',* was characteristic of the smaller irrigation systems and was usually subordinate to a more powerful official with wider jurisdiction. The *amīn al-mā' is* considerably less of a puzzle than the ṣāḥib al-sāqiya. The office persists in modern Islamic systems which have been studied; the derived Christian institution is well documented; and medieval sources are extant showing the office in transition.

Amīn means "reliable, trustworthy," and, as a noun, "guardian, keeper, superintendent," and was applied to a wide variety of officials to whom were entrusted functions of economic or administrative responsibility.[14] The derived Christian officer was known in Romance by the Valencian arabism alamí (*alamín* in Castilian)[15] or by the literal translation, *fiel* (as in the *fiel de pesos y medidas* or the *fiel de aguas* of Elche). The amīn differed from the cequier in that he was solely an administrative official with no criminal competence. He never initiated action against malefactors or exercised powers of summary justice.

The *amīn al-mā'* is the typical water officer of the Saharan oases and the Iberian alamí appears in places whose institutions strongly resemble them (notably Elche and Novelda). The amīn is concerned largely with the distribution of water and direction of turns and has often been associated with systems in which the sale of water has made the distribution so complicated that a special official is needed simply to keep track of all the transactions and ensure order in the turns.

Sir Clements Markham's description of the alamí of Novelda at mid-nineteenth century is one which accords completely with descriptions of Saharan amīns:

> The official, to whom is entrusted the difficult and confidential duties of managing the distribution and sale of the water, is nominated by the Municipality. He is called the *Alami,* a word of Arabic origin; and the post is one requiring so much special knowledge, that it is always given to the members of certain families, who have held it from time immemorial. . . . After the sale of the next night's *hilos,* those of the next day are sold, and the *Alami* then arranges the distribution of the water according to the number of hours it is to flow in this or that channel. . . .
>
> It will at once be seen that the duties of the *Alami* require constant attention, much care, and an intimate knowledge of the localities, for their satisfactory performance. He has to remember where the *hilos* of day and night irrigation have been delivered during the previous twenty-four hours, and which is the nearest among those sold on the current day, to combine the directions in which the water should flow with requisite volume, so that there may be sufficient in the places where each *azumbre* is to be delivered, adding to, or diverting from, the various channels, so as to secure these ends.[16] These complicated operations can only be satisfactorily performed by persons who have had long experience, and are withal quick-witted and intelligent.[17]

The similarity of Markham's alamí to Saharan amīns is striking. In Ghadames and Touat, where water is sold, the amīn sets and measures the hours of distribution of water. In Touzer, the intricacies of distribution at the beginning of the present century were known by one man only, the *amīn al-mā',* who had learned them from his father, who in turn had received the information from his ancestors.[18]

A medieval Romance document attests the existence of this

officer in Islamic Spain. The Moorish quarter of Elche, located in the Arrabal de San Juan, was irrigated from the Marchena Canal, a channel constructed in Christian times especially for the use of the *morería*. Nevertheless, it was administered autonomously by the Muslims themselves and the pertinent administrative officer was the alamí. In a document of 1435, Obaquer (Abū Bakr) Ismael, "*alamí*, or divider of the water, of the huerta of the Moors of the town of Elche during the past year," rendered an accounting of all the units of water "which were sold by me, the said *alamí*." [19] Similarly, in an account of the dues paid by the Moorish quarter in 1461, it was noted that "the queen possessed 43 units [of Marchena Canal water] which were sold each Friday by the *alamí* or *fiel* of the Moorish quarter (*morería*)." [20]

REGIONAL VARIATION

Proof of continuity in two administrative traditions would not authorize an assertion that irrigation administration was homogeneous throughout Al-Andalus. On the contrary, significant regional variations existed. The *acequiero* of sixteenth-century Granada, whose procedure was known at the time to be an Islamic inheritance, was quite distinct from his Valencian counterparts. In Granada the *acequiero* had jurisdiction over the mills and the tanning shed, as distinguished from Valencian regulations attributing these jurisdictions to the bailiff and the mustasaf. Moreover, the *acequiero,* though a municipal appointee, was paid by a weekly assessment of three *maravedís* per mill and received no other salary. The city was obliged to supplement this income with an annual stipend, since no one could be found to take office on the old terms.[21] The Valencian officials, on the other hand, either were paid annual salaries or received part of the fines they collected or both.

Furthermore, while both traditions studied here are characterized by appointment to an urban jurisdiction, there was also in Islamic law — especially Malikite law — a strong tradition of communal irrigation, characterized by autonomous administration. In Islamic law irrigation canals are the communal property of the individuals who contributed to their establishment. The commons alone have the right to use the water of their canal for irrigation and they rule the affairs of the canal. The co-irrigators set the turns, and no one of them may erect a mill or even build a bridge over the canal without the agreement of all.[22] Just as in Christian Spain, then, it is probable that in Al-Andalus centralization of irrigation administration was an independent variable. In some localities original communal or tribal autonomy may have been ceded to a higher authority; in others it may have persisted until the end of Islamic rule.

XI | Proportional Distribution and the Measuring of Water

Proportionality was the organizing factor of the water distribution systems of medieval eastern Spain. Each irrigator received water in proportion to the amount of land he held. The water to which he was entitled, moreover, was not a fixed amount per unit of land, but rather a proportional one which varied with the volume of the river. All irrigators shared in times of abundance and were equally deprived in times of drought. Likewise, the total debit of the river or spring was divided among the principal canals taking water from it, in proportion to the amount of land served by each canal. The water of each main canal was, in turn, subdivided proportionally among its branches. The proportional quotas for each carrier were expressed in an abstract measurement unit, usually the "thread" of water (fila in Catalan, *hila* or *hilo* in Castilian). A fila thus represented one share of the total amount of water in a given river (or sector of a river), spring, or canal.

The abstract unit fila was translated into the reality of volumes of water by the dimensions of physical structures — diversion dams, irrigation canals, and divisors — which divided the water that flowed in any channel proportionally among the branches into which that channel divided. In situations during which water was so short that these physical structures could not at the same time preserve the principle of proportionality *and* deliver water efficiently, equality was preserved by the institution of a system of turns called tanda, *dula,* and so on, whereby the irrigator, instead of drawing water at will, was obliged to irrigate in a specific order and could not irrigate again until his next turn.

One of the advantages of the proportional system was that equitable distribution could be assured without measure of time or of delivery orifice. When water was abundant the farmer took it at pleasure. When water was scarce a turn was imposed, and as the quantity of water available grew less, the interval between successive turns became greater. But no measure was needed in this eventuality, either. As long as the irrigator did not waste water (and there were penalties to ensure that he did not), he could only use so much; he had only a certain amount of land and could not overwater his crops without harming them. Therefore the only control needed to ensure equality of partition was to see that he could not water again until everyone else had had a turn.

Of the three components of the system of proportional distribution — the fila, the dam and divisors, and the turn — only the last two were palpable. Divisors and turns assured, or were designed to assure, that water from the river would reach the individual farmer in an amount proportional to the area of land which he had the right to irrigate. These complex arrangements (antedating the Reconquest) were preserved through custom and by ensuring that the dimensions of divisors and canals were not altered. But the fact that rights were often expressed in filas, the abstract and imaginary unit, has caused no end of confusion among students of Spanish irrigation.

In traditional irrigation systems that function on the basis of custom, and where proportionality is the basis of water distribution, irrigators may not always understand the rationale of the apportionment. E. R. Leach's observations of such a system of distribution in modern Ceylon — whose irrigation customs, like those of Spain, have developed over a period of more than one thousand years — are relevant to the Valencian system:

> It must be pointed out that from the Vel Vidâne's [irrigation officer] point of view the ratios [of distribution] are established by tradition; they are not something which he has to work out for himself. Moreover the ratios are not

> thought of as fractions but as *karahankota* [wooden divisors] of different sizes with the grooves measured in inches. . . .
> The complexity of the arrangement is itself relevant to my theme since such a system is virtually unalterable. Although the present generation of Pul Eliya villagers are not at all clear about the inner logic of it all they are keenly aware that the numerical formulae handed down from ancient times are very important. The general view seems to be: "We don't understand why things are arranged like this, but this is how they are, and we had better leave them alone." [1]

So it was that in medieval Valencia, when it was said that such a canal received so many filas of water from the river, the unit was not understood as being measurable; it was simply a formula for expressing proportional rights. The actual quantity of water represented by the amount of filas was secured by keeping unchanged the other arrangements: the system of turns and the physical dimensions of the distributory network. Nevertheless, to dismiss the fila simply as a ratio does not reach the heart of the historical problem; for there is evidence that it was originally in fact a *specific* ratio.

My analysis of the origin of the fila proceeds from one initial observation: the traditional quotas of filas of river water for different Spanish rivers, irrigation districts, or principal canals are typically expressed in multiples of twelve. Thus the Mijares, at the point where it serves the four towns of Castellón, Burriana, Villarreal, and Almassora, is said to carry sixty filas.[2] The river Sans at Játiva has twenty-four filas,[3] as has the river of Lorca.[4] The Royal Canal of Alcoy (serving Gandia and Oliva) takes forty-eight filas from the Serpis at the diversion dam of Alcoy.[5] The Moncada Canal of the Valencian huerta also carries forty-eight filas. These figures suggest that one fila is the equivalent of one hour of water, an assertion supported by the Marqués de Molin's definition of the *hilo* of Elche as "the twelfth part of the water which flows through the canal in the period of twelve hours." [6]

In cases where water is divided by time units as well as by filas, the equivalence of the hour and the fila is clear. There is such a twofold definition in the division of the water of the Mijares in 1346 (Table 18) by Prince Peter of Aragon, who de-

Table 18

Division of the Mijares River in 1346

Town (upstream to downstream)	Filas	Hours	Days
Villarreal	14	28	1 day 4 hours
Castellón	14.5	29	1 day 5 hours
Almassora	12.5	25	1 day 1 hour
Burriana	19	38	1 day 14 hours
Totals	60 filas = 120 hours = 5 days		

creed that in times of drought the water above the divisor of Villarreal should be considered as divisible into sixty filas, "or equal parts," of which fourteen were allotted to Villarreal, fourteen and a half to Castellón, twelve and a half to Almassora, and nineteen to Burriana. But should the water be so insufficient that not even one fila reached Almassora (the area served by the lower section of the canal of Castellón), then all the water of the river would flow into the canal of Villarreal for twenty-eight hours continuously and thereafter to Castellón for twenty-nine hours, Almassora for twenty-five, and Burriana for thirty-eight.[7]

Assuming that the Mijares fila was "doubled" in times of drought (the *dobla,* or *fila dobla,* meant that the water ran for twice as long as usual), the equivalence of filas and hours is clear: one fila would equal two hours. Thus in times of severe water shortage the four Mijares towns instituted a turn in which each received water in the same proportion as its basic right (expressed in filas). In times of abundant water all four could draw water from the river concurrently, but the proportions would be approximately the same because the capacity of each canal would determine the share each received. The statutory proportions were, of course, the basis of the size of the canals, reflecting the amount of land irrigated. Burriana, the farthest downstream of the four towns, irrigated more land than the others and therefore drew more water.

The fila, then, is a figure defining that proportion of the total debit which each canal takes (1) in times of abundance, according to its capacity, and (2) in times of drought, according to a commensurate number of hours. The traditional quotas of filas of the canals of the Valencian huerta present both a confirmation and a puzzle. Moncada and the seven lower canals received water in the amounts indicated in Table 19. The division of the lower canals into fourteens and tens suggests that these were originally paired, Moncada having two days' water then two lower canals having the next day, two the next, and so forth. If a fila was equivalent to an hour, then the traditional apportionment represented a turn of six days (or one week, if there was no irrigation on the Sabbath[8]). There still remains the problem of identifying the huerta's actual allotment of a hundred thirty-eight filas with an expected six-day total of one hundred forty-four hours. The discrepancy is resolved by completing the pairs: in order to preserve the logically indicated order, there must once have been another canal on the left bank of the Guadalaviar between Mestalla and Rascanya and paired with Favara, with an allotment of ten filas, as in Table 20.

Table 19

Fila allotments of the canals of the huerta of Valencia
(upstream to downstream)

Right bank		Filas	Left bank	
–		48	Moncada	(48)
Quart	(14)	24	Tormos	(10)
Mislata	(10)	24	Mestalla	(14)
Favara	(14)	14	–	
Rovella	(14)	28	Rascanya	(14)
Total		138 filas		

There is some doubt as to the correct allotment of Rovella.

Table 20

Original division of the Guadalaviar in the huerta of Valencia (upstream to downstream)

Right bank		Filas	Left bank	
–		48	Moncada	(48)
Quart	(14)	24	Tormos	(10)
Mislata	(10)	24	Mestalla	(14)
Favara	(14)	24	[Lost Canal]	(10)
Rovella	(10)	24	Rascanya	(14)
Total		144 filas		

Most of the sources (following Jaubert de Passa) say fourteen, but José Martínez Aloy states that the correct figure is ten.[9] Ten filas is what one would expect for Rovella, according to the logic of the pairing. If fourteen is the correct figure, however, then one must assume that Rovella's original ten filas were increased at such a time when the lost canal between Mestalla and Rascanya ceased to function autonomously. The main canals of the huerta, running parallel to each other along the river, often serve the same area and even pass over and under one another. These canal networks can merge and separate with considerable ease: for example, when the diversion dam of the Rovella Canal was destroyed in the mid-eighteenth century, Rovella's service area was supplied through the Favara system, which diverted Rovella's filas as well as its own through its diversion dam.[10] Rovella was then functionally a branch of Favara, but it received its customary quota of water and maintained its autonomy as a separate community of irrigators. At an earlier stage of development it might have merged with Favara completely, especially if the canals had been administered by a central authority. Still it is difficult to understand how the lost canal could have been absorbed into a neighboring system without that system's quota of filas being increased — without, that is to say, doing violence to the concept of proportionality. Most probably, the depopulation of its service area and the accompanying deterioration of the dam and channels would have impelled the remaining irrigators to attach themselves to a more prosperous, efficient delivery network.[11] In any case, the disappearance of the lost canal's diversion dam and absorption of its service area must have antedated the Reconquest; the neighboring canals are documented from the early fourteenth century, and any such major change would have attracted the attention of local authorities.

The true nature of the Valencian fila has puzzled scholars and engineers for generations. Since the seventeenth century attempts have been made to gauge it in hopes of discovering some uniform and concrete measure which would express the volume of water in terms of its flow through a known orifice at a constant velocity, in the same manner as other modern liquid measures.[12] Such efforts have of course been predestined to failure so long as the proportional concept was not understood.[13] The adoption of a fixed value for the fila in some areas (such as Murcia and Orihuela) and the continued use of the unspecific, proportional concept in others (Elche and Novelda) caused nineteenth-century scholars like Sir Clements Markham even more confusion.[14]

No less puzzling was the *pluma* (quill, feather), one hundred forty-four of which comprise a fila. Surely this measure too has the same rationale as the fila: just as a river was divided into so many days of water, expressed in filas equivalent to hours, the fila itself was subdivided according to the same principles. The *pluma* would be analogous to the fila in an internal turn, one *pluma* being equal to one hour of water in a channel carrying one fila. This would suggest that internal turns also were conducted on a weekly basis, with no irrigating on the Sabbath.

ISLAMIC AND SPANISH IRRIGATION MEASURES

The irrigation measurement unit based on hours is not unique. In widely separated Islamic countries it is standard. In southern Iraq the base unit of water measurement is the *waqt* (time), which means either the period from sunrise to sunset or that from sunset to sunrise — approximately twelve hours.[15]

In the Yemen (Shoʿūb) the unit is the *farḍ,* which equals twenty-four hours.[16] The twelve-hour base is also observable in the Ghūṭa of Damascus, although as in Valencia not all divisions of the river were easily reducible to it (see, for example, the division of the Barada River in the eighth century, Appendix 8). There the standard water unit is now the *qīrāṭ,* a proportional measure which, though it varies in value from place to place and canal to canal, has the basic connotation of one twenty-fourth. At Dārāya a *qīrāṭ* has the value of one half-hour of irrigation, although in most other places it means one hour. At Berzé, the value of the *qīrāṭ* is fixed by local option at one-fourth hour in dry years and one-half in wet.[17] This procedure is the same as that of "doubling the water" in Spain, whereby the amount of water is doubled and the amount of time halved.

The most striking characteristic of the Ghūṭa measurement system is that the river is not considered to hold a fixed sum of *qīrāṭs,* but at each point along the course where there is a derivation the debit is considered as divisible into twenty-four *qīrāṭs.* The various canals then draw from the river a proportional part of the debit at that point (the Yezid Canal diverts eighteen *qīrāṭs* — that is, eighteen twenty-fourths of the river's volume at its dam — Thūra fourteen, Bānās twenty-three or almost the entire volume at that point).[18] Each canal is also considered to have flowing in it twenty-four *qīrāṭs,* which are then divided proportionally by means of divisors (called *mezzaz*) both functionally and in form identical with those of Valencia.[19]

The relation between the concepts of division in the Valencian huerta and the Ghūṭa of Damascus can now be appreciated. The fila and the *qīrāṭ* are cognate measures. The Guadalaviar at each of four stages below the Moncada Canal was considered as having twenty-four units. The Syrian model was that which guided the Muslim irrigators of Valencia as well as of several nearby huertas, those of Castellón and Gandia in particular.[20]

The Syrian model was not the only one followed in the medieval kingdom. The measurement units of Elche are so similar to those of Shoʿūb that the irrigation system of Elche can truly be characterized as Yemenite. Both the *farḍ* and the *hilo*

Table 21

Water measurement units of Shoᶜūb (Yemen) and Elche (Spain)

Shoᶜūb			Elche		
Unit	Translation	Time value	Unit	Translation	Time value
Fard	share	24 hr.	Hilo	thread	12 hr.
Nuss fard	½ share	12 hr.	Medio hilo	½ thread	6 hr.
Rubᶜ fard	¼ share	4 hr.	Cuarta	¼	3 hr.
Thumun fard	⅛ share	3 hr.	Tomin (medieval)	⅛	–
Nuss thumun	½ eighth	1½ hr.	Media cuarta	½ quarter	1½ hr.
Rubᶜ thumun	¼ eighth	¾ hr.	–	–	–

Sources: Rossi, "Irrigazione nel Yemen," p. 352; Ibarra y Ruiz, *Riego de Elche,* p. 156.

alike have a twenty-four hour base (Table 21). The "quarter" apparently replaced the "eighth" as the standard subdivision of the *hilo,* due to a doubling process.[21] The Moorish quarter of Elche still used the eighth (*tomin*) in the fifteenth century;[22] it is also in evidence in Novelda, where the *hilo* is equal to twelve hours, and the *azumbre* to an hour and a half.[23] The measures of Lorca are similar. There, the *hila* is equivalent to twelve hours and is divided into *terciadas* (four hours), *cuartas* (three hours), *taullas* (one hour), and *jarros* (half-hour).[24]

PROPORTIONAL DISTRIBUTION: A DEVELOPMENTAL HYPOTHESIS

To people without mechanical clocks, time is like a distance traveled or a cycle of recurring events.[25] In the irrigation systems studied in this volume the first step toward institutionalization may have been the imposition of a specific cycle, or order, of distribution. Initially — that is, after demand for water became great enough to occasion the need for control — the river may have been divided so that all the water went to one canal for a specific amount of time and then to another. The greater the number of canals, however, the greater the hardship likely to fall on any one canal on the days when it does not have water. Furthermore, water could easily be wasted if on the day of its turn a given canal could not use all that was available. This initial stage in which the river itself was subjected to a turn was modified, therefore, so that instead of each canal receiving the total flow for a specific time, *continuous flow* was given to all the canals simultaneously but in such a way that the proportions of the original division were preserved. For this purpose carefully measured diversion dams and divisors must have been constructed.

The repeated divisions of the Guadalaviar in the fourteenth and fifteenth centuries (expressed in a variety of units including filas, mulnars, and days) seem to indicate an extended phase of transition from the concepts underlying the first stage to those underlying the second. The second, modified system is the more efficient, but it is harder to control than the first. Significantly, the greater the pressure for water in medieval Valencia, the more likely it was that arrangements for distribution would revert toward the archaic method of dividing the river by days. The division of the Mijares in 1346 combined both concepts. The significance of the term "thread" of water may be that it expressed this concept of continuous flow.[26]

XII | The Imprint of Islam upon the Terminology of Irrigation

The lexicon of irrigation in eastern Spain is characterized by two general features: semantic fluidity and regional variation. Definitions are often not strict and meanings frequently are vague. One word can apply to differing phenomena and, conversely, a single thing can be described by a variety of words.[1] Regional variation is a logical corollary to semantic fluidity: words vary in meaning from place to place. But within each locale irrigation terms are both semantically and chronologically stable.

Students of Spanish and Islamic irrigation terms alike have remarked the multiplicity of words describing canals. "The medieval texts of Spain abound in words and expressions relating to water carriers," wrote the philologist Eero Neuvonen. "It is often very difficult to establish a significant difference between one word and another."[2] In the texts studied by Neuvonen words such as *canal, calze, acueducto,* and *conducto de agua* are used alternately with each other indiscriminately. In like manner Etore Rossi reported that "in the region of Raimah (Yemen), water-courses are indicated with many synonyms: *gail, nahr, ʿain al-māʾ*."[3]

Medieval Valencian canal terminology was the poet's delight, as seen in Jaume Roig's description of the water flowing "per rius, braçals, conduyts, canals e çequioles, rolls e filloles."[4] Roig's catalog of canals is a mixture of literary terms of a general nature (for example, *conduyt,* conduit) and the specific terms of everyday irrigation practice in the Valencian huerta. The braçal was a major branch canal, second in size to the "mother canal" (*cequia mare*) from which it was derived. Next in size and order was the fila, then the *fillola,* and finally the *cequiola,* usually a small ditch within the bounds of an individual field and the property of its owner. The roll, an opening or conduit diverting water from the man canal into a secondary channel, is analogous to the *boquera,* or field turn-out, the opening or gate through which the cultivator diverts water from the irrigation system onto his own field.

The same or related terms may have different values in different places. In Gandia the *hilo* (the etymological equivalent of the Valencian *fila,* as mentioned above) is the functional equivalent of the Valencian *braçal.* One sizable channel (serving numerous irrigators) in medieval Castellón was called The Cequiol, hardly the equivalent of the insignificant *cequiola* of Valencia. The *hijuela* (etymological equivalent of *fillola*) of Orihuela is the same as the *hila* in the system of Alfeitami and the *regadera* of Murcia;[5] yet, all three systems draw water from the Segura River and are regulated by similar institutions.

The Segura vega is extraordinarily rich in irrigation vocabulary, particularly in terms for channels, owing to the double system of delivery and drainage canals. The former system (that of *aguas vivas*) is arranged from largest to smallest: *acequia madre* (or *mayor*), then *acequia menor* (or *arroba*), then *brazal,* and finally *hijuela* or *regadera.* The drainage (*aguas muertas*) hierarchy runs from *azarbe* (or *landrona* or *merancho*), the largest channel, to its diminutive, *azarbeta,* to *escorredor,* the smallest.[6] Thus the meaning of the terms is rarely generic, but depends upon the configuration of the particular system, the size of the canal and its place in the hierarchical network from main canal down to field, and the function of the system, whether for delivery or drainage.[7]

A similar pattern is evident in the words used to describe hydraulic wheels. Strictly speaking, a *noria* is a wheel moved by the force of the current alone, while a *sinia* is one moved by human or animal power: but only in Murcian usage is this distinction correctly made. In Catalonia and Valencia *sinia* displaced *noria* completely and was used to describe all hydraulic wheels. In other parts of Spain the common word for diversion dam, *azud* (and its variant *azuda*), came to be applied to the current wheel.[8]

At the base of this fluidity is regional variation in irrigation customs along ethnic lines. Terminology varied principally as the ethnic composition of the original Muslim irrigators and the Christian settlers varied. The place of arabisms in the picture is crucial. The fact that there were not enough terms in the Romance tongues of the conquerors to replace the multiplicity of definitions in a highly developed irrigation system made this sector of the lexicon particularly susceptible to linguistic borrowing; the fact that Arabic was unintelligible to most of the new settlers ensured that the borrowing would be loosely structured. Thus many areas of human thought and industry were described by two sets of terms. The Romance tongue of the Christians, together with Latin expressions of long acceptance in Catalonia and Castile, coexisted with words acquired from the Arabs. Irrigation was one of those areas of culture (warfare was another) where the arabisms tended to predominate.

That many of the concepts of irrigation were new to the Christians is indicated by the intrusion of vernacular arabisms into Latin documents in the thirteenth and early fourteenth centuries. This phenomenon is indicative of the lack of consensus concerning a given term (evidence of entrenched regional variation, in other words), and the lack of a Latin or Romance tradition to guide the scribe.[9]

REGIONAL AND LOCAL VARIATION

Most of the arabisms common throughout eastern Spain were present either in Castilian or Catalan by the thirteenth century. *Sinia* and *safareig*, both specialized hydraulic devices, were among the earliest peninsular arabisms, dating from the epoch of Islamic expansion (711 to the mid-eleventh century). Cequia and açut, the most common and typical of Valencian irrigation arabisms, date from the epoch of the great Christian conquests (mid-eleventh to the end of the twelfth century) and were therefore already in the language when the Valencian region was taken from the Muslims. *Alcaduf, aljub, azumbre, çabacequies,* and marjal are arabisms which passed into the Romance vernaculars in the course of the thirteenth century.[10] Most of the localized arabisms as well must have passed into the Christian vernacular in the late thirteenth century during the post-Conquest process of learning local customs from Muslim irrigators.

The lists that follow include all the important regional arabisms and most of the significant local ones; the contents of the lists are summarized in Tables 22 and 23.

Regional Irrigation Arabisms

1) açut (*çut, azut*), "diversion dam," Arabic *al-sudd;* the most common and widest spread arabism besides cequia. The Romance equivalent, *resclosa,* retains the sense of the original Arabic *sadda,* to close. The word has been used consistently throughout eastern Spain from the Reconquest to the present, and the Castilian form, *azud,* dates from the period of the great Christian conquests (1050 to the end of the twelfth century). Both the word *sudd* and the technology are typical of southern Arabia.[11]
2) *albellon,* "sewer, drain," Arabic *al-bālūʿa.* Usually an urban term in medieval Valencian documents, typically associated with *vall,* "sewer, moat." In Murcia, however, its oldest meaning was that of a permanent opening in a channel, and other references corroborate its association with irrigation.[12]
3) *alcaduf,* "bucket of a water wheel," Arabic *al-qādūs:* a very common term in Islamic irrigation, originally referring to the buckets of the Persian water wheel, by extension a measure of water and even (in Spain) a conduit. The *qādūs* is the common instrument for measuring water in the Saharan oases.[13]
4) *aljup* (*aljub*), "cistern," Arabic *al-jubb.*[14]
5) almenara, "return ditch," Arabic *al-manāhir,* pl. of *manhar,* "canal." [15]

6) *çabacequies* (*çabacequia* and, through influence of cequier, *çabacequier*), "master of the canals" (an irrigation official), Arabic ṣāḥib al-sāqiya.[16]
7) cequia, "irrigation canal," Arabic *sāqiya;* the most characteristic and widespread irrigation arabism. In thirteenth-century Castilian documents (Castilian *acequia*) it frequently has the meaning of a canal that supplies a mill, but in Valencia it means any kind of irrigation canal.[17] *Sāqiya* is the common word for irrigation canal throughout the Islamic world.
8) marjal (*almarjal, almargal*), "swamp," Arabic *marj,* "meadow." In Valencian documents it has no synonyms.[18]
9) *noria,* "hydraulic wheel," Arabic *nāʿūra* (from *naʿara,* to grunt); one of the earliest hydraulic arabisms in Spanish. The addition of the *i* to the primitive form *nora, annora,* is possibly explained by the influence of *acenia* (*sinia*) and *acequia.* The first document is of the twelfth century.[19]
10) *safareig,* "cistern," Arabic *ṣahrīj.*[20]
11) *sinia* (medieval Valencian *cenia*), "Persian wheel," Arabic *sāniya;* Castilian-Murcian *acenia, aceña.* In the Catalan zone *sinia* is understood as synonymous with *noria,* also an arabism. But strictly speaking a *sāniya* is driven by animal power, while the *nāʿūra* is moved by force of water alone. One of the earliest of irrigation arabisms, first documented in 945.[21]
12) tanda, "irrigation turn." Corominas suggests that an Arabic derivation from **tanzīm* (the root means to put in order) is preferable to one from the Latin *tanta.* The first appearance is in thirteenth-century Catalonia as the standard word for "irrigation turn," although in Valencia and Murcia some local variants (all arabisms) occur.[22]
13) *tarquim,* "silt." Corominas suggests a derivation from the Arabic **tarkīm,* from the root meaning "to pile up."[23]

Localized Arabisms: Valencia, Murcia, Lorca

1) *açarb,* "drainage canal" (Murcia), Arabic *al-zarb.*[24]
2) *ador,* "turn" (Gandia), Arabic *al-daur.* A feminine form

of *daur* is a common word for "turn" in modern Yemen.[25]

3) *albala,* "water-ticket" (Alicante), Arabic *al-barā'a.* The *albala* is a ticket representing a certain number of minutes of water on sale at the water auction at Alicante.[26]
4) *almahacen,* "water that remains unapportioned, for communal use" (Lorca), Arabic *al-maḥzan* (storehouse, deposit).[27]
5) *almatzem,* "divisor" (Gandia), Arabic *al-miqsam.* Not a common term in Spain but of wide diffusion in Islam, as far east as Persia.[28]
6) *azumbre,* a water measure (Novelda, Elche), Arabic *al-thumn,* "one-eighth."[29]
7) *dula,* "turn" (Gandia, Elche); also a water measure synonymous with *hila* (Alicante), Arabic *daula.* Another term found both in the Yemen and the Sahara.[30]
8) *jarique,* noun, or verb *jaricar,* "to unite various *hilas* of water" bought at auction in order to irrigate with a greater body of water and to transport it over a long distance with minimal loss (Lorca); Arabic *sharīk,* "a partner." An important reminiscence of communal irrigation practices among Muslim irrigators.[31]
9) *jarro,* a water measure (Lorca, Jumilla), Arabic *jarra,* "jug." In Lorca and Jumilla the *jarro* is equal to one-half hour of water. It takes its name from the clepsydra (water clock), a vessel with a hole in the bottom which was the standard time measure for irrigation in the North African oases and in southern Spain. The word is frequently found in thirteenth-century Romance texts, indicating its adoption after the conquest of Lorca and the Murcian region.[32]
10) *martava,* "turn" (Alicante, Novelda), Arabic *martaba.*[33]
11) *merancho,* "drainage ditch" (Murcia), Arabic *marj.*[34] The connection between drainage and marjals is clear. Drainage of marshland was an operation described largely in arabisms.
12) *rafa,* "canal check" (Murcia), Arabic *raf',* "raising up."[35] An interesting example of an arabism that describes a function different from that expressed by its common

Romance synonym, *parada.* The canal check (often simply a board placed across a canal) serves both to stop the flow (the meaning of *parada,* from *pararse,* to stop) in order to divert water and to raise the level (the meaning of *rafa*).

13) *sistar,* "divisor" (Vall de Segó). Corominas has suggested a derivation from the Latin *sistere* (to stop);[36] but this must be questioned because irrigation terms are usually precise as to function, and a *sistar* divides water (its more usual synonym is *partidor*) rather than stopping it. Since the Vall de Segó had a substantial Muslim population, it seems proper to seek a derivation from the Arabic root *sh-ṭ-r,* "to divide into two equal parts."

14) *tahulla,* a measure of land (Murcia) but also a measure of water equivalent to one hour (Lorca), Arabic *ṭahwila,* "field." [37]

The foregoing data permit several generalizations concerning the frequency and function of arabisms in the technical vocabulary of medieval irrigation: 1) The Arabic word and the arabism usually have the same meaning.[38] Agricultural terms especially are conservative by nature and tend to retain the original meaning so long as the use of the object is the same. 2) Romance synonyms are likely to be literal translations of the original Arabic meaning: thus *resclosa* is the equivalent of açut, from *sadda,* "to close"; partidor is the synonym of *almatzem,* from *qasama,* "to divide." 3) When all the synonyms describing a single phenomenon are arabisms, there is a strong presumption that the concept or technique was unknown or undeveloped outside of the Islamic orbit or, at any rate, was very intimately associated with the Islamic style of irrigating: for example, the words for turn (*dula, ador, martava*), for return ditch (almenara, *azarbe, merancho*), for cistern (*safareig, aljup*), and for water wheel (*sinia, noria*).

If these generalizations are applied to words of presumed but unproven Arabic etymology, it is possible to make some guesses merely on the basis of whether the concept was developed out-

Table 22

Regional irrigation arabisms

Arabism	Meaning	Major variants	Medieval Castilian	Toponyms	Medieval synonyms: Romance	Medieval synonyms: Arabism	Arabic	Meaning	Root
1. açut	diversion dam	*çut, azut*	*azud, açuda*	Burjassot (Valencia)	*resclosa, presa*	–	*as-sudd*	dam	to close
2. *albellon*	sewer, drain	–	*albollon*	El Albellon (Elche)	*vall*	–	*al-bālūʿa*	sewer, drain	–
3. *alcaduf*	bucket of a *noria*	–	*alcaduz, arcaduz*	Calduf (Gandia)	*fortera*	*jarro*	*al-qāds̄*	waterbucket, scoop	–
4. *aljup*	cistern	*aljub*	*algibe*	Aljupet (Elche), Port d'Aljup (Murcia)	–	*safareig*	*al-jubb*	cistern	–
5. almenara	return ditch	–	–	–	–	–	*al-manāhir*	canals	to flow, gush
6. *çabacequies*	an official, literally "master of the canals"	*çabacequia, çabacequier*	*sobrecequiero*	–	*sobrecequier*	–	ṣāḥib al-sāqiya	master of the canal	–
7. cequia	irrigation canal	–	*acequia*	Guadassequies (river)	*ayguaduyt*, fila	*arroba*	*sāqiya*	irrigation canal	to irrigate
8. marjal	swamp	*almarjal, almargal*	–	–	–	–	*marj*	meadow	–
9. *noria*	hydraulic wheel	–	*annora*	La Ñora (Murcia)	–	*cenia*	*nāʿūra*	hydraulic wheel	to grunt
10. *safareig*	cistern	–	*xafariz, zafariche*	Zarahiche (Murcia)	–	*aljup*	*ṣahrīj*	cistern	–
11. *sinia*	hydraulic wheel	*cenia*	*aceña*	Cenia (river), Aceña, Aceñas (many), Azana (Toledo), etc.	–	*noria*	*sāniya*	Persian wheel	–
12. tanda	turn	–	tanda	–	–	*ador, dula, martava*	**tanẓīm*	order	to put in order
13. *tarquim*	silt	–	*tarquin*	Tarquin (Murcia)	*fanch*	–	**tarkīm*	piling up	to pile up

Table 23

Localized irrigation arabisms

Arabism	Meaning	Place	Common synonym	Arabic	Meaning
1. *açarb*, *azarbe*	drainage canal	Murcia	*landrona*, *merancho*	*az-zarb*	canal, drain
2. *ador*	turn	Gandia	tanda, *dula*	*ad-daur*	rotation, turn
3. *albala*	water ticket	Alicante	–	*al-barā'a*	license
4. *almahacen*	undivided water for common use	Lorca	–	*al-maḥzan*	storehouse, deposit
5. *almatzem*	divisor	Gandia	partidor	*al-miqsām*	divider
6. *azumbre*	water unit	Novelda	–	*thumn*	⅛
7. *dula*	turn water unit	Gandia, Elche Alicante	tanda *hila*	*daula* *daula*	turn, rotation, alternation
8. *jarique*	sharer of water	Lorca	–	*sharīk*	partner
9. *jarro*	water measurement unit	Lorca	–	*jarra*	cup, jar
10. *martava*	turn	Alicante, Novelda	tanda, *dula*	*martaba*	rank; arrangement in regular sequence
11. *merancho*	drainage ditch	Murcia	*azarbe*	*marj*	meadow
12. *rafa*	canal check	Murcia	*parada*	*rafᶜ*	raising up
13. *sistar*	divisor	Vall de Segó	partidor	*shatr*, *shitra* (?)	partition; division in 2 equal parts
14. *tahulla*	measure of land water measurement unit	Murcia Lorca	*jovada*, *fanega* *caballeria*	*taḥwila* *taḥwila*	field, piece of land field, piece of land

side the Islamic framework. Thus tanda, in spite of phonetic difficulties in Corominas' derivation, is probably an arabism.[39] Tanda and *tanta* do not mean the same thing; *tanẓim* comes much closer.[40] The fact that all of its synonyms (*dula, ador, martava*) are arabisms suggests that the concept of the turn was unknown or not highly developed among Christians. Likewise *sistar* from *shaṭara,* in spite of the infrequently found change from *sh* to *s,* is attractive because it means the same as partidor. *Almatzem,* the local variant for partidor in Gandia, provides a parallel case. (Vall de Segó and Gandia were among the relatively few irrigated areas in the medieval kingdom of Valencia where there were significant numbers of Muslim irrigators.)

ARABIC THEMES IN ROMANCE VERNACULAR

The expressions for the typical irrigation land-use patterns, huerta and vega, both seem to bear the Islamic imprint. It is probable that huerta is a translation of the Arabic, *bustān,* both words meaning "garden."[41] Vega is of uncertain origin, but Arnald Steiger believes it to be an arabism.[42]

The irrigation system of Lorca preserves two interesting measures, *casa de agua* and *tiempo de agua,* both probably of Islamic ancestry. A royal privilege of September 23, 1268, provides that the water of Lorca is to be divided "communally, by days and by times" ("comunamente por dias y por tiempos"). The concept of a "time" of water recalls the Arabic *waqt,* used in the same way (as a unit of water) in Iraq.[43] Also in Lorca, a *casa* (house) is said to be twelve hours of water, each day being divided into two "houses." If this expression is in fact medieval it seems quite clearly to be a translation of *dār,* "house," which, in this case, is a dialectal pronunciation of *daur,* "turn." In the Saharan oasis of El Outaïa a *dār* is, similarly, twelve hours of water.[44]

The term expressing the standard unit of water measure in most of the kingdom of Valencia — the fila or *hilo,* "thread," of water — has proven enigmatic to generations of scholars: they have neither understood the rationale of the measurement nor have they explained what a "thread" has to do with water. The term "thread of water" was known in Andalusí Arabic. In a document of 1223 (619 H.) describing a water dispute between two towns near Morvedre, the measurement unit is the *khaiṭ,* or thread.[45] The only other pre-modern source known to me that corroborates this usage is Pedro de Alcalá's translation of *kait min mi* (thread of water) as *corriente venaje de agua* (source-current of water).[46] A modern Arabic-French dictionary published in Algiers, and thus reflecting Magribi usage, defines *khaiṭ* as "saignée à un canal d'irrigation, rigole d'arrosement" [47]: that is, not a measure, but the ditch itself. This usage corresponds to an alternative use of fila in Valencia with the meaning of ditch or canal, synonymous with cequia but perhaps connoting a smaller channel.

There is another example of the same term expressing both a liquid measure and a channel. In thirteenth-century Murcia *arrova,* usually a liquid measure (of olive oil, for example) also had the sense of a channel. Alfonso X of Castile ordered the *sobrecequiero* of Orihuela to look to the cleaning of the cequias, filas, *arrouas,* and *azarbes.*[48]

The medieval documentation, especially that of Castellón, provides numerous examples of important canals called filas: for example, "la fila dels dos hulls," the "files de la doberia" (that is, *adoberia,* tannery), and "les files de Na Orellana." [49] That the Arabic equivalent *khaiṭ* was used in the same way is clear from two toponyms previously misunderstood. In the town limits of Alberic there is a canal named Alfait with its origin in the same springs that give rise to the Riu dels Ulls.[50] Miguel Asín Palacios derived Alfait from *al-faiḍ,* "stream," from the verb "to overflow," so called because the head is augmented in times when run-off water infiltrates into it.[51] In view of the literalism displayed by medieval Spaniards — Arabs and Christians alike — in irrigation terminology, it is much simpler to derive Alfait from *al-khaiṭ,* "thread" (in the sense of fila, "ditch").[52] Naming a canal "The Canal" was by no means uncommon in the Valencian region.[53] An analogous development is the name of Faitanar, one of the main canals of the Valencian huerta; it is quite clearly derivable from Arabic, *khaiṭ al-nahr* ("thread of the river" in the sense of a canal diverted from the river), in spite of contrived efforts to find a Romance derivation for it.[54]

Several passages of the Repartimiento of Murcia allude to "irrigation of *alfayt*" (that is, "riego de çequia et dalffayat, se reguen dalfayt" [55]). The editor, possibly following Asín in associating *alfayt* with *al-faiḍ,* understands the term to be flood irrigation (*crecida*).[56] For reasons already indicated, a derivation from *faiḍ* is much less likely than from *khaiṭ;* therefore "riego de çequia et dalffayat" most likely means, simply, irrigation from a large channel and a smaller, secondary one.

HYDRAULIC PLACE NAMES IN VALENCIA AND MURCIA

Many of the arabisms discussed above also appear in place names. I shall not attempt to present a complete description of all toponyms having to do with irrigation; enumeration here is limited largely to arabisms as names of hydraulic appurtenances, such as canals and divisors, in order to give a graphic illustration of the Islamic imprint on irrigation.

Of the main canals of the huerta of Valencia two, in addition to Faitanar, are named with arabisms bearing hydraulic allusions. The Favara Canal derives its name from *al-fawwāra,* "spring." Nicolau Primitiu, basing his argument on a single variant, Alfara (which he believes is a shortened form of Alfafar) posits a Germanic derivation.[57] This is highly dubious reasoning, and the Arabic etymology gains further support from the presence of the form Favara in the Repartimiento and from the appearance of the primitive Arabic article Alfavara in a document of 1430.[58] Rascanya is also an arabism, from *rās,* "head," and *canya* (from *canna,* canal). Here, though, the canal takes its name from the place it irrigates, called Rascayna (*cayna,* the diminutive of *canna*) in the Repartimiento.[59]

A repeated canal name in the Valencian region is Algirós (also spelled Aljiros, Algeros, Algiroz). It is the name of one of the three major branches of the Mestalla system in Valencia and of an important canal near Alcira. Perhaps, too, the canal in Elche called Algorós is related etymologically.[60] The derivation is from *al-zurūb,* the plural of *zarb* (from *zariba,* to flow). But in spite of the plural form it has a singular meaning, "the canal," just as *almenara* is derived from the plural form *al-manāhir,* "the canals." Even though the canals in question are not now drainage channels, they are related etymologically to *azarbe* (from *al-zarb*) and may originally have served that purpose.[61]

Canals also take their names from other hydraulic expressions. One canal in Gandia is called Ador (turn), and a secondary channel of the Aljufia system in Murcia is called Zarahiche (cistern).[62] Similarly named are the divisors of Elche called Aljupet (little cistern) and El Albellon (drain).[63] Mills, too, frequently bear hydraulic names. The mill of Calduf (from *alcaduf,* by metathesis) on the Vernisa Canal in Gandia derived its name from the buckets of a water wheel.[64] The Murcian mill called Tarquin[65] must have been so named for good hydraulic reasons. Mills create backwaters, causing a silt deposit through a lowering in the velocity of the current.

The diversion dam and the canal — the two most important visible symbols of irrigation — lent their names to toponyms. Thus a place near Cullera was called El Azut, and the village of Burjassot (tower of the dam) in the Valencian huerta on the Moncada Canal similarly took its name from a dam.[66] The canal lent its name to a river, the Guadasequies (river of the canals).[67] The river Cenia, the northern boundary of the kingdom of Valencia, took its name from the *sāniya,* always a popular device in the province of Castellón.[68] Numerous other hydraulic arabisms, notably those derived from springs (*ʿayn, al-hama*) and wells (*b'ir*) dot the Spanish countryside.[69] One such, of more than passing interest, is Titaguas, a Valencian village near Chelva that takes its name from the Berber *tit,* "spring," plural *titáuan* or *titaguan,* easily becoming confused with *aguas,* which was added as a reinforcement.[70]

XIII | Image and Reality of Cultural Change

The vestigial arabisms and Islamic practices visible in late medieval Valencian irrigation form an ethnic mosaic corresponding to the various irrigated regions of the Islamic world which sent their sons to settle Al-Andalus. Two cultural types of water distribution systems can be broadly delineated: 1) a Syrian type, modeled after the Barada River of the Ghūṭa of Damascus (systems of Valencia, Castellón, and Gandia), characterized by proportional distribution of water; 2) a Yemenite type, typical of the small, oasislike huertas of the southern kingdom of Valencia (Alicante, Elche, Novelda), based on fixed time measurement units and associated with the sale of water.[1]

These types correspond as well to two rival systems of land tenure, one in which water rights are inseparable from the land, another in which the right to water may be sold separately from land. The second, which I call "Yemenite" on the basis of the irrigation vocabulary, has also been associated with the Saharan oases with which there is great institutional similarity. The evidence is suggestive not only of mixed settlement — both Yemenite and Berber — in eastern Al-Andalus but also of Yemenite influence upon Saharan irrigation practice.

Most administrative procedures and legal traditions are difficult to trace to precise localities in the Islamic world. The amīn was an official typical of the Saharan environment and was found in those areas of Spain where Yemenite measures were used; the ṣāḥib al-sāqiya, on the other hand, may have been indigenous to Al-Andalus. Many practices, too, were widely diffused through the homogenizing effect of Islamic Law, the Malikite school especially.

The processes of diffusion were varied. The transmission of certain terms and concepts — *sāqiya, sudd, sāniya,* and so on — was accomplished before the reconquest of the kingdom by nonformal mechanisms (such as the movement of Mozarabs northward) by which Islamic irrigation techniques became known in the Christian kingdoms. After the conquests were completed, transmission was accomplished both nonformally, as a result of contact between Muslim and Christian groups, and, more significantly, by formal processes of directed culture change whereby Christian officials sought deliberately to record and emulate irrigation customs of the Muslims. By another formal process certain elements, such as summary procedure in local irrigation justice, were included in the fueros and by that means diffused throughout the entire kingdom.[2]

The reconquest of the kingdom of Valencia was accomplished in the 1230's and early 1240's. James I, at the head of Aragonese and Catalan troops, captured Burriana in 1233, Valencia in 1238, and Alcira in 1242. At first, masses of Muslims remained in the subject territories. A series of revolts from around 1248, however, led to the expulsion of large numbers from the kingdom of Valencia and initiated a new phase of repopulation which, when terminated in 1270, resulted in the almost total demographic dominance of Christians in the major centers of the realm.[3] Of the large mass of Muslims who remained, most were now settled in the mountainous hinterland, most of which was unirrigated. In the huerta of Valencia no Muslims at all appear in the documentation from the mid-fourteenth century up to 1500. In Castellón, with one or two exceptions Muslim irrigators were limited to the village of Canet, on the southern limits of the huerta. Of the important towns, only Gandia, Játiva, and Elche[4] had significant numbers of Muslim irrigators who, in the latter two places, were largely restricted to discrete suburban settlements. Substantial numbers of Muslim irrigators remained in some secondary centers of regadíu, such as Benaguasil, Callosa d'En Sarriá, Chiva, Torres Torres, Vall de Segó, Vall de Uxó,[5] and numerous smaller, relatively unimportant villages and hamlets.[6]

With the exception of an occasional peculiarity (such as the use of an archaic measurement unit by the irrigators of the Moorish quarter of Elche), the documents do not reveal any striking differences in practice between Christian and Muslim irrigators in any given locale. The Christians seem to have absorbed local customs soon after settling the newly conquered territories and to have maintained them intact thereafter.

INITIAL DIVISIONS OF WATER AND LAND

The Repartimiento of Valencia is relatively poor in specific data concerning irrigation. Of the donations in the huerta of Valencia, about two dozen parcels fronted on unnamed canals, and an occasional field is said to be irrigated (*in regadivo*). Of the major canals, only Favara and the Algirós branch of Mestalla are mentioned by name.[7] Throughout the kingdom the method of *repartimiento* was the same. Old landholding patterns were not broken up, but newcomers were gradually settled upon the homesteads and fields abandoned by fleeing Muslims.[8] Since not all the defeated fled at once (many remained throughout the Middle Ages), reasonable continuity in agricultural and irrigation regime was assured in most places.

An early entry in the Repartimiento of Valencia (August 1237) proves that irrigated fields in the huerta were still held by Muslims at the same time that grants were being made to new settlers:

> To Bafiel, the king's physician,[9] some houses belonging to Abengeuir . . . and an estate[10] with its orchard contiguous to the same Saracen; and two large fields, with a road in between them, one of which is walled and bounded by three roads, and which fronts on the field of Abinanubil and that of Abdel Abenfulfel and that of his brothers and that of the sons of Abinbalapha, the smith; and another field which fronts on the road which goes to Rascayna [sic] [11] and on a road next to the said walled field and on the field

of Jahie Elluri, Saracen, and on a certain large irrigation canal,[12] and on country houses[13] up to the road of the canal.[14]

Given the interspersal of Muslim-held fields among those newly distributed to Christians, it is entirely natural to expect the continuation of the distribution arrangements then in effect. That this in fact was the case is seen in a formal act, by which customary information was deliberately sought of the Muslims and recorded, as in the division of the water of the Serpis River (Gandia):

> This is the division of the waters of the district of Gandia:
> *Anno domini* 1244, 18 Kalends August
> Because of disputes which the hundred settlers of Bayren[15] had with the settlers of the Benietos,[16] Don Peregrín de Otrosillo[17] ordered that the *çauaçequias* who used to apportion the water in the time of the Moors appear, and that the Moorish *çequieros* take an oath on pain of their persons and goods, and that they tell the truth about the waters, [that is], in what way they used to apportion them in the time of the Moors; and that those [*çequieros*] who apportioned [the water] in that way, according as the districts used to have the water in the time of the Moors, should in a like manner [continue to] give it to each district, in writing.
> And the Moors gave us [an account of] the waters by means of this [written instrument] just as it is here. And for this purpose, they went to the divisor[18] of Beniarjó and Benifla,[19] and they apportioned the waters for us in this manner, just as they had always been in the time of the Moors:
> Between the two of them, Benifla and Beniarjó ought to irrigate thirteen days and thirteen nights. But while Benifla is irrigating, Beniarjó ought not to irrigate.
> [An enumeration follows of extremely complex details for each divisor of the system. At the end of the inventory of divisors those who have participated in the inquest give witness:]
> Monday, August 4, in the presence of Martin d'Oblites,

> and Johan de Mora,[20] and Boniface, and Justo, and P. Oliver, and Bn. Sanauga, and Guillem Durant, and Bn. de Lloses, and the Saracens Avengamer, the qadi, and Gomar Auengomar, and Mahomet Abengayep.
>
> Witnesses: Juçeff Aucortinna, interpreter,[21] and the *çequiero,* Abrahim Arraguefl, and Abrahim Almocant, and Mahomat Abencalima, and Mahomet Ahuor, and Hamet Abendarraix. For the Christians, Lorenç Ruffa.[22]

It is noteworthy to find in the witness list a Muslim cequier, a qadi, and Christians who actually held irrigated land. The Gandia evidence — positive proof of the direct transmission of Islamic customary arrangements to the new settlers — must have been duplicated in other regions of Spain.[23]

THE ROLE OF THE FUEROS IN ASSURING CONTINUITY

A leitmotiv of thirteenth-century irrigation grants is the stipulation that the custom continue to be just as it was in the time of the Moors. The document which convinced Borrull of the fact of institutional continuity (and that which is most often cited) is from the fuero of James I, rubric xvi, "De servitut d'aygua," paragraph 35, granting to the settlers of the city and kingdom of Valencia "each and all the canals, large, middling, and small, freely, with spring waters, excepting the Royal Canal which goes to Puzol, from which canals and springs you may have water, and conductions of water, at all times, continually, by day and by night, so that you might irrigate from them and take water without any obligation, due or tribute, and that you might take those waters just as of old it was established and accustomed in the time of the Saracens."[24] The use of water was thus granted to a particular locality on the same basis as before the Reconquest. Similar provisions are to be found in privileges of James I to Morvedre, of Alfonso X of Castile to Alicante and Orihuela, and of Prince Manuel to Elche.[25] Usually the stipulation was general, but in one case Alfonso X indicated that cequiers should follow Moorish procedure in directing the flow of the water and in diverting it at the canal checks used by the Moors in their system of turns, "and in no other place." [26] That is, the configurations of the system down to the very points of diversion were not to be changed.

James I was also insistent on forbidding innovations in the existing works and was careful to declare that all repairs must be executed in conformity with "ancient mode, form, and state." [27] In ensuing years, repeated royal orders alluded to the provisions of the fuero, and frequent litigation was brought before the governor to denounce "innovations" (such as new canals or diversion dams) and have them destroyed. The "time of the Saracens," therefore, referred to a concrete moment in time—a reference point both for rights and for the physical structure of the system.

THE "TIME OF THE MOORS" AS LEGAL TRADITION

Priority of use was the principle of customary law upon which irrigation disputes were litigated in medieval Valencia. Each disputant tried to show that his right was sanctioned either by royal or seigneurial privilege or by continued use dating from time beyond the memory of living man. The appeal to "time immemorial" was common enough in the law of other European societies, where it was the intent to show that the custom in question antedated all surviving records.[28] The legal basis of the appeal was found in the Roman Law of Waters. Ulpian ruled: "Where a man has acquired a right of watercourse by lasting use and, so to speak, long possession, he is not required to bring evidence to show the legal title on which his right to the water was founded, such, for example, as a legacy or any other ground of title; he has a good right of action if he can show that he had the use of the watercourse, say for such and such a number of years, and that not by force or by stealth but by leave and license." [29] Proceedings were taken against anyone who "sought to disturb" (in Valencian legal parlance) the irrigator in his customary right to divert the water.

In Valencian documents the specific form taken by this procedure was the legal formula alleging possession of an irrigation right for "ten, twenty, thirty, fifty, one hundred years back and for so much time that the memory of men runneth not to the contrary." [30] Yet in spite of the formula, "time immemorial" was not an open-ended concept stretching back indefinitely, but it had a definite point of departure in the conquest from the Muslims. The medieval man had difficulty conceptualizing the establishment of a right in some vague, distant past, and so he preferred to set a specific date which would take on the meaning of time without memory in the procedure of establishing proof. This was the case in England, where, in quo warranto cases in the reign of Edward I (1272–1307), it was necessary to establish continuous use since before the time of King Richard I. "This was construed as meaning a seisin extending back beyond September 3, 1189 when Richard I was crowned and began his reign." [31] The efficient process of law — property law especially — seemed to require that a specific chronological limit be included in the concept of legal memory.

The conquest of James I was a natural limit because it was at this time that a great portion of the land was divided and granted to settlers by written instruments. Beyond the memory of living man the legal memory reached back to the date of the Reconquest, whenever it may have been, and before it to an undefined, foreshortened "time of the Moors" (*temps de moros, temps de sarrahins*).

Medieval Valencians were not interested in the specifics of the Islamic past of their realm. The only continuity felt with Al-Andalus, aside from traditions of Christian combat, was that perceived by the eye. The physical remains of Muslim settlement were at the core of what slight appreciation there was of an "Islamic heritage." Then, too, the terms of the original royal donations and charters granting the use of water for irrigation reinforced the disinclination to inquire too closely into the Islamic past. To prove that one's rights were among those covered generally in the fuero or specifically in a thirteenth-century charter or donation, all that was necessary was to adduce *any* proof of Islamic origin — the specific date was not important — or show continuity back to the original grant by alleging continuous use beyond the memory of men. Thus throughout the Middle Ages a play can be described between the legal tradition, recalling the words of James I, and an oral tradition based on testimony in court regarding the physical remains of the Islamic past.

The most common approach was a simple mixing of the "time immemorial" and the "time of the Moors" formulas. In a dispute of 1431 over irrigating turns in Vinalesa (on the Moncada Canal, Valencia), one side alleged that in times of water abundance the practice had been to irrigate "according to ancient custom, which is since the kingdom has been in Christian hands, down to the present time, the memory of man running not to the contrary." [32] The other side in the dispute (the syndic of the Moncada Canal) says that, for his part, it is the cequier who decides questions of division of water "according to ancient custom . . . just as it has been accustomed ever since the time of the Saracens." [33]

Similarly, the jurates of Quart heard that certain men irrigating from the Benacher Canal in 1421 wanted to change the old practice of dividing the water coming from the Canal of the Arches between the two canals of Benacher and Quart by raising the level of the divisor. This, and the closing of a certain fila were objectionable because the existing condition of these works was that which "they have been ever since the Canal of the Arches has existed, which is from the time of the Saracens and from the time much before this kingdom was conquered by the most excellent king, James the First, of glorious memory, continuously, down to the present." [34] Several years before, the bailiff of the Albufera had reported that the canals and springs there had been made of old and existed "already in the time of the Saracens." [35]

The vagueness of this method of asserting priority of rights is all the more glaring in briefs in which the formulas were mixed in confusing anachronisms, as in a dispute involving various branch canals of the Mislata Canal in 1454. The right in question was claimed from fifty years before to the present, from time immemorial, and "from the time when Saracens possessed the present city [of Valencia] and its district."[36] One of the parties commenced his brief with a full citation of the original grant of James I in the fuero cited above.

These stock legal appeals to the time of the Reconquest and before did not convince everyone, however. At least one lawyer, Dionis Dolit, who often appeared before the governor's court in fifteenth-century water disputes,[37] noted that there was a patent contradiction between the legal fiction that irrigation institutions and practices were fixed in a Moorish pattern and the everyday reality of irrigation practice. Representing the town of Burriana against Villarreal and Almassora in 1417, Dolit denied the pertinence of the Moorish past to present conflicts over the administration of water:

> It is not to the point, nor does it harm our case, as the contrary party alleges, that the landholders can set fines regarding water and apportion them as they please, and that the water and the canals ought to continue, be, and flow according as they were in the time of the Saracens . . . because it is certain that the residents, together with the town, set the fines, and change, augment, limit, and reduce them as it seems best to them. The circumstances demand this. And they, with the said town, change the canals and make the water run where it seems to them that it will be most to the profit and utility of the landholders, and at the least cost.[38]

Indeed, the realities of daily life, the constant modifications of regulations and physical plants alike which the practice of irrigation agriculture required, clearly precluded the inalterable fixedness of the Valencian irrigation system in the form encountered by the first settlers.

So it is that, for the purpose of establishing specific and valid links with the Islamic past, the historian cannot put too much faith (as Borrull did) in the legal tradition of the fuero alone. But alongside the often-cited formulas, documents recorded a strong oral tradition, drawing sustenance not from charters or royal privileges, but from the visible remains of Islamic settlement.

In medieval Valencia certain methods of construction were commonly identifiable as "Moorish," and works so identified were assumed to have antedated the Reconquest.[39] "Moorish" construction was recognizable both by the technique of building and by the material used. In a complaint brought by the city of Valencia against the lord of the Valley of Chelva in 1413, the latter sought to refute the charge that the canal of Facuquer was newly built and contrary to ancient custom by showing that the diversion dam and canal alike were of obvious Moorish construction. A visual inspection not only could reveal no sign of new construction, but it was clear that in many places the canal was "constructed and worked from very strong and ancient mortar, and in other parts . . . from right old rough-stone, which work is clearly seen to be completely Moorish and worked in the time of the Moors. And so the said canal from the said old dam up to an old and ruined mill near a depopulated place called Facuquer is made and constructed, as has been said, [in an] old and Moorish [fashion.]" [40]

Conversely, it was equally apparent when an old right was claimed for what was obviously a modern work. Two millers on the Algirós branch of the Mestalla Canal in 1452 demanded certain rights on the basis of ancient practice. The lawyer for the commons of the canal, Berenguer Cardona, argued that their plea was inadmissible because the mills in question clearly were not ancient. Let us suppose, he argued, that the mills in question

> are very ancient mills from the time of the Moors, and *re ipsa* there are evident in each of them Moorish constructions which could be recognized as such by anyone seeing them, built in the time of the Moors before Valencia belonged to the Christians, and that the said constructions are held by those seeing them as Moorish, ancient, and built in the time that Valencia belonged to Moors.
>
> It would follow, therefore, if the said mills are from the time of the Moors, before Valencia was Christian, that in the time they were built they were granted the water, and it is their property, and they are in peaceful possession of having and using the said amounts of water of the Algirós Canal.

But, continued Cardona, one of his opponents cannot possibly allege such rights for his mill because

> it is a new mill, and its millhouse and all its works are new, that is, they were clearly made one hundred years ago or less, and there is nothing ancient about it, nor does it have any construction from the time of the Moors, but rather anyone who understands building would hold, and does hold, the construction and building of the said fulling mill as the typical work of Christian times and from one hundred years ago and less, or a little more.[41]

Because Moorish construction was the best evidence to cite in support of a claim of priority, it was adduced to corroborate rights alleged to have dated from the time of the Reconquest. In 1310 the new owners of a village called Tales claimed that they were inheritors of the rights (including those of irrigation) granted by James I to the original three hundred settlers of the place. The owners, "justly and with reason, have by the ancient diversion dams of Saracen times . . . the water which passes through the district of Tales and flows in the said dams." [42] The right, that is, came from the king; but Moorish diversion dams were the surest proof that the works were actually those cited in the charter.

As late as the seventeenth century, distinctive technology was still considered trustworthy historical evidence of the Islamic past. In 1609 the city of Orihuela mentioned in a letter to the King that "in the district of the city of Murcia . . . in our Segura River, there is built a diversion dam, made in the time of the Moors with such industry and art that with the water which passes over it this city and the villages and towns of the region are sustained, because it is made in the form of a spoon." [43] This striking and totally unusual, nonindigenous method of construction was, as late as the seventeenth century, positive proof that the authors of the splendid sluice were the Moors of olden days.

THE LIMITS OF CULTIVATION

Valencian tradition, reflected in the common law, regarded the huerta as fixed in the limits imposed upon it in the time of the Moors. In this concept the limits of Moorish cultivation (well known from the donations of James I) were held as defining the optimum service area for the city's water resources. Strictures in the fueros against building new canals or introducing innovations in the irrigation network served to prevent any major alterations in the traditional balance between water supply and irrigated area. Thus the jurates of Valencia complained to Peter IV in 1342 that, according to the fueros, the water of a public river ought to be divided in relation to the size of the possession irrigated; but, subsequent to the enactment of that law, new lands had been brought under cultivation that were taking water from the Guadalaviar "to the prejudice of the lands of the huerta of Valencia which already in the time of the Saracens were cultivated and had sufficiency of water from the river to irrigate them." It was not just, the jurates argued, for lands which had not customarily been irrigated to have abundance of water while lands cultivated since olden times lost their crops. The jurates sought the right to divide the huerta's water as they saw fit — in this case, favoring those irrigators with the most ancient rights. The King agreed with the jurates.[44]

In 1368, in order to preserve the customary limits of the huerta, Peter ordered compliance with a previous stricture against widening the Moncada Canal;[45] and in 1413, as a further example, the city asked the governor to destroy all new works on the river.[46] Nevertheless, in spite of the weight of tradition, the huerta did expand. In the first part of this study two cycles of expansion and contraction were described. The first, from 1238 to 1321, was characterized by 1) expansion of the huerta throughout the thirteenth and into the fourteenth century; 2) the halting of expansion by a catastrophic drought; 3) a legal solution, imposed by the King and entailing an accommodation with the Pueblos Castillos, to the problems arising from acute pressure for water; 4) enforced destruction of new works and a return to the "Islamic image" of the huerta. The second cycle, extending from 1321 to 1413, was also marked by several decades of expansion, capped by reclamation of the Marjals of Valencia at the end of the fourteenth century. In this case, too, expansion was terminated by a long drought, during the course of which renewed pressure from the Pueblos Castillos was met by a political solution imposed by the city of Valencia.

Several interrelated factors — climatic, demographic, and economic — governed the expansion of irrigation cultivation. The water resources of the huerta varied so widely from year to year and from decade to decade that the potentialities for expansion must have appeared quite different to different generations. In drought-free years men would think about expansion; when the water level fell, contraction and restriction were inevitable. All these factors converged and were most graphically tested in the marginal areas of cultivation, areas which were fruitful in good years and perhaps not even worth holding in bad: the Marjals of Valencia were typical of such regions. In the 1380's and 1390's the Marjals seemed ripe for cultivation. The last quarter of the fourteenth century was nearly free from drought and was at the same time a period of rising population (the city had about four thousand seven hundred hearths in 1355 but eight thousand in 1418).[47] If the settlement plans were successful — as they seem to have been — the added population would surely have contributed to huerta-wide pressure for water under the drought conditions of 1412–1415.

It has been observed above that drought brought a quickening of communal activities; the mechanisms of control were tightened in order to attain the maximum possible efficiency in distributing a contracting water supply. On the regional level drought forced reassessment and sometimes realignment of the balance of power among all the irrigators of the watershed. Drought also caused reassessment of public policy toward expansion of cultivation into marginal areas. It curtailed expansion and at the same time it awakened interest in water diversion plans.

In measuring conflicting pressures and objectives of its irrigation-based agricultural economy, Valencian society had always to account for its basic hydrological potentiality. The tradition of the Islamic bounds of the huerta, by providing both a reference to past experience and a legal mechanism to brake expansion whenever it was detrimental to the common good, served reality well.

Appendixes
Bibliography
Notes
Index

Appendix 1 | Municipal Provisions for Road Maintenance, 1396

The state of the roads in irrigated areas was so precarious that the city council had to intervene to ensure their repair:

Concerning the Repair of Roads[1]

Since in past councils mention was made of the befouling of the roads of the huerta of the said city [Valencia], at times through the fault of irrigators, and at times through the fault of cequiers, it was decided that it would be an expedient and most useful thing that . . . a good person be appointed to have single care and special charge of this. Because those to whom it pertains ordinarily (the cequiers have usually taken the fines for spilling and flooding of water) do not take care of it any longer.

And since the mustasaf — an officer to whom it pertains, according to fuero, to guard and prevent the roads and highways from deteriorating — is busy with other affairs and does not take care of it, nor either do the *obrer* or *sotsobrer*,[2] to whom pertain the repair of roads, take care of it, it was decided in past councils to find a good man who would undertake such a chore without salary from the general funds of the city, but only from the emoluments and fines coming from this matter, the present council . . . in order to remove exasperation and harm from travelers and to save the great expenses which the said *obrer* has to make each year in good repairs, deliberately and unanimously provides that the jurates seek out and find . . . a person fitting to hold the commission in this matter . . . which commission can be made by the mustasaf or others, according to what they decide is best.

Appendix 2 | Irrigation of Vineyards

Contrary to the practice in Spain today, vineyards were often irrigated in medieval Valencia, as shown in the account of a fifteenth-century agricultural worker:

> Account of that which En Miquel Giner owes me, En Bernat Ebri, of that which I have spent from my own pocket in caring for his vineyards in the present year of 1438.[1]

First, he owes for pruning (*podar*) and cultivating (*exarmentar*) 4 irrigated vineyards (*vinyes de la orta*), and it appears in detail just as he and I had already figured it out	4£	19s	
Item, for pruning the unirrigated vineyard (*vinya del secá*), for 29 men at the rate of 3s. a day	4£	7s	
Item, for irrigating the four vineyards of the huerta, at 4s. each		16s	
Item, for cultivating the "heir's vineyard" (*la vinya del hereu*), 18 men at the rate of 3s. 2d.	3£	1s	
Item, for cultivating the widow's vineyard, 19 workdays at 3s.	2£	17s	
Item, for the vineyard of *el morestrell* for cultivating, 19 days at 3s. 6d.	3£	7s	
Item, for the vineyard of *el mancurro,* 19 men, at 3s. 6d., which are	3£	6s	6d
Item, for cultivating the unirrigated vineyard, as long as the season lasts, there were 29 workdays, at the rate of 3s.	4£	7s	
Item, for pruning (*sporgar*) all the vineyards	1£	10s	

Item, for repairing the road used by the cequiers		7s	6d
Item, for irrigating the "heir's vineyard"		4s	
Item, for hoeing (*manchegar*) 14 workdays at 3s.	2£	6s	

All what I have paid out for him totals, 31 lliures 3 sous 6 diners.[2]

Appendix 3 | Crop Priorities in Time of Drought, 1376

In time of drought, the town officials of Valencia exercised their authority in order to establish a uniform irrigation policy over the entire huerta. They enforced the turns with the upstream towns and established crop priorities for all the canals while taking care not to interfere with the actual internal administration of any one of them.

Of the Form of Irrigating Crops[1]

By reason of the great droughts which have long continued and persist at present, the water of the Guadalaviar River is diminished to such an extent that the turns set up for the diversion dams and canals which draw from the said river do not suffice, nor can they suffice, the irrigation of all the fields of their systems, especially because of the rice being grown there, [which] by its nature has frequent need of water for irrigation.

For this reason, conflicts and strife have arisen among the irrigators of each canal, and scandal could now ensue. Therefore the [Valencia] council, deciding and agreeing to put an end to conflicts and scandal and to give norm and form to the cequiers of the said canals in regard to the distribution or partition of the water of the canal of each one of them for the days of their turns during the present time of need, provides and ordains and holds as good that each one of the said cequiers give and authorize . . . the water of his canal, from first to last [in priority], first to bread grains and vegetables, next to grapevines, next to fallow fields about to be sown, and last to rice, according to the greater need of each of these things. . . . Likewise, [if] . . . any canal in its turn cannot supply all of its irrigation district and has to deprive [some of it], let it deprive the last of the said things, according to what is lowest in the order.

Appendix 4 | Regulations of the Water Company of Benidorm[1]

According to one method of financing the building of an irrigation canal in medieval Valencia a company was formed, the members of which subscribed equal — or nearly equal — shares. This company, organized to construct an irrigation canal in Benidorm sometime during the first quarter of the fifteenth century, included on its membership list such local dignitaries as the Vicar of Polop and Johan de Luna and some contractors including the leveler Garcia Mezquita.

The following points were agreed upon by the members of the whole company:

First, it was decided and held as good that all the things done and still to be done ought to be arranged and discussed by the honorable Johan de Luna, Guillem Ramon Çaportella, Johan de Vilanova the Younger, Jacme Gomiz, and Garcia Mezquita; that all should pass under the consideration of the above mentioned; and that all the other partners be held to authorize whatever the above shall do.

Likewise, no person may be admitted to the company without the wish and consent of the above deputies.

Likewise, it was decided by all the partners that if by chance any member of the said company, for any cause, manner or reason, tries or plans to do wrong or commit any fraud deemed harmful to the company, in such a case — if it be proven — he who wishes to do such a thing . . . will lose however much he may have invested in the company and shall be expelled therefrom.

Likewise, it was decided that when the work was brought to a perfect end, all the partners would be obliged to deal justly with Garcia Mezquita's share, just as they might do in case of their own, doing there [on behalf of] Mezquita that which each would do with his own property.

Likewise, Garcia Mezquita, as master, is to receive as salary for each working day . . . 6s.

Likewise, the elder son of the said master, for the same reason, 3s.

Likewise, the younger son of the said master, in the form above mentioned, 2s.

Likewise, it was decided and held as good that Johan de Vilanova the Younger be treasurer and administrator and that he keep account of receipts and expenditures of money for the duration of the said work, and of the scribe who keeps the book of receipts and expenditures and of the workdays; let him have as salary 330s.

Likewise, it was decided that Guillem Ramon Çaportella be the master mason of the work and that he have as salary for each working day, 2s.

Likewise, it was decided that all the partners be held to put in the first sack of money to be made up, for each 100 florins, 15 florins.

Likewise, that all the partners be held to give to the treasurer the above mentioned portion [of his share] whenever he might ask for it, and thereafter however much may be needed for the work will be levied accordingly by the elected deputies.

The below-written are the partners of the work of the canal:

	[Share]
First, the honorable en Johan de Luna	200 florins
Likewise, en Guillem Ramon Çaportella	100 fl.
Likewise, en Johan de Vilanova, Elder and Younger	200 fl.
Likewise, en Jacme Iuarç, the Younger	100 fl.
Likewise, en Jacme Gomiz	100 fl.
Likewise, n'Anthoni Armengol	100 fl.
Likewise, en Garcia Mezquita, master	100 fl.
Likewise, Anthoni Mezquita	100 fl.
Likewise, en Macia Catala	100 fl.
Likewise, en Guillem Oliuer	100 fl.
Likewise, en Marti Ferrer	100 fl.
Likewise, en Jacme Vives, vicar of Polop	100 fl.
Likewise, en Jacme Fira	50 fl.
[Total	1450 florins]

Appendix 5 | Rainfall and Temperature in Valencia

a) Average Monthly Rainfall (in mm.) in Valencia, 1920–1954

	1920–1929	1930–1939	1940–1949	1950–1954
January	9.7	30.3	36.0	15.8
February	64.3	26.5	40.6	8.1
March	22.9	27.6	25.9	45.1
April	16.1	22.1	34.0	47.8
May	28.0	40.4	25.9	10.2
June	46.4	43.8	12.2	15.0
July	6.1	11.6	7.2	11.6
August	7.5	15.6	28.7	32.1
September	37.6	52.3	59.5	53.4
October	44.8	49.2	46.7	92.7
November	71.7	32.1	22.9	16.9
December	28.3	33.1	47.9	39.3

b) Average Monthly Temperature (in °C) in Valencia, 1920–1954

	1920–1929	1930–1939	1940–1949	1950–1954
January	10.0	10.4	10.0	9.6
February	11.1	10.8	11.1	10.7
March	13.1	13.0	13.1	13.0
April	14.8	15.3	15.3	14.3
May	18.1	17.9	17.8	17.9
June	21.8	21.7	21.4	21.9
July	24.3	24.6	23.8	24.3
August	24.7	25.0	24.4	24.7
September	22.2	22.6	22.4	22.5
October	18.9	18.6	18.5	18.7
November	13.4	14.1	14.3	15.4
December	10.3	11.3	10.6	12.0

Source: Instituto Nacional de Estadística, *Reseña estadística de la provincia de Valencia* (Madrid, 1957), pp. 35–36.

Appendix 6 | Droughts and Floods in Medieval Valencia

Table A lists all the references to droughts in Valencia between 1313 and 1450 which are known to me. Most of the references are from the series of *Manuals de Consell* (MC) and *Cartas Misivas* (CM) in the Municipal Archives of Valencia (AMV). Noted are 1) general indications of the state of drought as indicated either by the word *seccada* or by allusions to the lack of water generally or to rainfall; 2) specific allusions to shortness of river or canal water; 3) indication, if any, of the temporal extent of the drought; 4) repercussions of the drought, as indicated by the following symbols:

D = water diversion plan
G = grain shortage, usually because of nonfunctioning of mills
H = harvest endangered
P = public prayer or procession for rain
T = setting of turns among various towns and canals on the river

Table B lists references to floods and torrential rains between 1321 and 1487, noting 1) the general term expressing the phenomenon (e.g., *diluvi*) or specific allusions to high water in the river and 2) the locus of the flooding and its effects, if any.

Floods are also indicated in Table A, in chronological sequence, by cross-references to Table B.

Table A

Droughts in Valencia (1313–1449)

No.	Year	Day/month	Source	Description of climate	Description of river water	Indication of time	Ramifications of drought		Comment
1.	1313	21 August	AMV/MC 1, fols. 91r–93r	seguitat	minva daygues en lo riu	(seguitat) lan pasat e del present es	H	T	–
2.	1321	25 May	Branchat, *Tratado de derechos*, III, 204	siccitatem temporis ex raritate aquarum pluvialium procedentem	aquae rivi de Godalaviar defluentes per cequias ad dictam Civitatem et ejus terminos sunt adeo diminutae	–	–	T	Not enough water in huerta for irrigation and mills; turn arranged with Pueblos Castillos
3.	1321	5 July	Branchat, III, 206	indigentia aquarum (. . . immensa necessitate)	–	–	–	T	Specifically sets turn decreed in (2) at 4 days for Pueblos, 4 for city
	1321	16 October	Flood no. 1]						
	[1328	28 September	Flood no. 2]						
4.	1335	28 August	AMV/CM 1, f. 41	–	gran mirva daygua del flum de godalaviar/ cequies de la orta son mirves daygues/[never had been] semblant mirva daygues en lo dit flum	–	H	T	Crops of the huerta in danger of being lost if God and good friends don't come to the rescue
	[1340	6 November	Flood no. 3]						
5.	1341	21 March	AMV/MC 4, f. 171v	mirva daygues pluvials en lo regne de Valencia	–	–	G	–	–
6.	1343	2 January	AMV/MC 4, f. 285r	–	–	–	–	P	–
7.	1345	6 May	AMV/MC 4, f. 486v	–	–	–	G	–	Because women wear long skirts, God has withheld rain, so that the grain (*blats*) cannot ripen
8.	1351–1352	–	Santamaria, *Aportación*, 82	–	–	–	–	–	Lowering of taxes; increase in duty on salt and meat

9.	1355	2 January	AMV/MC 12, f. 52	gran e evident necessitat daygues/ gran segdat	-	(gran segdat) que a present es	-	-	Inspection of all the canals ordered
	[1356	-	Flood no. 4]						
10.	1358	25 February	Branchat, III, 308–309	maximam siccitatem	-	praeteritorum et praesentis temporum (maximam siccitatem)	-	T	Half the huerta dry
	[1358	17 August	Flood no. 5]						
11.	1368	31 July	AMV/MC 15, f. 8r	gran secada/poquea de la aygua	(river water is) molt mirvada	-	-	T	Water so low that only 1 or 2 of the *cequies sobiranes* have enough water, and city's mills can't mill
12.	1371	-	Santamaria, 82	-	-	-	-	P	Outbreak of plague
13.	1372	31 May	AMV/MC 16, f. 90; N. Primitiu, "Molineria valenciana," 730.	-	-	-	-	-	Harvest small owing to droughts of past years
14.	1374	30 May	AMV/CM 3, n.p.	sequedat	-	-	G	-	Grain shortage owing to drought, mildew (? —neula) and theft
15.	1374	14 July	AMV/MC 16, f. 207v	seccada	necessitat de les aygues del riu	(seccada) e necessitat del temps present	-	-	Re. rights of Puig (i.e. downstream) on cequia of Moncada
16.	1374	2 October	AMV/MC 16, f. 225r	seccades	-	temps de seccades segons en lo present temp alcuns altres ja passats	-	D	Júcar diversion project
17.	1375	3 October	AMV/MC 17, f. 20r	seccades	aygua del riu, la qual per grans e longues seccades es m diminuida	-	-	D	Júcar project
18.	1376	11 February	AMV/MC 17, f. 38v	seccades	(see time column)	(same wording as no. 16 except) . . . longues seccades *era e es* molt diminuida	-	D	Júcar project

Table A (*continued*)

No.	Year	Day/month	Source	Description of climate	Description of river water	Indication of time	Ramifications of drought		Comment
19.	1376	26 March	AMV/MC 17, fols. 47v–48r	la gran sequedat	la gran minua daygua que es en lo dit riu de godalaviar	(la gran sequedat) que es estada de lonch temps a ença e es de present en la terra	–	D	Júcar project
20.	1376	18 July	AMV/MC 17, f. 62r	grans seccades	aygua diminuida	(grans seccades) que longament son estades e son de present	T	–	Water diminished so much that cequies cannot supply the whole area of their rech
	[1378	16 June	Flood no. 6]						
21.	1384	4 March	AMV/MC 18, f. 26v; Primitiu, 730	secada per fretura de pluja	–	–	H	–	. . . en tant que lo blat, maiorment dels secans, son o estan en perill de perdicio
22.	1393	12 August	AMV/CM 5, f. 215r	–	fretura daygua/ necessitat	–	H	T	fretura . . . en la nostra orta
23.	1400	24 July	AMV/MC 22, f. 120v	spesses secades	gran penuria daygues	per frequentacio de (spesses seccades)	H	D	Júcar project; not enough water for mills; crops lost
24.	1401	18 May	AMV/CM 7, n.p.	–	fretura daygues	–	–	T	Whole kingdom affected by drought
25.	1401	15 July	AMV/CM 7, n.p.	–	distret daygues	–	–	T	–
26.	1401	19 July	AMV/CM 7, n.p.	–	fretura daygues	–	G	–	Mills have no water
27.	1401	23 July	AMV/CM 7, n.p.	–	esterelitat e penuria de les aygues	–	–	D	Júcar project
28.	1403	12 January	AMV/MC 22, f. 229r	tanta mirva daygua per les frequentes secades	–	frequentes (secades)	–	–	Water in Albufera low with harm to fish; Cequia d'En Fluvia ought to be extended so it flows into Albufera
	[1403	–	Flood no. 7]						

29.	1404	16 January	Branchat, III, 210–213	ob pluviarum nimiam raritatem	–	–	G	–	–
30.	1406	25 October	AMV/MC 23, f. 92v	–	poquea de laygua (in Guadalaviar)	–	–	D	Júcar project
	[1406	3 November	Flood no. 8]						
	[1410	22 January	Flood no. 9]						
31.	1412	10 October	AMV/CM 11, f. 101r	–	poquea de aygua (in Guadalaviar)	–	–	–	Water needed to float wood downstream
32.	1413	19 April	AMV/CM 11, f. 207r	–	fretura e distret daygues	–	G	T	–
33.	1413	5 May	AMV/CM 11, fols. 210v–211r	–	minua de laygua/poquea de laygua del riu	–	–	–	Mills needed to mill city's grain
34.	1413	3 June	AMV/MC 25, f. 196r	gran secada	penuria daygua	en aquest temps	–	–	Re. Riu de Xelva
35.	1413	13 July	AMV/CM 12, f. 14v	–	gran destret daygues en lo nostre riu	–	–	T	–
36.	1413	14 July	AMV/CM 12, f. 16r	–	fretura de aygues . . . en lo nostre riu	–	G	–	–
37.	1413	14 July	AMV/CM 12, f. 16r	–	nosaltres siam constituits en gran penuria de aygues/pocha aygua (in Júcar)	–	G	–	–
38.	1413	24 July	AMV/CM 12, f. 23v	–	penuria daygua	–	G	–	–
39.	1413	28 July	AMV/CM 12, f. 24v	–	gran fretura daygues	–	G	–	But great abundance of water in barony of Oliva
40.	1413	19 August	AMV/MC 25, f. 258v	–	gran fretura e sterelitat que de present eren daygues en lo riu de godalaviar	de present	H	T	Pueblos Castillos affair; not enough water for milling, harvest lost
41.	1413	20 August	AMV/CM 12, f. 35vff.	seccada en totes aquestes parts	(river is so) minue que memoria de homens no fo vist semblant	–	G	–	Letter to King
	[1413	14 October	Flood no. 10]						
42.	1413	22 November	AMV/CM 12, f. 83r	–	gran penuria daygues	–	G	–	–
43.	1414	7 August	AMV/CM 12, f. 172r	–	gran fretura daygua	–	–	T	–

Table A (*continued*)

No.	Year	Day/month	Source	Description of climate	Description of river water	Indication of time	Ramifications of drought		Comment
44.	1414	7 August	AMV/CM 12, f. 172v	–	gran fretura daygua	–	G	T	–
45.	1414	8 August	AMV/CM 12, f. 173v	–	–	en los anys prop passats	G	–	Ref. is to gran penuria e carestia de blats en los anys prop passats
46.	1415	3 April	ARV/Gob., 2211, 6th hand, f. 2	–	fretura daygua	–	–	–	In upstream-downstream dispute it is requested that water be shut off for cleaning the canal
47.	1415	6 May	AMV/CM 13, f. 3r	–	destret de la aygua	–	G	T	People cannot find bread in the stores
48.	1415	17 July	AMV/CM 13, f. 11r	–	poques aygues (in canals)	–	G	T	Mills cannot mill
49.	1415 [1415 [1415 [1416 [1417	9 August 24 September 19 October 3 April 27 March	AMV/CM 13, f. 13r Flood no. 11] Flood no. 12] Flood no. 13] Flood no. 14]	–	poques aygues (in canals)	–	G	T	–
50.	1420	10 April	ARV/Gob., 2224, 5th hand, f. 20	–	(great lack of water)	–	–	–	In Morvedre: general meeting to take action
51.	1421	12 November	ARV/Gob., 2227, 16th hand, f. 27	sequada	–	de present [the land is secá] cinc o ses mesos ença (town depopulated)	–	–	In Chiva: sequada de la font (of Chiva). 90–100 *moros* reduced to 24 in last 5–6 months. Because of drought they have neither sown nor harvested their crops
	[1427 [1428	25 October 6 October	Floods nos. 15, 16] Flood no. 17]						
52.	1432	12 April	AMV/MC 29, terç libre, f. 193v	gran fretura de pluja en aquest regne	–	–	–	P	–
53.	1435 [1436	10 March 5 June	AMV/MC 30, f. 296 Flood no. 18]	fretura (of rain)	–	–	–	P	–

54.	1443	1 May	AMV/CM 20, n.p.	(God) nos ha privat de pluia del cel	–	de lonch temps ença	–	P	Owing to lack of rain the land cannot fructify nor give forth the harvests (*splets*) entrusted to it
55.	1444	24 April	AMV/CM 20, n.p.	(same as no. 54)	–	–	–	P	(same as no. 54)
56.	1444	11 August	AMV/CM 20, n.p.	(same as no. 54)	–	–	–	P	(same as no. 54)
57.	1445	24 July	AMV/MC 33, f. 125r	secada	–	–	G	–	Moncada Canal ordered surveyed "in order to know how much water might come here in order to mill the flour (*farines*), owing to the drought"
58.	1445	August	AMV/CM 20, n.p.	(same as no. 54)	–	–	–	P	(same as no. 54)
59.	1447	27 November	AMV/MC 34, f. 35r	–	–	–	–	P	–
60.	1449	6 May	ARV/Gob., 2277, 2nd hand, f. 46v, 7th hand, f. 36r	gran sequedat, gran fretura	fretura de aygues	tres mesos ença	H	T	Cequia of Moncada; people of Riff canal (at tail) complain
61.	1449	(before 31 May)	AMV/MC 34, f. 196r	–	–	–	–	P	–

Table B

Floods in Valencia, 1321–1487

No.	Year	Day/month	Source	Hydrological data	Location and effects
1.	1321	16, 24 October	F. Almela y Vives, *Las riadas del Turia* (Valencia, 1957), p. 18.	abundantes lluvias, aguas pluviales	Destruction of houses, bridges
2.	1328	28 September	Almela, *Riadas*, 19–23; ARV/Protoc. 4151, n.p.	dulubi	Valencia and huerta, Roteros, Alcira, Foios
3.	1340	6 November	Almela, 23; Santamaria, 81	–	Procession of *Lignum Crucis* from Valencia cathedral
4.	1356	–	Santamaria, 82	avenida del Turia	–
5.	1358	17 August	Almela, 23–26	floods of Turia and excessive rain	Inundation of city, ruin of houses
6.	1377 [1378	Summer 16 June]	AMV/MC 17, f. 168v; Primitiu, 731	tempestat de pedra com de gran diluvi daygues pluvials (in past summer, i.e., 1377)	Huerta and district of city: grape harvest lost owing to greatest hail storm in memory
7.	1403	–	Almela, 27	–	–
8.	1406	3 November	AMV/MC 23, f. 169v	gran diluvi daygues	Lack of grain; dams and canals so damaged that mills not working. Valencia: Júcar overflowed 22 October (Almela, 27)
9.	1410	22 January	AMC/LC 17, n.f.	–	Castellón: river very full; almenares to be raised
10.	1413	14 October	AMV/CM 12, f. 62v	gran multitut daygua que en lo riu es venguda	Valencia: grain shortage; diversion dams broken; drought continues nonetheless
11.	1415	24 September	ARV/Gob., 2213, 31st hand, 3r–9v	gran diluvi daygues . . . les pluges grans	Valencia (Chirivella and Mestalla, Favara, Faitanar canals): many vineyards and canals destroyed; time reference not clear

12.	1415	19 October	ARV/Gob., 2212, 24th hand, fols. 6r–8v	heavy rains	Valencia: rains cause problem with city sewers, which hadn't been cleaned
13.	(1416	3 April)	ARV/Gob., 2214, 9th hand, f. 12	–	Huerta of Valencia: ref. to floods of "several years ago" (date is of doc., not flood)
14.	1417	27 March 16 April	ARV/Gob., 2219, 7th hand, fols. 5r–6r	aygues pluviales	Huerta of Valencia (Puig)
15.	1427	25 October	ARV/Gob., 2238, 11th hand, fols. 13r–14v; Almela, 28–30	grans multituts daygues celestials	Valencia: recently (*prop passades*)
16.	1427	–	ARV/Gob., 2252, 11th hand, f. 29	–	Ref. in a 1434 doc. to "floods of 1427"
17.	1428	6 October	AMV/MC 29, f. 34r	–	Mortar works to be done in Mislata Canal because of rain-water damage
18.	1436	5 June	ARV/Gob., 2257, 5th hand, f. 2v	–	Valencia: flooding (when not clear) destroys diversion dams
19.	1453	15 January	AMV/MC 35, 5th book, f. 255r	–	Valencia: procession in order that the waters might cease (*cessassen les aygues*)
20.	1462	1 February	AMV/CM 25, f. 58r	Tantes e continues pluges . . .	Valencia: so much rain that land is in peril and people cannot sow; prayer for cessation of rain
21.	1462	24 April	AMV/CM 25, f. 74r	Tantes e continues pluges . . .	Valencia: prayer for cessation
22.	1475	17–20 November	Almela, 30–31	(lluvias espantables) "as if the sky were open and the world was going to end"	Whole kingdom: rains last until January 1476
23.	1487	28 October	Almela, 31	–	Valencia: river destroys bridges, floods city

Appendix 7 | Plans for Hydraulic Wheel in Valencia to Irrigate Garden of the Lonja, 1529

The council of Valencia often discussed hydraulic problems of interest to the city. A proposal for irrigating the garden of the Commodities Exchange (Lonja) in 1529[1] included plans for the construction of a *sāniya:*

> All the magnificent jurates, the treasurer, and the syndic take note that the garden (*ort*) of the Lonja is very dry and ruined because it has no water for irrigation, which redounds to the great blame and shame of the city inasmuch as the Lonja is so beautiful, famous, and important and a thing which so ennobles the city. For this reason the magnificent jurates who held office in times gone by . . . wished to provide for the needs of the said garden and made a great expenditure on a certain work which was begun in hopes of drawing water from the Favara Canal at the bridge of Les Mealles[2] and diverting it to the garden of the Lonja in order to make a fountain which would irrigate and embellish the said garden. This expenditure was totally useless and unfruitful because it was impossible to make the water come that way.
>
> Taking into consideration this difficulty and the needs of the said garden . . . the jurates, treasurer, and syndic, desiring the retention of the said garden for the great embellishment which it provides the Lonja (in which not only merchants do business but also many knights and honorable persons, not only of the said

city but also others who come here from diverse parts of the world) . . . have decided to make a hydraulic wheel *(cenia)* in a place set off from the said garden and even to make a fountain in the middle which might honor and beautify both it and the Lonja.

And now at present there stands vacant a house belonging to the widow of Anthoni Aliaga, the merchant, which is adjoining to and part of the [house] constructed on a dead-end street[3] which used to be between the said house and another house belonging to the father of the said Anthoni Aliaga, which street is behind the house of Pere Lobet, the notary, which is in front of the garden of the Lonja, the which house of the said widow, together with the said street, is most handy and suited to the making of such a hydraulic device so that from there water might come along the said street in front of the house of Pere Lobet with very little work and expense . . . to the middle of the garden where the said fountain could be made, from which, by means of the said device . . . water could flow (*rajar*) all day in the said fountain and irrigate the garden whenever necessary.

And in this manner the garden will be preserved and the Lonja ennobled and beautified and the city much decorated and honored.

Therefore the magnificent jurates, treasurer, and syndic, moved by these considerations, provide that the house of the widow ot Anthoni Aliaga be bought by the city for the price agreed upon by the magnificent treasurer and syndic, to whom they give power to discuss and agree upon the price as it seems best to them. They provide that the house, together with the said street, be seized by the city in order to make the wheel and its works, and that the fountain be made in the garden of the Lonja in the form abovestated, and that the expense which it will entail, not only in the price of the house but also in the making of the wheel and fountain mentioned above, be paid by the administrator of the Lonja.

Once the wheel is completed and the space which will be necessary for its functioning marked off and set aside, they provide that that which is left over . . . be sold by the city in order to make up all which is possible of the said expense.

Appendix 8 | Division of the Barada River in 125 H. (742/743 A.D.)

The river of the Ghūṭa of Damascus was divided, both physically and institutionally, in a manner similar to the Guadalaviar and others rivers of eastern Spain. As a result of an appeal from the irrigators, the Caliph Hishām decreed a proportional division in the year 125 H.:[1]

Then the people of the Barada complained to Hishām b. ʿAbd al-Malik[2] about the insufficiency of the water. He ordered al-Qāsim b. Ziyād[3] to apportion[4] the canals for them; and he apportioned them:

He allotted to the Yazīd Canal: 16 parts.[5]

He allotted to al-Ghaur al-Kabīr: 10 parts.

And to al-Ghaur al-Ṣaghīr: 5 parts.

And to the Dāriyā Canal: 16 parts.

He allotted to the Thūra Canal 42 parts. In it at that time 14 siphons were functioning, and upon it there was not a [single] mill.

And to the Qīniyyah Canal: 11 parts.

And to the Bānās Canal: 30 parts. One part was carried in it to Yazīd b. Abi Maryam,[6] client of Sahl b. al-Ḥanẓaliya; and after him 3 parts were carried to al-Faḍl b. Sālih al-Hāshimī.[7]

And to the Majdūl Canal: 12 parts.

And to the Dāʿīya Canal: 13 parts.

And to the Ḥayuwa Canal — it is (also called) the Canal of al-Zuluf(?) — 12 parts.

And to the Upper Tūma Canal: 5 parts.

And to the Lower Tūma Canal: 4 parts.
And to the Canal of al-Zābūn: 4 parts.
And to the Canal of al-Malak: 4 parts.[8]
Al-Qanawāt[9] was not apportioned at that time, its 2 sides running full.

Select Bibliography

Abbreviations facing p. 1

MANUSCRIPT COLLECTIONS

ALCIRA

Archivo Municipal

Llibres de Consells, 45–48.

Nivelacio feta per Micer Guillem del Rey y Joan Salvador, experts nomenats per la Real Audiencia instancia de la Vila de Alzira [1592], 066/1.

BARCELONA

Archivo de la Corona de Aragón

Consejo de Aragón, legajo 615.

CASTELLÓN DE LA PLANA

Archivo Municipal

Libres de Consell, 8–11, 14–20, 22–23, 27–28, 30–32, 35. The council books for the second half of the fifteenth century are not numbered. When both books and folios are unnumbered, the date alone provides adequate reference because the meetings were recorded in chronological order.

Libres de Values, years 1462, 1468.

Pergaminos (uncatalogued), Oct. 18, 1325; May 7, 1347; Apr. 7, 1400.

ELCHE

Archivo Municipal

Actas de Consell (vols. unnumbered), years 1420, 1528, 1531.

MADRID

Archivo Histórico Nacional

Ordenes Militares. Orden de Montesa (documentos particulares), pergaminos 626, 653, 690. Orden de Montesa, libro 542c.

VALENCIA

Archivo de la Acequia de Robella
Legajo 1, no. 1.
Archivo de la Catedral
Actas Capitulares, 3579.
Notals de Bonanat Monar, 3518.
Pergaminos 3924, 5597, 7429.
Archivo Municipal
Cartas Misivas [Lletres Misives], 1(1334)–33(1501).
Cartas Reales, 7, 8.
Libros de Sotsobreria de Muros y Valladares, 2.
Manuals de Consell, 1(1313)–50(1500).
Notals de Domingo Juan, 1.
Protocolos de Jaime Dezpla, 2, 25.
Archivo del Patriarca
Protocolos de Francesc Vilba, 1684.
Archivo del Reino
Fondos en depósito, Pergaminos, caja 6, no. 2.
Gobernación, 2188–2378 (Litium).
Procesos de Madrid, letra *S*, no. 429.
Protocolos, 2864, 2974, 3009, 3253, 4151.
Real, 272.
Real Acuerdo, 70.
Real Patrimonio.
Bailía, 1430.
Maestre Racional, legajos 482, 9852.
Varia, libros 126, 402.

GENERAL

Alcover, Antoni M. *Diccionari català-valencià-balear.* 10 vols. Palma: M. Alcover, 1930–1962.

Almela y Vives, Francisco. *Las riadas del Turia (1321–1949).* Valencia: Ayuntamiento de Valencia, 1957.

——— *Valencia y su reino.* Valencia: Ediciones Mariola, 1965.

Altamira, Rafael. *Derecho consuetudinario y economía popular de la provincia de Alicante.* Madrid: Asilo de Huérfanos, 1905.

——— "Mercado de agua para riego en la huerta de Alicante y en otras localidades de la península y Canarias," *Derecho consuetudinario y economía popular de España,* ed. Joaquín Costa. 2 vols. Barcelona: Manuel Soler, 1902. Volume II, pp. 136–164, 441–447.

Andreu Valls, Guillermo. *Archivos de la provincia de Castellón inventariados por D. Luis Revest Corzo en 1925.* Castellón de la Plana: Sociedad Castellonense de Cultura, 1965.

Archivos de las instituciones valencianas: exposición conmemorativa de la inauguración del nuevo edificio del Archivo del Reino de Valencia. Valencia: Dirección General de Archivos y Bibliotecas, 1965.

Asín Palacios, Miguel. *Contribución a la toponimia árabe de España.* Madrid: CSIC, 1944.

Aureum opus regalium priuilegiorum ciuitatis et regni Valentiae cum historia cristianissimi regis Jacobi ipsius primi conquistatoris. Valencia, 1515.

Aymard, Maurice. *Irrigations du midi de l'Espagne.* Paris, 1864.

Azofra, Manuel María. *Memoria sobre la exacta medición del agua corriente por medio de muelas, filas y plumas.* Valencia, 1844.

Baradez, Jean. *Fossatum africae.* Paris, 1949.

Bataller Bataller, Adelina. "La expulsión de los moriscos: su repercusión en la propiedad y la población en la zona de los riegos del Vernisa," *Saitabi* (Valencia), 10 (1960), 81–100.

Bellver, Melchor, and Vicente del Cacho. *Influencia que ejerció la dominación de los árabes en la agricultura, industria y comercio de la provincia de Castellón de la Plana. Castellón,* 1889.

Beneyto i Pérez, Joan. "La propietat predial i l'aigua de rec: sobre una hipòtesi d'evolució històrica," ACCV, 4 (1931), 123–126.

Blasco Ibañez, Vicente. *La Barraca.* Valencia, 1898.

Bofarull y Mascaró, Próspero de. See *Colección de documentos inéditos. . . .*

Borrull y Vilanova. Francisco Xavier. *Tratado de la distribución de las aguas del río Turia.* Valencia, 1831.

Branchat, Vicente. *Tratado de los derechos y regalías que corresponden al real patrimonio en el reyno de Valencia y de la jurisdicción del intendente como subrogado en lugar del antiguo bayle general.* 3 vols. Valencia, 1784–1786.

Brunhes, Jean. *L'Irrigation: ses conditions géographiques, ses modes et son organisation dans la Péninsule Ibérique et dans l'Afrique du Nord.* Paris: C. Naud, 1902.

Bruno, Henri. *Contribution a l'étude du régime des eaux en droit musulman.* Paris: Arthur Rousseau, 1913.

Bunting, Brian T. *The Geography of Soil.* London: Hutchinson, 1965.

Burns, Robert Ignatius, S.J. *The Crusader Kingdom of Valencia.* 2 vols. Cambridge, Mass.: Harvard University Press, 1967.

——— "Irrigation Taxes in Early Mudéjar Valencia," *Speculum,* 44 (1969), 560–567.

Camarena Mahiques, José. *Historia del distrito de Gandia.* Gandia: Ediciones de Fomento de Agricultura, Industria y Comercio de Gandia, 1965.

——— See *Colección de documentos para la historia de Gandia y su comarca.*

Carbó, Juan B. "El Archivo Municipal de Castellón," *BSCC,* 3 (1922), 85–

88, 169–173, 266–269, 344–347. Irrigation documents are discussed on p. 172.

Caro Baroja, Julio. "Norias, azudas, aceñas," *RDTP,* 10 (1954), 29–160. An exhaustive study of the diffusion of current-powered wheels in the Iberian Peninsula.

——— "Sobre cigüeñales y otros ingenios para elevar agua," *Revista de Guimarães,* 65 (1955), 161–181.

——— "Sobre la historia de la noria de tiro," *RDTP,* 11 (1955), 15–79. The history and diffusion of animal-powered hydraulic wheels in Spain.

Carpentier, Etienne. *Les réglements et les tribunaux des eaux dans les provinces du sud-est de l'Espagne (Murcie et Valence).* Paris: L. Larose et L. Tenin, 1912.

Casas Torres, José Manuel. *La vivienda y los núcleos de población rurales de la huerta de Valencia.* Madrid: CSIC, 1944.

Catálogo del Archivo Municipal de la Ciudad de Alcira, primera parte. Alcira: Publicaciones del Ayuntamiento, 1961.

Catálogo de la exposición bibliográfica y de recuerdos históricos de los riegos en el Reino de Valencia celebrada por la Sociedad "Lo Rat Penat" (separado del libro del III Congreso Nacional de Riegos). Valencia: Hijo de F. Vives Mora, 1923.

Catálogo de la exposición de derecho histórico del Reino de Valencia. Valencia: Miguel Laguarda, 1955.

Cavanilles, Antonio Joseph. *Observaciones sobre la historia natural, geografía, agricultura, población y frutos del Reyno de Valencia.* 2 vols. Madrid, 1795–1797. 2nd ed. Zaragoza: CSIC, 1958. I have used the second edition but, like the editors, have followed the pagination of the first.

Chabás, Roque. *Distribución de las aguas en 1244 y donaciones del término de Gandia por D. Jaime I.* Valencia, 1898. Document of an inquest held by order of the King to ascertain the local irrigation procedures from Muslim officials.

Chabret, Antonio. *Sagunto: su historia y sus monumentos.* 2 vols. Barcelona, 1888.

Chaytor, H. J. *A History of Aragon and Catalonia.* London: Methuen, 1933.

Colección diplomática de Jaime I, el Conquistador. 3 vols. Ed. Ambrosio Huici Miranda. Valencia: Hijo de F. Vives Mora, 1916–1920.

Colección de documentos inéditos del Archivo General de la Corona de Aragón. 41 vols. Ed. Próspero Bofarull y Mascaró *et al.* Barcelona: Imprenta del Archivo, 1847–1910. Volume XI (pp. 143–656) contains the text of the *Repartimiento de Valencia.*

Colección de documentos para la historia de Gandia y su comarca. 3 parts. Ed. José Camarena Mahiques. Gandia: Casa Ferrer, 1959–1961.

Colin, G. S. "La noria marocaine et les machines hydrauliques dans le monde arabe," *Hespéris*, 14 (1932), 22–61.

——— "L'origine des norias de Fès," *Hespéris*, 16 (1933), 156–157. A note on the diffusion of hydraulic techniques from Al-Andalus to North Africa.

Corominas, J. *Diccionario crítico etimológico de la lengua española*. 4 vols. Madrid: Gredos, 1954.

Costa, Joaquín. *Colectivismo agrario en España*. Madrid, 1898.

Cressey, George B. "Qanats, Karez, and Foggaras," *Geographical Review*, 48 (1958), 27–44.

Cueco Adrian, José M. "La Font de la Vall de Segó," *Generalitat* (Valencia), no. 3 (June 1963), pp. 83–88.

——— *La Font de la Vall de Segó*. Valencia: Ayuntamiento de Valencia, 1965. History of a small Spanish irrigation system from the twelfth century to the present.

——— "Un documento árabe inédito: la historia de los aprovechamientos de agua para el riego," *Levante* (Valencia), Aug. 1, 1965. Description of the discovery of a thirteenth-century Arabic document concerning an irrigation dispute (ARV, Procesos de Madrid, letra *S*, no. 429).

Danvila, Augusto. "Las ruinas de Pallantia," *Almanaque "Las Provincias,"* 1890, pp. 209–214. Description of Roman aqueducts.

Danvila y Collado, Manuel. *La germanía de Valencia*. Madrid, 1884.

Despois, J. "Development of Land Use in Northern Africa," *A History of Land Use in Arid Regions*, ed. L. Dudley Stamp. Paris: UNESCO, 1961. Pages 219–237.

Dias, Jorge, and Fernando Galhano. *Aparelhos de elevar a água de rega: contribuição para o estudo de regadio em Portugal*. Oporto: Junta de Província do Douro-Litoral, 1953. The diffusion of hydraulic lifting devices in Portugal.

Diaz Cassou, Pedro. *La huerta de Murcia*. Madrid, 1887. A study of the historical development of the Murcian huerta and its irrigation system by a perceptive local historian.

——— *Ordenanzas y costumbres de la huerta de Murcia*. Madrid, 1889.

Diaz Nieto, Ignacio, and Carlos Arrieta Alvarez. *Ensayo de bibliografía en materia de aguas*. Madrid: Dirección General de Obras Hidraulicas, 1964. A hastily compiled and poorly edited bibliography of contemporary and historical hydraulic material with too many errors and imprecisions to be used without a great deal of care.

Dolz Moros, José. "Antecedentes, peculiaridades y datos curiosos sobre la Junta de Aguas de La Plana," *Buris-ana* (Burriana), vol. 10, no. 103 (1967), 16–17.

Doñate Sebastiá, José María. "Riegos romanos del Mijares," *Archivo de prehistoria levantina* (Valencia), 11 (1966), 203–214.

Dozy, R. *Supplément aux dictionnaires arabes.* 2nd ed. 2 vols. Leiden: E. J. Brill, 1927.

Dozy, R., and W. H. Engelmann. *Glossaire des mots espagnols et portugais dérivés de l'arabe.* Leiden, 1869.

Eguílaz y Yanguas, Leopoldo de. *Glosario etimológico de las palabras españolas de origen oriental.* Granada, 1886.

Eiximenis, Francesc. *Regiment de la cosa publica.* Ed. Daniel Molins de Rei. Barcelona: Els Nostres Clasics, 1927.

English, Paul Ward. "Qanats in the Old World," *Proceedings of the American Philosophical Society,* 112 (1968), 170–181.

Escoolano, Gaspar. *Décadas de la historia de la insigne e coronada ciudad y reino de Valencia,* 2 vols. Valencia, 1878–1880 [1610–1611].

Exposición de distintas actuaciones del Tribunal de las Aguas de la vega de Valencia. Valencia: M. Guillot Aguilar, 1944.

Fernea, Robert A. "Conflict in Irrigation," *Comparative Studies in Society and History,* 6 (1963–1964), 76–83.

——— *Shaykh and Effendi.* Cambridge, Mass.: Harvard University Press, 1969.

Fletcher Valls, Domingo. "Acueductos romanos en término de Ribarroja del Turia (Valencia)," in *Notas sobre la antigüedad de la agricultura y el regadío en tierras valencianas.*

Font, Joan. *Discurso breve con que se prueva le possibilidad de sacar agua del Rio Xucar, para los llanos de Quart, Liria, Morviedro, y otros.* Valencia, 1628.

Font y Rius, José M. "La reconquista y repoblación de Levante y Murcia," in José M. Lacarra *et al., La reconquista española y la repoblación del país.* Zaragoza: CSIC, 1951. Pages 85–126.

Fontavella González, Vicente. *La huerta de Gandia.* Zaragoza: CSIC, 1952.

Forbes, R. J. *Studies in Ancient Technology.* 9 vols. Leiden: E. J. Brill, 1955–1964. (2nd rev. ed., 1964——.)

Fori antiqui valentiae. Ed. Manuel Dualde Serrano. Madrid-Valencia: CSIC, 1950–1967.

Fori regni valentiae. Valencia, 1548.

Foster, Alice. *The Geographic Structure of the Vega of Valencia.* Chicago: University of Chicago Libraries, 1936.

Franquet y Bertrán, Cirilo. *Ensayo sobre el origen, espíritu y progresos de la legislación de las aguas.* 2 vols. Madrid, 1864.

Fröhling, Margret. *Die Bewässerungslandschaften an der spanischen Mittelmeerküste.* Munich: Selbstverlag des Instituts für Geographie und Länderkunde und der Geographischen Kommission für Westfalen, 1965 (Westfälische Geographische Studien, 17). The physical and cultural geography of eastern Spanish irrigation.

Fuster, Joan. *Nosaltres, els valencians.* 2nd ed. Barcelona: Edicions 62, 1964.

Garcia, Arcadio. "La actuación procesal del mustaçaf," *Ausa* (Vich), 2 (1955–1956), 301–310.

Garrido Atienza, Miguel. *Los alquezares de Santafe.* Granada, 1893. Institutional history of irrigation from the Genil River, including a contemporary Spanish translation of a fifteenth-century Arabic document concerning division of the Genil's water.

Geertz, Clifford. "Form and Variation in Balinese Village Structure," *American Anthropologist,* 61 (1959), 991–1012.

Geografía general del Reino de Valencia, Ed. F. Carreras y Candi. 5 vols. Barcelona: Alberto Martín, 1920–1927.

Gimeno Michavila, Vicente. *La Rambla de la Viuda.* Castellón: Sociedad Castellonense de Cultura, 1935. History of irrigation from an intermittent stream near Castellón, with a documental appendix.

——— "Los riegos de la comarca de la Plana," *BSCC,* 19 (1944), 139–145; 20 (1944), 17–32, 205–222.

Giner Boira, Vicente. "La Sang de les Santes Escudelleres," *Las Provincias* (Valencia), July 20, 1965.

——— *El Tribunal de las Aguas de la vega de Valencia.* Valencia: Tipografía Moderna, 1960.

——— See Ros Marín, "El asesor jurídico. . . ."

Glick, Thomas F. "Abogados de aguas en la edad media valenciana," *Las Provincias* (Valencia), June 22, 1967.

——— "Dos documentos medievales referentes al Tribunal de las Aguas," *BSCC,* 43 (1967), 81–84.

——— "Levels and Levelers: Surveying Irrigation Canals in Medieval Valencia," *Technology and Culture,* 9 (1968), 165–180.

——— "Medieval Irrigation Clocks," *Technology and Culture,* 10 (1969), 424–428.

Glick, Thomas F., and Oriol Pi-Sunyer. "Acculturation as an Explanatory Concept in Spanish History," *Comparative Studies in Society and History,* 11 (1969), 136–154.

Goblot, Henri. "Dans l'ancien Iran, les techniques de l'eau et la grande histoire," *Annales,* 18 (1963), 499–520.

Gómez Serrano, Nicolás Primitivo [Nicolau Primitiu]. "D'Arqueologia: excavacions de Valencia," *ACCV,* 2 (1929), 75–96, 154–169; 3 (1930), 57–69, 205–212; 4 (1931), 57–72, 137–160; 5 (1932), 1–28.

——— "Contribució al estudi de la molinería valenciana mijeval," *III Congreso de Historia de la Corona de Aragón,* 2 vols. Valencia, 1923. Volume II, pp. 695–766.

——— "Salterio toponímico: Paiporta y Faitanar," *Almanaque "Las Provincias,"* 1925, pp. 199–201.

González Palencia, Angel. "Documentos árabes del Cenete," *Al-Andalus,* 5 (1940), 301–382.

——— "Notas sobre el régimen de riegos en la región de Veruela en los siglos XII y XIII," *Al-Andalus*, 10 (1945), 79–88.

Gray, Robert F. *The Sonjo of Tanganyika: An Anthropological Study of an Irrigation-based Society*. London: Oxford University Press, 1963.

Guillén, A. *El Tribunal de Aguas de Valencia y los modernos jurados de riego*. Valencia: Domenech, 1920.

Hall, William Hamilton. *Irrigation Development; History, Customs, Laws, and Administrative Systems Relating to Irrigation, Watercourses, and Waters in France, Italy, and Spain*. Sacramento, Calif., 1886.

Hamilton, Earl J. *Money, Prices, and Wages in Valencia, Aragon, and Navarre, 1351–1500*. Cambridge, Mass.: Harvard University Press, 1936.

Ḥimyarī, Muḥammad ibn ʿAbd al-Munʿim, al-. *La péninsule ibérique au moyen-âge d'après le Kitāb ar-Rawḍ al-mi ʿṭār fī ḥabar al-akṭār d'Ibn ʿAbd al-Munʿim al-Ḥimyarī*. Ed. and trans. E. Lévi-Provençal. Leiden: E. J. Brill, 1938.

Història del país valencià. Vol. I. Barcelona: Edicions 62, 1965.

Houston, J. M. "Land Use and Society in the Plain of Valencia," in *Geographical Essays in Memory of Alan G. Ogilvie*, ed. R. Miller and J. Wreford Watson. London: Nelson, 1959. Pages 166–194.

——— "Urban Geography of Valencia: The Regional Development of a Huerta City," *Institute of British Geographers: Transactions and Papers*, 1949, pp. 17–35.

——— *The Western Mediterranean World*. London: Longmans, Green, 1964.

Huici Miranda, Ambrosio. "Dos acequieros musulmanes, reyes de Valencia," *Las Provincias* (Valencia), May 21, 1961.

——— See *Colección dipolmática de Jaime I, el Conquistador*.

Humlum, J. "Underjordiske vandingskanaler: kareze, qanat, foggara," *Kulturgeografi*, 90 (1965), 81–131.

Hunter, Louis C. "Water-mills in Southern Europe," *Technology and Culture*, 8 (1967), 446–466.

Ibarra y Ruiz, Pedro. *Estudio acerca de la institución del riego de Elche*. Madrid: Jaime Ratés, 1914. Historical study including numerous documents.

Imamuddin, S. M. *Some Aspects of the Socio-economic and Cultural History of Muslim Spain, 711–1492* A.D. Leiden: E. J. Brill, 1965.

Jaldero, Francisco de Paula. *Monografía y planos de los antiquísimos canales existentes en el término de Ribarroja que vienen de Villamarchante y se pierden en Manises*. Valencia, 1853.

[James I, the Conqueror, king of Aragon.] *Chronica ó comentaris del gloriosissim é invictissim rey en Jacme primer, rey Daragó, de Mallorques é de Valencia, compte de Barcelona é de Montpesler*. E. Mariano Aguiló y Fuster. Barcelona: Alvar Verdaguer, [1873] 1905.

Jaubert de Passa, François-Jacques. *Canales de riego de Cataluña y Reino de Valencia.* 2 vols. Trans. Juan Fiol. Valencia, 1844. Volume II contains *Reglamentos y ordenanzas de las principales acequias del Reino de Valencia,* separately paginated.

——— *Voyage en Espagne dans les années 1816, 1817, 1818, 1819, ou recherches sur les arrosages.* 2 vols. Paris, 1823.

Jordana de Pozas, Luis. *Ensayo de una bibliografía española de aguas y riegos (separado del libro del III Congreso Nacional de Riegos).* Valencia: Hijo de F. Vives Mora, 1923.

Kunow, Paul. *El clima de Valencia y Baleares.* Valencia: Institución Alfonso el Magnánimo, 1966.

La Blanchère, M.–R. de. "L'Aménagement de l'eau et l'installation rurale dans l'Afrique ancienne," *Nouvelles archives des missions scientifiques,* 7 (1897), 1–109.

——— "Fossa," in *Dictionnaire des antiquités grecques et romaines,* 5 vols., ed. Charles Daremberg and Edmund Saglio. Paris: 1877–1919. Volume II, pt. 2, pp. 1321–1333. The most concise discussion of ancient irrigation.

Lapeyre, Henri. "Les archives de Valence," *Caravelle: cahiers du monde hispanique et luso-brésilien* (Toulouse), 6 (1966), 54–71.

Latour Brotons, José. *Antecedentes de la primitiva ley de aguas.* Madrid: Ilustre Colegio de Abogados de Elche, 1955.

——— "La distribución de las aguas en la huerta de Valencia," *Astrea* (Valladolid), 2nd semester, 1957, pp. 1–8.

Laufer, B. "The Noria or Persian Wheel," *Oriental Studies in Honour of Cursetji Erachji Pavry.* London: Oxford University Press, 1933. Pages 238–250.

Leach, E. R. "Hydraulic Society in Ceylon," *Past and Present,* no. 15 (April 1959), 2–26.

——— *Pul Eliya: A Village in Ceylon.* Cambridge: Cambridge University Press, 1961.

Lévi-Provençal, E. *L'Espagne musulmane au Xème siècle.* Paris: Larose, 1932.

Libre de ordinacions de la vila de Castelló de la Plana. Ed. Luis Revest y Corzo. Castellón: Sociedad Castellonense de Cultura, 1957.

Llauradó, Andrés. *Tratado de aguas y riegos.* 2 vols. Madrid, 1884.

Llorca Rodríguez, José. "Romanidad de los riegos de la huerta de Valencia," in *Notas sobre la antigüedad de la agricultura y el regadío en tierras valencianas.* Describes the recent discovery of Roman sluice gates in Valencia.

Lop, Joseph. *De la institucio, gouern politich y juridich, observancies, costums, rentes i obligaciones dels officials de les ill.^s fabriques de murs e valls y nova dita del Riu de la Insigne, Leal y Coronada Ciutat de Valencia.* Valencia, 1675.

Manrique, M. "La jovada valenciana," *BSCC*, 15 (1934), 155–166.

Marco Baidal, Jaime. *El Turia y el hombre ribereño*. Valencia: Mari Montañana, 1960.

Markham, Clements R. *Report on the Irrigation of Eastern Spain*. London, 1867.

Martínez Aloy, José. *Provincia de Valencia*, in *Geografía general del Reino de Valencia*.

Martínez Bara, José Antonio. "El Tribunal de las Aguas de Valencia de 1775 a 1780," *Martínez Ferrando, archivero: miscelánea de estudios dedicados a su memoria*. Madrid: Asociación Nacional de Bibliotecarios, Archiveros y Arqueólogos, 1968. Pages 515–533.

Martínez Morellá, Vicente. *Cartas de los reyes de Castilla a Orihuela, 1265–1295*. Alicante: Artes Gráficas Alicante, 1954.

——— *Guia del Archivo Municipal de Alicante*. 4 parts. Alicante: Artes Gráficas Alicante, 1954–1957.

——— *Privilegios y provisiones de Fernando, el católico, a Alicante*. Alicante: Artes Gráficas Alicante, 1951.

Millon, René. "Variations in Social Responses to the Practice of Irrigation Agriculture," in *Civilizations in Desert Lands*, ed. Richard V. Woodbury. University of Utah Anthropological Papers, no. 62 (December 1962), pp. 56–88.

Millon, René, Clara Hall, and May Díaz. "Conflict in the Modern Teotihuacan Irrigation System," *Comparative Studies in Society and History*, 6 (1961–1962), 404–521.

Molins, Marqués de. See Roca de Togores.

Momblanch, Francisco de Paula. *Historia de la Albufera de Valencia*. Valencia: Ayuntamiento de Valencia, 1960.

Moncrieff, Colin Campbell Scott. *Irrigation in Southern Europe*. London, 1868.

Muñoz Pérez, José, and Juan Benito Arranz. *Guia bibliográfica para una geografía agraria de España*. Madrid: CSIC, 1961. Invaluable bibliography of agrarian geography and history.

Muntaner, Ramón. *Crònica*. 9 vols. in 2. Barcelona: Editorial Barcino, 1927–1952.

Musso y Fontes, J. *Historia de los riegos de Lorca, de los ríos Castril y Guardal*. Murcia, 1847.

Neuvonen, Eero K. *Los arabismos del español en el siglo XIII*. Helsinki: Society of Finnish Literature, 1941.

Nicolau Primitiu. See Gómez Serrano.

Notas sobre la antigüedad de la agricultura y el regadío en tierras valencianas. Valencia: Ortizá, 1964.

Oliver Asín, Jaime. *Historia del nombre "Madrid."* Madrid: CSIC, 1959. A comparative study of the *qanat* system of Madrid.

Oliver Moros, J. M. "Las aguas del Mijares: distribución de las mismas entre las comunidades que componen la Junta de Aguas de la Plana; cesión posterior de Burriana a los regantes de Nules," *Buris-ana* (Burriana), vol. 10, 103 (1967), 18–19.

Querol Faus, Fina. *La vida valenciana en el siglo XV: un eco de Jaume Roig*. Valencia: Institución Alfonso el Magnánimo, 1963.

Reglamentos y ordenanzas de las principales acequias del Reino de Valencia. See F.-J. Jaubert de Passa.

Repartimiento de Murcia. Ed. Juan Torres Fontes. Madrid: CSIC, 1960.

Repartimiento de Valencia. See *Colección de documentos inéditos. . . .*

Revest y Corzo, Luis. See G. Andreu Valls and *Libre de ordinacions. . . .*

Ribera, Julián. *Lo científico en la historia.* Madrid: P. Apalategui, 1906.

——— *De historia arábigo-valenciana.* Valencia: Hijo de F. Vives Mora, 1925.

——— "Mons parturiens," *El archivo* (Denia), 2 (1887–1888), 232–235.

——— *Orígines del justicia de Aragón.* Zaragoza, 1897.

——— "El sistema de riegos en la huerta valenciana no es obra de los árabes," *Disertaciones y opúsculos,* 2 vols. Madrid, 1928. Volume II, pp. 309–313.

Roca de Togores, Mariano. "Antigüedades de Elche," *Discursos académicos.* Madrid, 1890. Volume I, pp. 261–395.

Roca Traver, Francisco A. "Cuestiones de demografía medieval," *Hispania,* 13 (1953), 1–20.

Roig, Jaume. *Spill o libre de consells,* ed. R. Miquel y Planas. 2 vols. Barcelona: Biblioteca Catalana, 1929–1950.

Ros Marín, R. "El asesor jurídico del Tribunal de las Aguas," *Las Provincias* (Valencia), Nov. 12, 1960. An interview with Vicente Giner Boira.

Rossi, Etore. "Note sull'irrigazione, l'agricoltura e le stagioni nel Yemen," *Oriente moderno* (Rome), 33 (1953), 349–360.

Ruiz-Funes Garcia, Mariano. *Derecho consuetudinario y economía popular de la provincia de Murcia.* Madrid: Jaime Ratés, 1916.

Russell, J. C. "The Medieval Monedatge of Aragon and Valencia," *Proceedings of the American Philosophical Society,* 106 (1962), 483–504.

Sánchez Adell, José. "Las murallas medievales de Castellón," *BSCC,* 28 (1952), 44–77.

Sanchis Guarner, Manuel. "Epoca musulmana," in *Història del país valencià.*

——— *Introducció a la història lingüística de Valencia.* Valencia: Institución Alfonso el Magnánimo, 1950.

Santamaria Arandez, Alvaro. *Aportación al estudio de la economía de Valencia durante el siglo XV.* Valencia: Institución Alfonso el Magnánimo, 1966.

San Valero Aparisi, Julián. "El compesino valenciano y su obra," in *Notas sobre la antigüedad. . . .*

Sarthou Carreres, Carlos. *Datos para la historia de Játiva.* 3 vols. Játiva: Sucesora de Bellver, 1933–1935.

———*Provincia de Castellón* in *Geografía general del Reino de Valencia.*

Schulten, Adolf. *Geografía y etnografía antiguas de la península ibérica.* Vol. II. Madrid: CSIC, 1963.

Semple, Ellen Churchill. *The Geography of the Mediterranean Region: Its Relation to Ancient History.* New York: H. Holt, 1931.

Serjeant, R. B. "Some Irrigation Systems in Ḥaḍramawt," *Bulletin of the School of Oriental and African Studies* (London), 27 (1964), 33–76.

Sevillano Colom, Francisco. *Valencia urbana medieval a través del oficio de Mustaçaf.* Valencia: CSIC, 1957.

Solignac, Marcel. "Recherches sur les installations hydrauliques de Kairouan et des steppes tunisiennes du VIIe au XIe siècle," *Annales de l'Institut des Etudes Orientales* (Algiers), 10 (1952), 5–273; 11(1953), 60–170.

——— "Travaux hydrauliques hafsides de Tunis," *Revue africaine,* 79 (1939), 517–580.

Steiger, Arnald. *Contribución a la fonética del hispano-árabe.* Madrid, 1932. *Revista de filología española, anejo XVII.*

Tarradell, Miquel. "Prehistòria i antiguitat," in *Història del país valencià.*

Tasso Yzquierdo, Rafael. *Algunos datos sobre la historia, descripción y actuación de la Acequia Real del Júcar.* 2nd ed. Valencia: Ortizá, 1964.

Thoumin, Richard. *Géographie humaine de la Syrie centrale.* Tours, 1936.

——— "Notes sur l'aménagement et la distribution des eaux à Damas et dans sa Ghouta," *Bulletin d'études orientales,* 4 (1934) [offprint].

Toledo Girau, José. *Las aguas de riego en la historia de Valldigna.* Castellón: Sociedad Castellonense de Cultura, 1958. History of small irrigation community from the middle ages through the nineteenth century.

Tormo, Rosa R. Troncoso de. "El Archivo del Reino de Valencia," *Boletín de información municipal* (Valencia), vol. 14, no. 52 (1966), 4–51.

Torres Balbás, Leopoldo. "La Abolafia de Córdoba y la gran noria toledana," *Al-Andalus,* 7 (1942), 461–469.

——— "Los contornos de las ciudades hispanomusulmanas," *Al-Andalus,* 15 (1950), 437–486.

——— "Las norias fluviales en España," *Al-Andalus,* 5 (1940), 195–208.

Torres Fontes, Juan. *Medidas de superficie y de valoración en el Repartimiento de Murica.* Murica: CSIC, 1959.

——— See *Repartimiento de Murcia.*

Traver Garcia, Benito. *Historia de Villarreal.* Villarreal: Juan Botella, 1909.

Traver Tomás, Vicente. *Antigüedades de Castellón de la Plana.* Castellón: Ayuntamiento de Castellón, 1958.

Tresse, René. "L'Irrigation dans la Ghouta de Damas," *Revue des études islamiques,* 3 (1929), 459–573.

Troll, Carl. "Qanat–Bewässerung in der Alten und Neuen Welt," *Mittei-*

lungen der Oesterreichischen Geographischen Gessellschaft, 105 (1963), 314–330.
Tyan, Emile. *Histoire de l'organisation judiciaire en pays d'Islam.* 2nd ed. Leiden: E. J. Brill, 1960.
ʿUdhrī, Aḥmad ibn ʿUmar ibn Anas, al-. *Fragmentos geográfico-históricos de al-Masālik alā gamīʿ al-Mamālik.* Ed. ʿAbd al-ʿAzīz al-Ahwānī. Madrid: Instituto de Estudios Islámicos, 1965.
Valls David, Rafael. *Pallantia, vulgo, Valencia la Vieja.* Vinaroz: Antonio Fernández, 1902. A study of Roman and pre-Roman irrigation in Valencia.
Valls Taberner, Fernando. *Los privilegios de Alfonso X a la ciudad de Murcia.* Barcelona: Tipografía Católica Casals, 1923.
Vicent Cortina, Vincente. *Bibliografía geográfica del Reino de Valencia.* Zaragoza: CSIC, 1964.
Wittfogel, Karl A. "Developmental Aspects of Hydraulic Societies," in *Irrigation Civilizations: A Comparative Study,* ed. J. H. Steward. Washington, D.C.: Pan American Union, 1955.
——— *Oriental Despotism.* New Haven: Yale University Press, 1957.

Notes

Abbreviations facing p. 1

INTRODUCTION

1. Clements Robert Markham (1830–1916) was selected by the Secretary of State for India to collect cinchona specimens in Peru in 1860 and superintend their subsequent acclimatization and cultivation in India. This work made him aware of the desirability of irrigating previously unirrigated farmland, especially in Madura district. From 1867 to 1877 he headed the geographical department of the India office; his *Report on the Irrigation of Eastern Spain* (London, 1867) was printed by order of the Secretary of State for India. Sir Colin Scott Moncrieff (1836–1916) served in the irrigation department of the northwest provinces of India and as chief engineer in Burma before being transferred to Egypt. There he was undersecretary for public works (1883–1892) and was responsible for the engineering of the New Nile Barrage. In 1901–1903 he served on the Indian Irrigation Commission. His book *Irrigation in Southern Europe* (London, 1868) was subtitled "the report of a tour of inspection of the irrigation works of France, Spain, and Italy undertaken in 1867–1868 for the Government of India." For biographical details, see *Geographical Journal*, 47 (1916), 481. Roberts' book, *Irrigation in Spain* (London, 1867) was likewise prefaced with an allusion to the attention which irrigation was attracting at the time "particularly in connexion with our possessions in India."

2. Aymard was an Ingenieur des Ports et de Chausées stationed in Algiers, and a knight of the Legion of Honor; his book *Irrigations du midi de l'Espagne* was published in Paris in 1864. François-Jacques Jaubert de Passa (1785–1856) was a government functionary and scholar who specialized in the history of irrigation and the language and history of Catalonia. His observations on Spanish irrigation were recorded in *Voyage en Espagne dans les années 1816, 1817, 1818, 1819, ou recherches sur les arrosages*, 2 vols. (Paris, 1823). (See Chapter VIII, nn. 17–19, below.) Two substantial works on French and ancient irrigation complement Jaubert's work on Spain: *De l'arrosage dans le département des Pyrénées-Orientales et des droits de arrosants sur les eaux* (Paris, 1848) and *Recherches sur les arrosages chez les peuples anciens*, 4 vols. (Paris, 1846–1847). He was elected a correspondent of the French Academy of Sciences in 1824 (section of rural

economy). (See *Nouvelle biographie générale* [Paris, 1861]. XXVI, 403; J. Mattes, "Notice biographique de François Jaubert de Passa," *Bulletin de la Société Agricole, Scientifique et Littéraire des Pyrénées-Orientales,* 11 [1858], 426–437).

3. William Hamilton Hall, *Irrigation Development: History, Customs, Laws, and Administrative Systems Relating to Irrigation, Watercourses, and Waters in France, Italy, and Spain* (Sacramento, 1886).

4. Jean Brunhes, *L'Irrigation: ses conditions géographiques, ses modes et son organisation . . .* (Paris, 1902); Joaquín Costa, *Colectivismo agrario en España* (Madrid, 1898).

5. This polemic is the subject of Chapter VIII.

6. The city of Valencia has now spread over about one-fifth of the traditional huerta (J. M. Houston, *The Western Mediterranean World* [London, 1964], p. 282). On the relation of the city and huerta in the nineteenth century, see Houston, "Urban Geography of Valencia: The Regional Development of a Huerta City," *Institute of British Geographers: Transactions and Papers,* 1949, p. 33, and Joan Fuster, *Nosaltres, els valencians,* 2nd ed. (Barcelona, 1964), pp. 197–201. A few regions having substantial numbers of Morisco irrigators were an exception to the general long-term demographic stability of huerta areas; for repercussions to the expulsion of Moriscos from the huerta of Gandia in 1609, see Adelina Bataller Bataller, "La expulsión de los moriscos: su repercusión en la propiedad y la población en la zona de los riegos del Vernisa," *Saitabi* (Valencia), 10 (1960), 81–100.

7. On the decline of mulberries and the introduction of new fertilizers, see José Manuel Casas Torres, *La vivienda rural y los núcleos de población rurales de la huerta de Valencia* (Madrid, 1944), pp. 46–48. (Hereafter cited as *Huerta de Valencia.*) For the orange revolution, see Fuster, *Nosaltres els valencians,* p. 195, and Vicente Fontavella, *La huerta de Gandia* (Zaragoza, 1952), pp. 187–189.

8. See especially the much-cited chap. iv of *La Barraca* (1898). An English translation of Batiste's trial by the Tribunal of Waters can be found in Seymour Resnick and Jeanne Pasmantle, eds., *Highlights of Spanish Literature* (New York: Ungar, 1963), pp. 400–405. I have used the Editorial Planeta edition (Barcelona, 1958).

9. *Oriental Despotism* (New Haven, 1957); and "Developmental Aspects of Hydraulic Societies" in J. H. Steward, ed., *Irrigation Civilizations: A Comparative Study* (Washington, D.C., 1955). A helpful summary of Wittfogel's thesis (from an anthropological point of view) can be found in Robert F. Gray, *The Sonjo of Tanganyika . . .* (London, 1963), pp. 2–8. A discussion of the historiographical ramifications of Wittfogel's theories is in Chapter VIII, below.

10. See René Millon, "Variations in Social Responses to the Practice of Irrigation Agriculture," in Richard B. Woodbury, ed., *Civilizations in*

Desert Lands, University of Utah Anthropological Papers, no. 62 (December 1962), pp. 56–88; findings are summarized from anthropological studies of irrigation societies in Tanganyika, Ceylon, Japan, Iraq, Bali, South Arabia, and Mexico.

11. On the positive role of conflict in communal life, see Herbert J. Gans, *The Levittowners* (New York: Pantheon, 1967), pp. 142–143; on the relationship of conflicts to rule-making, see Carl J. Friedrich, *Man and His Government* (New York: McGraw-Hill, 1963), chaps. xxiv and xxv.

12. See Eliot Dismore Chapple and Carleton Stevens Coon, *Principles of Anthropology* (New York: Holt, Rinehart, 1942), p. 185: "This business of irrigation involves cooperative action . . . because the irrigation water is shared by all men, and hence all are equally responsible for the maintenance of the system."

13. René Millon, Clara Hall, and May Díaz, "Conflict in the Modern Teotihuacan Irrigation System," *Comparative Studies in Society and History,* 6 (1961–62), 513.

14. Alan R. Beals and Bernard J. Siegel, *Divisiveness and Social Conflict* (Stanford: Stanford University Press, 1966), pp. 91–93.

PART ONE: THE IRRIGATION COMMUNITIES OF MEDIEVAL VALENCIA

CHAPTER I. THE HUERTA ENVIRONMENT

1. Geographers characterize vegas and huertas by their red or brown alluvial soil, with high (2 to 4 percent) iron content and groundwater at about 2 m. (Brian T. Bunting, *The Geography of Soil* [London, 1965], p. 120). Houston (*Western Mediterranean World,* p. 277) alludes to the "oriental density" of huerta population.

2. On marjals, see Alice Foster, *The Geographic Structure of the Vega of Valencia* (Chicago, 1936), p. 33; Fontavella, *Huerta de Gandia,* pp. 90–91; and Francisco de Paula Momblanch, *Historia de la Albufera de Valencia* (Valencia, 1960). On the swampy, pestiferous nature of Mediterranean lowlands in general, see Fernand Braudel, *El mediterraneo y el mundo mediterraneo en la época de Felipe II,* 2 vols. (Mexico: Fondo de Cultura Económica, 1953), I, 55–60.

3. On various aspects of the interdependence of town and huerta, see Robert I. Burns, *The Crusader Kingdom of Valencia,* 2 vols. (Cambridge, Mass., 1967), I, 151, and Houston, "Urban Geography of Valencia," p. 22.

4. At least 118 such in the Valencian region (Houston, *Western Mediterranean World,* pp. 277–278).

5. See Brunhes, *L'Irrigation,* chap. ii; Rafael Altamira, "Mercado de agua para riego en la huerta de Alicante y en otras localidades de la

península y Canarias," in Joaquín Costa, ed., *Derecho consuetudinario de España,* 2 vols. (Barcelona, 1902), II, 136–164, 441–447; Mariano Ruiz-Funes Garcia, *Derecho consuetudinario y economía popular de la provincia de Murcia* (Madrid, 1916), pp. 165–192.

6. *Repartimiento de Valencia,* ed. Próspero de Bofarull y Mascaró, *Colección de documentos inéditos del Archivo General de la Corona de Aragón,* 41 vols. (Barcelona, 1847–1910), IX, 349.

All translations are mine except where otherwise indicated.

7. See below, Chap. II.

8. Later, specific rights in an irrigation turn were sometimes specified in grants (usually in places where water was sold by the hour), as in a grant of 1708 of "a piece of land . . . with one-half hour of water" ("un tros de terra . . . ab miga hora de aygua de la damunt"); Joan Beneyto i Pérez, "La propietat predial i l'aigua de rec," *ACCV,* 4 (1931), 125 n. 2.

9. For example, a sale of land in the district of Moncada (Oct. 15, 1305) "cum cequiis aquis ad rigandum" (AHN, Montesa, Documentos particulares, no. 626).

10. Manuel Dualde Serrano, ed., *Fori antiqui valentiae* (Madrid-Valencia, 1950–1967), p. 280.

11. ARV, Gobernación, 2227, 16th hand, fol. 27 (Nov. 12, 1421). Similar cases arose with regard to nonuse by mills. If a mill was decrepit and had been inoperative for years, should the owner still have to contribute to the maintenance of the canal? See ARV, Gobernación, 2300, 2nd hand, fol. 7v. (Feb. 20, 1461), where the syndic of the Rascanya Canal argues that the miller still must pay because as owner of the mill he still has the right to use the water, whether or not he actually does so. The fueros did provide for a change in status if an irrigation source had been dry for 10 years or more (*Fori antiqui valentiae,* p. 75).

12. In an agreement between the master of the Order of Montesa and the men of Moncada and Massarrojos, concluded May 17, 1403, entry fees were set at 90 sous per cafizate of regadiu and 30s. per jovate of secá. With the jovate equivalent to 6 cafizates, irrigated land was worth 18 times more than unirrigated land at this time. (But the tax rate — *sens* — was set at 7s. per jovate of secá and per cafizate of regadiu, indicating a multiple of only 6) (AHN, Montesa, libro 542 c, fol. 144r).

13. Variants of this complaint: "regave per loch hon no tenie anpriu, regava per lla hon no deu, sorregava per lla hon no ha regadiu, regava per loch que no a regadiu." The illegal conversion of unirrigated land to irrigated has been a constant concern. In a letter to *El Mercantil Valenciano,* Feb. 20, 1878, "an irrigator" writes that people without water rights had been taking water from the Júcar Canal and converting secano to huerta. The *acequiero* should have fined these people but had not.

14. See M. Manrique, "La jovada valenciana," *BSCC,* 15 (1934), 156–166.

Until the conquest of Valencia the jovate had twelve cafizates, and the cafizate twelve fanecates. The king reduced the measures when he discovered that he had granted more land than existed (*ibid.*, pp. 164–165). *Fanegada* is from Arabic, *fanīqa*, "sack" (Eero K. Neuvonen, *Los arabismos del español en el siglo XIII* [Helsinki, 1941], pp. 129–130), and *cafizada* from Arabic, *qafīz*, a dry measure (*ibid.*, p. 75), indications originally of the amount of grain which could be sown on a given area. *Jovada* is related to the Castilian *yugada*, the amount of land which can be plowed by a yoke of oxen in one day.

15. *Servicio nacional de concentración parcelaria, 1953–1963*, 2 vols. (Madrid: Ministerio de Agricultura, 1963), I, 20–21, cuadro no. 2.

16. ARV, Gobernación, 2268, 4th hand, fol. 8r (May 1, 1442).

17. ARV, Gobernación, 2272, 2nd hand, fol. 29r (June 13, 1446).

18. ARV, Gobernación, 2307, 32nd hand, fols. 18r–19v (1462).

19. See Fontavella (*Huerta de Gandia*, pp. 103–104), who points out the extreme irregularity of *secano* fields in comparison with the orderliness of the irrigated huerta. See his parcel map of Beniopa (p. 106) and compare with the aerial photo of rice land (opp. p. 212), in which the fields tend to be extremely elongated. Cf. Houston, *Western Mediterranean World*, p. 133, fig. 49.

20. The *casita* or *casa de labor* is relatively modern (see Fontavella, *Huerta de Gandia*, pp. 314–315). On the *barraca*, see Casas Torres, *Huerta de Valencia*, pp. 79–139; J. M. Houston, "Land Use and Society in the Plain of Valencia," in R. Miller and J. Wreford Watson, eds., *Geographical Essays in Memory of Alan G. Ogilvie* (London, 1959), p. 191; on the *alqueria*, Casas Torres, *Huerta de Valencia*, pp. 141–168; Fontavella, *Huerta de Gandia*, pp. 310–311. Pere Garcia, in 1462, owned "one cafizate of vineyard with a *barraca*, fronting on the Mislata Canal" ("una cafiçada de vinya ab la barraqua afronta ab cequia de mizlata") (ARV, Gobernación, 2307, 32nd hand, fol. 19v).

21. ARV, Protocolos, 3009, n.p. (July 28, 1427): "prout afrontatur ab una parte cum quadam cequia nuncupata de Na Rovella et ab alia parte cum vinea Bernardi Rodrigo, braçallo medio."

22. ARV, Gobernación, 2306, 23rd hand, fol. 25r (July 31, 1462).

23. Pedro Diaz Cassou, *Ordenanzas y costumbres de la huerta de Murcia* (Madrid, 1889), p. 21, n. 2.

24. James I. *Crònica*, ed. M. Aguiló y Fuster (Barcelona, 1873), chap. cclviii: "e encara no sabem si han los camps regats, e per les cequies, e porien caure alguns, e pendre gran mal . . . mas que a la nuyt fariem cercar homens de valor, e que guardarien sils camps eren regats o no, e si nols hauien regats que vinguessen a nos, e que teniem per bo ques faes la brocada." There are numerous references to the strategic ramifications of huerta fighting in the *Chronicle*: in Mallorca James recounted that (chap.

lxvii) "we put the Aragonese on one side and the Catalans on the other, and the canal was in between" ("metem los aragonesos d'una part e els cathalans de l'altra, e la cequia era en mig").

25. Muntaner, *Crònica,* 9 vols. in 2 (Barcelona, 1927–1952), chap. xiii: "Segurament que, si no fossen les cequies qui eren entre amdues les hosts, que'l dit senyor infant haguera brocat sobre ells; mas les cequies e les aygues eren tan grans al mig d'ells, que nou pogueren fer."

26. *Fori antiqui valentiae,* p. 280 (Latin). Catalan version, *Fori regni valentiae* (Valencia, 1548): "E sil hereter quan regara, o no regara, o daltra manera gitara laygua en les vies, o en les carreres pach cinch sols, e restituesque a aquells qui hauran soffert lo dan, e a aquells qui passaran per les vies, o per les carreres lo dan que hauran sofert per lo gotament daquelles aygues."

27. Ruiz-Funes, *Derecho consuetudinario de Murcia,* p. 144. A 1,000-*maravedí* fine was set by the council (n.d.).

28. Archivo del Patriarca de Valencia, Protoc. Francesc Vilba, no. 1684 (May 29, 1435), hereafter cited Benacher-Faitanar Ordinances, 1435: "Item que si algun regant o en altra manera sorregara camps o sendes publiques pach per cascuna vegada que contrafara X. sous pagadors la mitat al dit cequier e laltra mitat al acusador e ultra aço sia tengut a tot lo dan que per lo dit sorregament se seguira axi als vehins com als altres."

29. The duties of cequiers are discussed in Chap. II.

30. The roads to Morvedre, Alboraia, Moncada, Liria, Burjassot, Quart, Chirivella and Alaquàs, Játiva, Picassent, Torrent, Russafa, and the road encircling the city, according to Joseph Lop, *De la institució . . . de les fabriques . . . de murs e valls* (Valencia, 1675), pp. 292–334.

31. The council (*consell*), a popularly elected body whose purpose was to advise the jurates and otherwise deliberate the affairs of the city, was comprised of 4 members elected from each of the 12 parishes and another 4 from each guild. The 6 jurates (*jurats*), elected annually by the council, were the executive and administrative officers of the city. See Manuel Danvila y Collado, *La germanía de Valencia* (Madrid, 1884), pp. 434–438.

32. AMV, Manuals de Consell, 26, fol. 217r: "los dits camins sien derruits e los escorredors riblerts per les aygues qui los regants frontalers o qui han possessions prop dels camins lancen en aquells dits camins e escorredors."

33. AMC, Libres de Consell, 16, n.p. (Aug. 11, 1408).

34. ARV, Real, 272, fol. 143r (May 22, 1450): "la punicio se deia fer per los cequiers aquells empero la cessen fer perque son hereters mateix e lauradors participants en lo abus."

35. Provision for public space on the banks of canals is an inheritance from Islamic law. Legal scholars differed, however, on the width of the *ḥarīm* (forbidden space), specifying between 25 and 50 cubits. Abū Ḥanīfa said that the *ḥarīm* of a canal was constituted by the space occupied by the mud thrown upon the banks. The practice of the modern Ghūṭa of

Damascus is that the *ḥarīm* should be equal on each bank to the width of the canal (Henri Bruno, *Contribution a l'étude du régime des eaux en droit musulman* [Paris, 1913], pp. 56–58; al-Māwardi, *Les statuts gouvernementaux* [*al-Aḥkām al-Sulṭāniya*], trans. E. Fagnan [Algiers: Adolphe Jourdan, 1915], pp. 390–392; René Tresse, "L'Irrigation dans la Ghouta de Damas," *Revue des études islamiques,* 3 [1929], 477). In Castellón, there had to be a space 4 palms wide on one side of the canal and 2 on the other in which plants or trees could not be planted (*Libre de ordinacions de la vila de Castelló de la Plana,* ed. Luis Revest y Corzo [Castellón, 1957], p. 67). A Murcian ordinance of Oct. 3, 1476, specified a space of 10 palms on the banks of main canals, 5 palms on secondary channels (Diaz Cassou, *Ordenanzas y costumbres,* p. 65).

36. Lop, *Murs e valls,* p. 298: "Lo cami de Alboraya es fillola del cami Real de Molvedre."

37. AMV, Manuals de Consell, 32, primer libre, fol. 28v (Oct. 15, 1438), concerning the hereters of a *carrero* (lane).

38. See Lop, *Murs e valls,* pp. 292–340, for indications of specific bridges, including those over the canals of Faitanar, Favara, Rascanya, and Rovella.

39. In AMV, Libros de Sotsobreria de Muros y Valladares, 2, fol. 155r (Dec. 24, 1390), the *sotsobrer* lists his duties in the Marjals as overseeing the making of canals and "repairs of roads, bridges, and pontoons" (*ponts e pontons*). On May 22, 1391 (*ibid.*, 4, n.p.) he paid 336 sous 1 diner to Jacme Demenso, a carpenter, for "repairing, making, and remaking . . . two bridges of mortar and rubble which are over the Favara Canal near the *alqueria* of Guerau de Castellvert" ("dos ponts de reble e dargamassa que son sobre le cequia de Favara").

40. Lop, *Murs e valls,* pp. 255–256: "en tots los puestos, que travessen per damunt los cequiols, pera passar a les heretats facen ponts. Per ço, que per escusarse los particulars de fer dits ponts rebleixen dits cequiols, y les aygues no poden tenir eixida."

41. AMC, Libres de Consell, 20, n.p. (Aug. 5, 1414): "Item lo honorable consell acorda quel pont appellat den Bona fos enderroquat et de nou adobat e refet de bona amplaria dels peus de quexer a quexer e que los dits peus no hisquen mes quels limits del quexer de la cequia." For another bridge over the cequia maior of Castellón, near the river, which was on the point of falling down and causing trouble to pedestrians, see Libres de Consell, 18, n.p. (June 1, 1411).

42. See Chapple and Coon, *Principles of Anthropology,* p. 184.

43. The *Libre de ordinacions de la villa de Castelló de la Plana* contains numerous medieval and early modern cattle regulations.

44. See *ibid.*, pp. 68–72, "Del bovalar."

45. AMV, Manuals de Consell, 3, fol. 98v (July 28, 1334). The bounds of the huerta: "del Riusech de Catarroya aença e devalla tro a la mar e puja tro a la cequia de Quart e de Moncada e devalla tro a la creu que

es posada en lo cami de Murvedre apud Meliana e devalla tro a la mar." Compare with similar versions in Manuals de Consell, 1, fol. 172r (Sept. 26, 1321) and 13, 2nd fasc., fol. 28r (Aug. 21, 1357).

46. AMV, Cartas Misivas, 7, n.p. (Aug. 20, 1400): "de tenir en la dita orta e de pexer e pasturar en aquella sens fer tala e dan, vint cabres e i cabro tant solament." Compare similar license to Anthoni Esteve to pasture a sheep in the huerta (June 14, 1401).

47. AMV, Manuals de Consell, 4, fol. 88v. See also Cartas Misivas, 1, fol. 54r (Jan. 8, 1335), wherein the city writes to the Archbishop of Zaragoza that one Lop de Luna was trying to prejudice the rights of both king and city by pasturing beasts in Paterna — i.e., within the limits of the huerta — contrary to the fueros of the city.

48. AMV, Manuals de Consell, 20, fol. 296r (Apr. 1, 1396):

> alcuns havents . . . possessions en les marjals . . . en los lochs de roçafa e dalfofar . . . prenien gran dan e senyaladament en desidia e tarda de la part de la universitat de la dita ciutat per cessar o tardar de fer ponts e pontons de les cequies de les dites marjals per fretura dels quals ponts e pontons los bous e altres besties majors e menors pasturants en les dites marjals traversaven per les dites cequies e braçals e enrunaven e destrovien aquelles embargants lo cas e les aygues de manera que aquelles regolfaven en lurs possessions e aquelles fahien tornar al primer estament marjalench.

49. AMV, Manuals de Consell, 23, fol. 48v (July 16, 1406): "al present dia la dita cequia era stada enrunada e malmenada per besties e per bestiars."

50. AMV, Manuals de Consell, 33, fol. 92r (Apr. 30, 1445): "Es veritat que en la dita cequia per beneffici de aquella eren necessaris alguns ponts per los quals poguessen passar los bous vaques e altres besties e bestiars per ço no enrunassen aquella." Similar concerns occupied the council of Castellón with regard to its marjals. See disposition regarding the New Canal near Benicasim (AMC, Libres de Consell, 16, n.p., Jan. 17, 1409): where there were no *caxers*, beasts could pass only in designated places and sheep could graze only in special areas set off by the city. On Aug. 5, 1436 (*ibid.*, 35, n.p.), the council noted that because of the coming and going of beasts the ditches of the Marjals were broken down causing stagnation of the water.

51. Hereters of Algirós (Mestalla) were reimbursed for damages caused them "axi en lo alfalç com en los altres splets" (AMV, Manuals de Consell, 34, fol. 265r, Jan. 14, 1450).

52. ARV, Gobernación, 2262, 1st hand, fol. 35r (Feb. 17, 1438). Regarding Oliva and Nules: no water was getting through from the Mijares; hence the *triga* (a castilianism?), *blats*, and crops (*splets*) in general were endangered. With regard to the general use of *blat* as "cereal," see ARV,

Gobernación, 2262, 2nd hand, fol. 40r (July 1, 1438): Jacme Prats held irrigated land in Massamagrell (Moncada Canal), which he held from the lord of that place at a rent of five measures (*almuts*) of grain, "that is, wheat and barley" ("un troç de terra campa situat en terme del dit loch [of Massamagrell] en lo regadiu tengut . . . a cens de v almuts de blat, ço es de forment e dordi."

53. Compare Francesc Eiximenis' list of Valencian grains, vegetables, and spices (*Regiment de la cosa publica,* ed. D. Molins de Rey [Barcelona: Els Nostres Classics, 1927], p. 25): "[Valencia] abunda en diversitats de grans, axí com de forment, ordi, mill, panís d'acça, avena, espelta, tramella, faves, ciurons, llentilles, fesols, pèsols, tremussos, arròs, alquena, adcerco, pastell, comí, batafalua, alcaraulla."

54. The Jesuit Pere Gil, in his *Geografia de Catalunya,* wrote that "what most commonly causes bad harvests [of wheat] is the lack of water, drought, particularly in the months of April and May"; Josep Iglésies, *Pere Gil, S.I. (1551–1622): i la seva Geografia de Catalunya* [Barcelona, 1949], p. 294.

55. *Libre de ordinacions de la vila de Castelló de le Plana,* p. 111 ("Com totes les vinyes axí de sequa com de la orta son bovalar").

56. ARV, Gobernación, 2262, 7th hand, fol. 17v (Aug. 28, 1438): "E dix que per tant com ell dit ts. les ha regades e es stat pagat del dit rech." See also ARV, Gobernación, 2286, 11th hand, fol. 15r (1455): "Item . . . lo dit en Salvador Prats, menor de dies, en l'any mil CCCCXXXXIII comença cavar, podar, magençar, regar, veremar, dar tot recapte continuament en la dita vinya." The vineyard in question was in the district of Museros.

57. See ARV, Gobernación, 2219, 3rd hand, fol. 3r (Jan. 28, 1417), *garroferes* (carobs) in the huerta of Penchisa; and Gobernación, 2241, 6th hand, fol. 39r., regarding fields owned by some Muslims in Torres Torres: "un troç de figueral en lo sequa. . . . Item, un troç de garoferal e figuerol situat en lo dit terme franch qui te conffronta ab la cequia chiqua e ab vinya de Albalati."

58. Vicente Branchat, *Tratado de los derechos y regalías que corresponden al real patrimonio en el reyno de Valencia,* 3 vols. (Valencia, 1784–1786), III, 277–279 (May 8, 1479).

59. AMC, Llibres de Consell, 17 n.p. (Aug. 18, 1409). Payment to Ferrando the water-carrier for 3 loads of water which he carried to irrigate the palace orange grove. ("Item an fferrando ayguader per iii carregues de aygua que porta per a regar lo teronger del palau.")

60. ARV, Gobernación, 2267, 11th hand, fol. 27v (Aug. 6, 1441), regarding an obstruction in the canal of orange and apple trees.

61. ARV, Gobernación, 2273, 14th hand, fols. 4v–9r (May 2, 1447). References to trees planted on the banks of the canal including oranges, mulberries, pomegranates, two kinds of peaches (*berxiquers* and *presiguers*) and willows.

62. "La dita barqua ab la qual lo dit Sr. Rey per hauer plaer ab lo inclit D. Johan frare del dit Senyor a jugar les tarongades en la albufera del dit Sr. Rey"; ARV, doc. of 1417 cited by Antoni M. Alcover, *Diccionari català-valencià-balear,* 10 vols. [Palma, 1930–1962]), X, 160.

63. Muntaner, *Crònica,* chap. clv: "Vaérets batallas de taronges, que del regne de València n'avien feyt venir més de cinquanta càrregues."

64. AMV, Manuals de Consell, 17, fol. 62r (July 18, 1376), "De forma de regar los esplets": The water to be given first to "paniços e ortalices, enapres a les vinyes, enapres als guarets per a sembrar, e derrerament als arroços." Compare order of priorities in the Ghūṭa: 1) fruit trees, 2) grapes, 3) fodder, 4) rice (Tresse, "L'Irrigation," p. 539).

65. For modern rotations, see Fontavella, *Huerta de Gandia,* pp. 165–166.

CHAPTER II. IRRIGATION COMMUNITIES AND THEIR ADMINISTRATION

1. If a man belonged to two communities by virtue of holding land in the rech of each, he could hold office in only one. See Ordinances of the Favara Canal (1701), chap. vi: "Hereter de Rovella no tinga ofici en la comuna" (*Reglamentos y ordenanzas de las principales acequias del Reino de Valencia,* in Jaubert de Passa, *Canales de riego de Cataluña y reino de Valencia,* 2 vols. [Valencia, 1844], II, 182. The eighteenth-century ordinances of all the huerta canals are reproduced in Jaubert de Passa, *ibid.* Those of Moncada are in vol. I, pp. 158–379; the rest are in vol. II in a separately paginated section entitled *Reglamentos y ordenanzas.*) For a medieval example, Bernat Frexa, cequier and subsyndic of the Benacher Canal in 1430 (ARV, Gobernación, 2242, 14th hand, fol. 13), was involved in a dispute the year before by virtue of his right to irrigate from a certain ditch *(fila)* of the Mislata system (ARV, Gobernación, 2239, 13th hand, fols. 19r–20v, Oct. 21, 1429).

2. Castellón Fine-Book (1486) (see Chap. III, n. 7, below), May 16, Apr. 17, Sept. 28: "Regave per la ciquia que no tenie anpriu; regave per loch hon no tenie anpriu; regave per la cequia que no ha regadiu."

3. Galceran Bou stated in 1492 that he was the owner of a rice mill "situated on the canal once of [the] Mislata [system] and now of Favara" ("situat e constituit en la cequia olim de Mislata e hara de Favara"); Jaubert de Passa, *Reglamentos y ordenanzas,* p. 235, Aug. 22, 1492.

4. See Costa, *Colectivismo agrario,* p. 538, on relation of canal and town jurisdictions. The autonomous irrigation community of Valencia strongly resembles the irrigation society *(subak)* of Bali, which is wholly separate from hamlet organization (Clifford Geertz, "Form and Variation in Balinese Village Structure," *American Anthropologist,* 61 [1959], 995).

5. By virtue of the fueros (rubric XVI, no. 35), all canals were considered the common property of the irrigators except for the Moncada Canal ("the Royal Canal which goes to Puçol," in the fuero's description), which was reserved for the king. By a privilege of May 9, 1268, however, James I gave this canal also to its irrigators (Branchat, *Tratado de derechos,* III, 300–302). The distinction between lower and upper canals of the huerta is made in a document of Aug. 21, 1313 (AMV, Manuals de Consell, 1, fols. 91r–93r).

6. In more recent times there was a land test for eligibility for commons membership (the nineteenth-century ordinances of the Tormos Canal, e.g., required that a proprietor own either 2 fanecates of land or a mill in order to vote in the general meeting; *Reglamentos y ordenanzas,* p. 72). I know of no such test in medieval times.

7. "Hereditari e hauentes hereditates e possessiones que rigantur ex aquis decurrentibus per cequias de Beneger [sic] et Faytanar productis ex riuo de Guadalaviar . . ." (Benacher-Faitanar Ordinances, 1435).

8. One important distinction between cultivators and millers, however, was that millers were not allowed to serve as officers of the commons. See chap. lxiv of the Ordinances of Moncada, from Sept. 21 1553: "That no miller might be *acequiero,* and if by chance they should elect one, let him then be removed" (Jaubert de Passa, *Canales de riego,* I, 192). Compare ordinances of Favara; *Reglamentos y ordenanzas,* p. 182.

9. "Lo sindich cequier moliners e hereters de la dita cequia de rascanya se ajustaren per tenir consell" (ARV, Gobernación, 2228, 14th hand, fol. 19r, Sept. 16, 1422).

10. The meeting of Benacher and Faitanar in 1435 to draw up regulations (see above, Chap. I, n. 28 and II, n. 7) was held "in quadam claustra monasterii beati Francisci." The commons of Rovella met on Apr. 5, 1496, "in the small cloister of the monastery and convent of the Blessed Saint Francis of the said city of Valencia" (Archivo de la Acequia de Robella, legajo 1, no. 1). The Franciscan monastery was just outside the city walls, at the point of their southeast salient, on the road to Russafa; see Burns, *Crusader Kingdom,* p. 200.

11. Monestir de la Santa Trinitat; see the *parlement* or *ajust* of June 22, 1437 (ARV, Gobernación, 2260, 8th hand, fol. 16). A general meeting of the Rascanya commons on Apr. 17, 1607, was held "en lo porche de la iglesia y hospital de Sent Llaser" on the Morvedre road outside the walls of the city, next to the monastery of St. Julian ("Acte de ajust general de la cequia y comuna de Rascanya del any 1607," ARV, Varias). See Burns, *Crusader Kingdom,* pp. 241–242, for these buildings.

12. "De licencia nobilis gubernatoris regni valencie" (Benacher-Faitanar Ordinances, 1435). The governor was the chief royal officer of the kingdom, the medieval antecedent of the viceroy.

13. ARV, Gobernación, 2249, 3rd hand, fol. 36r (Apr. 24, 1433): "Diu . . . que en la orta de la ciutat de Valencia ha una sequia entre les altres appellada vulgarment de tormos la aygua de la qual continuament serueix a regar certs quarters de la dita orta en aquella partida situats, e cascun any en cert temps los ereters e tinents teres [sic] en la dita partida se ajusten e fan eleccio de certes persones necesaries per al regiment de les aygues de la dita cequia; fan e costumen fer eleccio de deputats, clavari, e cequier o regidor de la cequia, e colector, donant a cascun de aquells cert carech e regiment." In the nineteenth century Tormos met only every third year, on Oct. 28 (*Reglamentos y ordenanzas*, p. 71).

14. ARV, Gobernación, 2262, 1st hand, fol. 40r (Mar. 8, 1438). The meeting was held Nov. 21, 1436.

15. ARV, Gobernación, 2272, 3rd hand, fol. 34r (Oct. 5, 1446). Because of "a certain work, begun by the hereters of Rovella," of an undescribed nature, the river was breaking down the channel of Favara. The cost of the repairs was estimated at 2,000 florins. The meeting was needed to provide for the "indempnitat e necessitat e negocis de la dita cequia."

16. The consent of the hereters was a common theme, and those accused of making an "innovation" in the system inevitably claimed that they had such consent. A certain Johan Moya, accused of building a conduit over the canal of Na Axera (Rasacanya system) in contravention of the rights of some irrigators, was said by a friendly witness to have "caused the said conduit to be made with the will and consent of the hereters, as is well-known" ("la dita canal feu fer lo dit en Moya segons fama de voluntat e consentiment dels hereters"); ARV, Gobernación, 2211 9th hand, fol. 8r, Apr. 16, 1415.

17. ARV, Gobernación, 2242, 14th hand, fol. 13, in which the crida is reproduced. Reasons for low attendance must have varied (e.g., fear of extra expenses, general satisfaction with the distribution arrangements, recognition that improvement or change would be difficult because of entrenched internal opposition or insuperable physical obstacles). Compare James Hudson, *Irrigation Water Use in the Utah Valley, Utah* (Chicago: Department of Geography, University of Chicago, 1962), p. 119. An aspect of the medieval *ajust* still discernible is the choice of Sunday as meeting day. The annual meeting (in 2 sessions) of the Community of Irrigators of the Rovella Canal in 1966 took place on Sundays Apr. 10 and 24; it was held on the premises of the Real Sociedad Económica de Amigos del País rather than at a monastery, and the crida was effected by newspaper (*Las Provincias*, Apr. 2, 1966).

18. ARV, Gobernación, 2242, 14th hand, fol. 41 (Dec. 12, 1430).

19. Jaubert de Passa, *Canales de riego,* I, 148. In the seventeenth century Rascanya chose 4 *elets*, representing four sectors of its service area, and another who represented the mill owners; they also elected a syndic and a cequier (see n. 11 above).

20. ARV, Gobernación, 2209, 27th hand, fol. 47r (Nov. 28, 1414).

21. In the nineteenth century the Acequiero Real was named by the 12 syndics (Jaubert de Passa, *Canales de riego*, I, 149).

22. ARV, Gobernación, 2272, 3rd hand, fol. 28r (Sept. 27, 1446).

23. *Fori antiqui valentiae*, p. 279: "Savacequie vadant super aquas et cequias totius termini Valentie . . ." (indicating that the *çabacequier* had huerta-wide jurisdiction).

24. The jurates' role as *sobrecequier* of the huerta is discussed in Chapter VI.

25. ARV, Gobernación, 2188, 15th hand of 1404, n.p. (loose sheets after fol. 48). Brackets indicate lacunae or illegible words: "Segons fur e privilegis del regne cascuna cequia del terme de la ciutat haie [. . .] son cequier e aquell conex e determena les questions que son de hereter e hereter de [cascuna] cequia e a mundar la dita cequia e fer altres coses tocants aquella cequia [e . . .] totes les cequies del terme de la ciutat han e deuen haure çabacequies los quals deuen [anar] sobre les aygues e cequies e tot lo terme de la ciutat e aquests son los jurats [de la] ciutat." Some of the wording is taken from the fuero, rubric *De cequiers;* see n. 23 above. The governor, the document continues, is *sobrecequier* of "canals of nobles and knights which are outside of the district of the city," and over the Moncada Canal he shares this power with the royal bailiff. Compare the petition addressed to the King in 1510 (*Fori regni valentiae*, fol 91r): "com per furs e privilegis del dit vostre regne de Valencia, la conexença, de les cequies e aygues del riu de Guadalaviar pertanga als jurats de la dita ciutat, excepto [sic] la ciquia real appellada de Moncada."

26. *Fori regni valentiae*, fol. 248r (1283): "Item concedimus quod supercequiarius de su officio in perpetuum sit eiectus, et unusquisque ex cequiariis itatur, et utiposit secundum quod erat antiqitus consuetum."

27. AMV, Cartas Misivas, fol. 11r (July 15, 1415). Another letter to Rochafort (Cartas Misivas, 12, fol. 151v [June 19, 1414] also concerned arrangements with the Pueblos Castillos.) José Martínez Aloy (*Provincia de Valencia* [Barcelona, n.d.], 297) states, on the basis of a privilege of James II in 1326 (*Aureum opus* [Valencia, 1515] privilege 156), that "in 1326 there was still an ancient jurisdiction of *acequiero* of the city of Valencia" "questionum cequiarum ciuitatis Valentiae et termini sui et aquarum ejusdem expectat ad cequiarum ciuitatis eiusdem: qui eas audire et decidere debet cum consilio provisorum ad ipsas cequias deputatorum").

28. *Fori regni valentiae*, fols, 247v–248r.

29. "Per la força ab la cort o sens la cort."

30. See privilege of Peter I, Dec. 1, 1283 (*Aureum opus*, fol. 33, chap. xxii): "We decree . . . under the penalty of sixty sous that the cequiers of Moncada and Favara or others not divert so much water through their canals to the prejudice of the hereters."

31. This was true, although probably less so, of *sobrecequiers*, where the

daily operations would be handled by the cequiers of the individual canals. An ordinance of 1276 stated that the *sobrecequier* of Orihuela "had to be a man well acquainted with the ways of the huerta." "El sobrecequiero ha de ser un ome bien sabidor de las cosas de la huerta," quoted in Etienne Carpentier, *Les réglements et les tribunaux des eaux dans les provinces du sud-est de l'Espagne* (Paris, 1912), p. 90. On June 25, 1295, Fernando IV decreed that the council of Orihuela should elect each year as *sobrecequier* "a knight or prominent man" ("un cauallero o un home bueno de vuestro lugar"), which indicates a different principle from that of the llaurador-cequier; Vicente Martínez Morellá, *Cartas de los reyes de Castilla a Orihuela, 1265–1295* (Alicante, 1954), p. 83. In Alicante, where the *sobrecequier* was a municipal officer of high status (equal in rank to the treasurer), Antoni Martínez, holder of the office for 1492, was nevertheless a llaurador; Martínez Morellá, *Privilegios y provisiones de Fernando el Católico a Alicante,* (Alicante, 1951), pp. 28–30. 34.

32. ARV. Gobernación, 2189, lst fasc., fol. 42r (Sept. 18, 1404): "lo dit cequier de Favara e sos veedors . . . son homens pagesos e grossos e no saben legir."

33. *Ibid.,* fol. 42v: "[by the privilege of James II, Barcelona, May 30, 1309] lo cequier de prima voluntate vel ex officio pot e deu pendre en assessor un jurista ab consell."

34. ARV, Gobernación, 2308, 8th hand, fol. 9v (May 6, 1463):

> Item diu que quant algu es fet cequier de Tormos e de altra cequia de la dita huerta de Valencia es acostumat que la dita cequia se subasta publicament davant lo comu dels hereters de aquella qui son ajustats o la maior part de aquells e per lo corredor publich se subasta publicament ab los capitols de la cequia e lo qui trau la cequia ha [i.e., a] millor mercat e per menor preu aquell . . . trau la dita cequia ab lo carrech dels capitols de aquella e segons los dits capitols aquell qui trau la dita cequia es fet cequier de la dita cequia e es tengut de tenir seruar e complir los capitols de la dita cequia.

35. *Ibid.,* 2nd hand, fol. 30v (Mar. 22, 1463).

36. ARV, Gobernación, 2255, 25th hand, fol. 39r (July 6, 1435): "en lorenç lobregat, cequier e o arrendador de la aygua de la cequia de chiluella."

37. In the fourteenth- and fifteenth-century documents I have examined, cequiatge is always a maintenance charge assessed by the community on its members, or by a town on irrigators, for the upkeep of the canal. Robert I. Burns, in "Irrigation Taxes in Early Mudéjar Valencia," *Speculum,* 44 (1969), 560–567, suggests that in the thirteenth century, directly after the Reconquest, cequiatge may have been considered a regalian tax collected by kings and lords as a rent. He assumes that the term *alfarda* — known to

have been an irrigation tax in the Ebro Valley but whose meaning in Valencian usage is not at all unequivocal — was the equivalent of cequiatge. If Burns is correct, he describes a transitional phase during which lords who knew little about irrigation appropriated the maintenance charge for themselves. It is in the nature of things that very soon the charge would have to be returned to the irrigators unless the canals were to deteriorate. This perhaps explains what happened in the case of the Moncada Canal system of Valencia and why James I only held it for 30 years (see above, n. 5).

38. ARV, Gobernación, 2241, 9th hand, fol. 34r (Aug. 11, 1430): "En Jacme Marti de Paterna, laurador, cequier, e açuter . . . de la cequia de Mizlata."

39. ARV, Governación, 2260, 8th hand, fol. 16r (June 22, 1437). Johan Dano was elected lieutenant of Jacme Gil, cequier, and charged with the direction of cleaning a certain section of the canal.

40. ARV, Gobernación, 2201, 8th hand, fol. 3r (Mar. 30, 1413).

41. AMC, Libres de Consell, n.p. (Dec. 16, 1472): "e hun altre que sie familiar del dit cequier."

42. For the etymology of *çabacequies,* see Chap. XII, n. 16, below.

43. (Benacher–Faitanar) ARV, Gobernación, 2253, 6th hand, fol. 39v (Apr. 29, 1435). (Moncada) ARV, Gobernación, 2277, 7th hand, fol. 36 (May 17, 1449): "los cequiers e çabacequiers de la dita cequia de muncada hagen carrech e sien sollicits de regir e guardar les aygues e de notificar los fraus que troben al cequier maior." In modern times the Acequiero Real has been aided by two "lieutenants called cavacequias [sic]" and three guards (Jaubert de Passa, *Canales de riego,* I, 152).

44. ARV, Gobernación, 2202, 11th hand, fol. 1r (May 2, 1413).

45. Anthoni Prats, a llaurador of Valencia, was *partidor e guarda* of Benacher-Faitanar in 1435 (ARV, Gobernación, 2255, 25th hand, fol. 36r). In 1441 he became cequier (*ibid.,* 2267, 11th hand, fol. 25r).

46. Benacher-Faitanar Ordinances, 1435: "Item que lo dit cequier haja a donar cascun any homens convenients per guardians e partidors de la dita aygua de les dites cequies, un per los sobirans e altre per los jusans, cascu de son terme aconeguda del dit comu. . . . Item que los dits partidors e guardians de les dites cequies sien tenguts de manifestar als dits cequier e senyor de la heretat al qual sera leuada laygua lo dia que lo sera atorgada qui lay haura leuada dins V. dies."

47. ARV, Protocolos, 2864, n.p. (April 1421).

48. ARV, Gobernación, 2208, 19th hand, fol. 11r (Aug. 1, 1414): "en hora prima ell parti de la aygua e dona aquella an Jacme Fuster."

49. The *veedor* was a typical guild official; each guild had 2. Pedro Diaz Cassou (*La huerta de Murcia* [Madrid, 1887], p. 220) recognized the similarity between guild and irrigation community organization.

50. ARV, Gobernación, 2211, 8th hand, fol. 48r (May 25, 1415): Jacme

Gisbert, cequier of Favara, decided with the advice of his inspectors ("ab consell de sos veedors") against the miller, Anthoni Reverdit, in a dispute concerning the latter's responsibilities in maintaining his canal frontage. Also ARV, Gobernación, 2267, 11th hand, fol. 26r (Aug. 6, 1441): a dispute between 2 irrigators of Chirivella was decided by Anthoni Prats, cequier of Benacher–Faitanar, "with the counsel of Pere Colomer and Pere Garcia, cultivators, and inspectors of the said canals" ("lauradors vehedors de les dites cequies").

51. See ARV, Gobernación, 2249, 3rd hand, fol. 36r (Apr. 24, 1433). At the annual meeting of the commons of the Tormos Canal the hereters decided upon the quantity of money "necessary to sustain and clean and repair the said canal and its diversion dam," and "by common will" they imposed on each hereter a tax per cafizate of land. The collector was elected at this meeting to collect the amount agreed upon from each hereter and to give an accounting of his receipts to the commons and/or the deputies and the cequier. In this particular case Nadal Tarrega had been elected collector and for four years had collected the cequiatge, which had been set at 1 sou 2 diners per cafizate. He had been told to collect from 257 jovates of land known to be irrigated at the time of his election; but in the process of collection he had found that in reality there were 25 jovates and 4 cafizates of irrigated land in excess of the original figure, the receipts of which he had kept for himself. Compare ARV, Gobernación, 2236, fol. 42r (Jan. 31, 1426). Bernat Sala, treasurer of the commons of Benacher-Faitanar, had collected money for the purpose of cleaning and repairs of these canals; 2 men were ordered appointed to hear his accounts.

52. ARV, Gobernación, 2251, 24th hand, fol. 29r (Apr. 24, 1433). Bernat Sablit of Valencia testified that he had owned an *alqueria* in Campanar for the past 11 or 12 years, during which time he had known no other collector than Tarrega. Pere Rog (fol. 32r), who said that he remembered back 45 or 50 years, also stated that Tarrega had been collector for 11 years.

53. ARV, Gobernación, 2262, 7th hand, fol. 14v (July 28, 1438): "En Guillem Cabot laurador habitador en lo loch de puçol. E dix que de xxx anys ença en lany present exclusive ell ts. es stat sabacequier de la cequia de muncada." Compare Aparici's testimony, fol. 16r.

54. AMC, Libres de Consell, 14, Mar. 14, 1405: "Item lo dit honrat consell a suplicacio del honrat en johan darenys cequier maior de la dita vila volch e comana al dit cequier que façe registrar ço que fall de les ordenacions e sentencies del seu libre ab les quals lo cequier se reig e deu regir per fer justicia e egualment a les gens segons les dites sentencies e ordenacions de la dita cequia."

55. AMC, Libres de Consell, 20, Aug. 5, 1414. A discussion at the same meeting regarding a dispute with Alamassora precipitated this order.

56. AMC, Libres de Consell, 22, July 12, 1416:

> Libre de les terres del regadiu de la cequia maior. Lo dit honrat consell per be profit e utilitat de la cosa publica ordena volch e mana que libre sia ordenat e fet de tots aquells que tenen terres en lo regadiu de la cequia maior per ço que daci anant lo cequiatge sie clar car daci a enrere lo dit cequiatge es estat molt torbat e quey sie fet libre nou quels jurats fan fer crida que tot hom que haie terres en lo regadiu de la cequia maior dins certs dies per aquells dits jurats declaradors porten lurs cedules als dits honrats jurats e scriva del consell de tots los bens quen tinguen en lo regadiu de la dita cequia maior sots certa pena per aquells ordonadora e imposadora als contrafahents e quen sie fet libre per la manera ques pertany.

57. See AMC, Libres de Consell, meetings of Oct. 30, 1416; July 2, July 16, Dec. 8, 1419.

58. AMC, Libres de Consell, 32, Jan. 7, 1432: "Lo libre del cequiatge. Item lo honorable consell comana als honorables jurats que sobre lo fet del libre del cequiatge lo qual sta molt torbat e enfrasquat e es gran dan de la vila com hic haia moltes faltes."

59. See Chap. III, n. 7.

60. AMC, "Compadoras de en Pere Bosch de la administracio de scurar la sequia maior en lany 1578."

61. See, for example, the ordinances of the Mislata Canal (1751), chaps. x–xiii (*Reglamentos y ordenanzas,* pp. 40–42); chap. iv (pp. 36–37) provides for a scribe (*síndico escribano*) and for archives.

62. ARV, Gobernación, 2190, 38th hand, fol. 3r (Dec. 18, 1404):

> E en senyal de la dita nunciacio de novella obra e per comprovar la possessio de la dita vila de Algezira e universitat de aquella pres en les sues mans tres pedres e tira . . . aquelles en la dita cequia nova e tirant e gitant aquelles la una daça laltre dalla en presencia de mi dit notari e dels testimonis deius scrits e tirant e gitant les dites pedres per la fet deuant specificada dix: Yo giti aci en aquesta obra en presencia de vosaltres nunciant novella obra aquestes tres pedres una pres laltre e per conservar la possessio de la dita vila e universitat de aquella en e per aquella forma manera e condicio que mills puxa valer segons fur e bona raho.

The Cequia del Canó (or Canyo), Canal of the Siphon, was described as "an aqueduct or canal which flows from the River Júcar and passes through a subterranean conduit [i.e., a siphon] which is beneath the river" ("el ayguaduyt o cequia qui es passa del riu de xuquar e passa per lo cano subterraneu que es davall lo riu"); ARV, Gobernación, 2190, 37th hand, fol. 26v.

63. AMC, Libres de Consell, 7, Dec. 26, 1385; see also elections of same series, same date, for 1379 (Berenguer Serra), 1382 (Guillem Company), 1384 (Bernat Cavaller), 1386 (Bernat Andreu), 1391 (Guillem Miquel), etc.

64. AMC, Libres de Consell, for the cited dates.

65. Throughout the fifteenth century the *moços* received a yearly salary of 60s. (AMC, Libres de Consell, Jan. 18, 1414, A. Codina and P. Belloch, guards; May 31, 1473, Johan Pelega and Johan Sales, guards).

66. Pedro Ibarra y Ruiz, *Estudio acerca de la institución del riego de Elche* (Madrid, 1914), p. 98; Martínez Morellá, *Cartas de los reyes de Castilla a Orihuela,* pp. 58–60 (privilege of Alfonso X, May 14, 1275). Note that although the cequier of Castellón had only minor functionaries below him, the *sobrecequier* of Orihuela had jurisdiction over *acequieros* of individual canals. Thus Alfonso X's charge to the *sobrecequier* Pedro Zapatero, was: "E mando que ande e cate sobre las acequias e las açarbes. E sobre los acequieros que vos y pusieredes e sobre todas las otras cosas que pertenecen en endreçar las eguas de Orihuela e de su termino." In an earlier privilege (Apr. 28, 1272) Alfonso had authorized the officials of Murcia to appoint two *zequieros,* one for the Christians, another for the Muslims (whose settlement was apparently concentrated on certain canals); Diaz Cassou, *La huerta de Murcia,* pp. 158–160.

67. Archivo de la Corona de Aragón, Consejo de Aragón, leg. 615, Memorial de los regantes de Orihuela (Sept. 16, 1625). I am indebted to James Casey for bringing this document to my attention.

68. Archivo de la Catedral de Valencia, Notals de Bonanat Monar, Libro 3518, fol. 167²r (1386): "una gran partida deius los lochs de Roçafa e d'Alfofar e daltres fos e sia tornada marjalença e erma." Also AMV, Manuals de Consell, 32, 2nd bk., fol. 197r (Apr. 22, 1444): "los almarjals que son davall los lochs de Ruçaffa e Alfafar."

69. The *veedors* were originally chosen by the municipal officers in the same manner as the cequier. By 1439, however, the *veedors* were llauradors who were apparently elected by the hereters of the Marjals (AMV, Manuals de Consell, 32, 1st bk., fol. 61v, Apr. 20, 1439).

70. ACV, Notals de Bonanat Monar, Libro 3518, fol. 168v:

> Item quels justicia civil jurats e consell de la dita ciutat perpetualment a bienni o a trienni o a aquells anys que mills los sera vist eligen e meten cequier e administrador e tenidor en condret de les dites cequies, braçals, escorredors, ponts, e pontons, e dos promens veedors e consellers a aquell lo qual ab consell dels dits veedors haia plena jurisdiccio e poder de e sobre les dites cequies, braçals, escorredors, ponts, e pontons, e hereters daquelles e aquells, segons per fur e privilegis cascun cequier ha sobre sa cequia e hereters daquella. E si alcu de juhi o acte del dit cequier e administrador se sentra [sic] agreujat quey puxa recorrer als dits justicia e jurats e no a altri.

71. AMV, Manuals de Consell, 30, fol. 61r (May 11, 1433), also fols. 88r. 181.

72. AMV, Manuals de Consell, 31, fol. 47r (Nov. 5, 1435). The disposition of June 30 is cited in *ibid.*, 32, primer libre, fol. 45v, and 37, 9th bk., fol. 32v.

73. AMV, Manuals de Consell, 32, primer libre, fol. 45v (Feb. 10, 1439). Mariana (or Meriana) held 7 cafizates of fields and vineyards within the limits of the Marjals. In 1442 he was involved in litigation with the deputies of the Favara Canal over a drainage canal (*escorredor*) bordering his land, one end of which joined the rech of Favara, the other end the "water of Melilla" (a place near Russafa, mentioned in the Repartimiento of Valencia as Malilla).

74. Activities of Mariana 1439–May 5, 1441: Mar. 5, 1439 — annual 100 £ allotted by the council for cleaning the canals of the Marjals ordered paid to Mariana (AMV, Manuals de Consell, 32, primer libre, fols. 48v–49r); May 27, 1439 — additional funds paid him for *treballs del scurament* (*ibid.*, fol. 65r); Dec. 24, 1439 — Mariana ordered to clean the Cequia del Grau, as well as the basin where the oaken boat for the galleys is anchored ("e les basses en ques cou la fusta de carrasca per a les galees;" *ibid.*, fol. 132v); May 5, 1441 — he rendered account of expenses on New Canal of the Marjals (*ibid.*, fol. 248r).

75. AMV, Manuals de Consell, 32, segon libre, fol. 53v: "en Marti de Mariana, cequier de les marjals . . . era stat segons ell dehia per rahon del dit ofici denunciat per alguns hereters los quals li havien fet proces essent en la dita ciutat la senyora reyna per la qual raho havia fet moltes despeses en defendres."

76. Oct. 4, 1442, and Apr. 12, 1443 (AMV, Manuals de Consell, 32, segon libre, fols. 82r, 123r).

77. AMV, Manuals de Consell, 33, fol. 152r (Nov. 20, 1445).

78. AMV, Manuals de Consell, 34, fol. 22v (Aug. 25, 1447): "Attenents que en les marjals de la dita ciutat noy havia cequier ne guarda alguna."

79. Activities of Mariana Aug. 26, 1447–Oct. 8, 1451: Aug. 26, Oct. 19, Nov. 27, 1447 — Marti worked on cleaning the Grau canals (AMV, Manuals de Consell, 34, fols. 23v, 32v, 35r); the fact that he began work the day following his appointment suggests great urgency in his hiring; Nov. 27, 1447 — he received his salary (*ibid.*, fol. 43v); May 11, 1448 — salary paid (*ibid.*, fol. 90r); May 30, 1448 — he directed cleaning of the *cequia de la traversa* of the Canal of Castelló d'En Arrufat (*ibid.*, fol. 102v); Dec. 20, 1448 — salary paid (*ibid.*, fol. 148); Feb. 15, 1449 — he was in charge of cleaning the city's share of the Cequia del Cavaller in the Marjals (*ibid.*, fol. 168v); May 26, 1449 — work on the Font del Grau and the New Canal (*ibid.*, fols. 192r, 239v–240r); Dec. 16, 1449 — salary paid (*ibid.*, fol. 252r); Jan. 14, 1450 — his work on the Grau has led to the ruin

of a neighboring field (*ibid.*, fol. 265r); Feb. 19, 1450 — jurates authorize payment to Mariana of 59s. 6d. for cleaning the New Canal, a job which took 5 days (*ibid.*, fol. 271v); May 8, 1451 — Mariana directs cleaning of the Cequia Morisca (*ibid.*, 35, 5th bk., fol. 87r); Sept. 2, Sept. 6, Oct. 8, 1451 — work on the New Canal continues (*ibid.*, fols. 136r, 136v, 142).

80. AMV, Manuals de Consell, 35, 5th bk., fol. 166v (Jan. 28, 1452): "continuament ne havien infinides clamors dels hereters de les dites marjals de les grans . . . senrahons quels fahia molt sovint en tal forma que nol podien comportar en alguna manera."

81. AMV, Manuals de Consell, 35, 5th bk., fols. 162r, 182r: "[The jurates have] remogut e revocat an marti de mariana del offici de cequier de les almarjals e de qualsevol altra administracio que tinga de la ciutat en qualsevol manera." On May 27 he was given that part of his salary due him up to the day of his dismissal ("tro al dia que fon revocat") (*ibid.*, fols. 187v–188r).

82. Activities of Mojoli, July 20, 1452–Oct. 25, 1454: July 20, 1452 — cleaning canals below that of Castelló in the Marjals (AMV, Manuals de Consell, 35, 5th bk., fol. 221r); Aug. 14, 1452 — cleaning canals on the road to the Grau (*ibid.*, fol. 277r); Sept. 9, 1452 — cleaning the Cequia Morisca and the Cequia del Cavaller (*ibid.*, fol. 232v); Aug. 25, 1453 and Jan. 16, 1454 — more cleaning in the Marjals (*ibid.*, 6th bk., fols. 27r, 62v); Oct. 25, 1454 — Mojoli removed from office, *sens infamia* (*ibid.*, fol. 151r). Activities of Mariana, Nov. 20, 1454–Feb. 15, 1459: Nov. 20, 1454, and March 27, 1455 — work in the Marjals (*ibid.*, fols. 153v, 186v); Sept. 2, 1457 — he undertakes a trip to Tortosa on city business (*ibid.*, 36, 8th bk., fol. 22r); Feb. 15, 1459 — work at Grau (*ibid.*, fol. 154v).

83. Oct. 8 and 14, 1460 (AMV, Manuals de Consell, 37, 9th bk., fols. 31v, 32v).

84. AMV, Manuals de Consell, 37, 9th bk., fol. 53v (Jan. 29, 1461): work on the New Canal.

85. *Ibid.*, 10th bk., fol. 102v (Dec. 17, 1463).

86. AMV, Manuals de Consell, 38, 11th bk., fol. 31v (Sept. 30, 1465):

> Crida de les marjals. Ara oiats que us fan saber los molts honorables justicia e jurats de la ciutat de Valencia que tots e qualsevol hereters de les marjals scorrents en la cequia appellada del riu o de mossen perot mercader e benifet nes pren haure que per dimarts primer vinent o en aquella setmana si en davant en johan alffonso elet en administrador e regidor per los diputats de les marjals de la dita cequia e los qui treballar hi volran en erbejar e scurar en la dita cequia lo sera pres en compte de la taixa quels es stada feta o fahedora e si treballar noy volran seran logats homens per a les dites coses a despesa dels dits hereters.

87. *Ibid.*, fol. 54r (Feb. 11, 1466).

88. *Ibid.,* 12th bk., fol. 117v (May 19, 1470): "que an marti de mariana cequier de les marjals sien donats e pagats . . . cent sous . . . per obs de comprar hun roci [to replace] hun altre roci que casualment li es stat mort en les dites marjals anant a aquelles per exercir lo dit offici de cequier."

89. A frequent theme in the documents; see AMC, Libres de Consell, 34, Jan. 15, 1436: the *escorredors* of the town were so badly cleaned that the swamps were filling with water and giving rise to disease. Alfonso X of Castile, in a privilege to the city of Murcia, ordered that each year the jurates were to select two men "to clean the major drainage channels of the huerta so that it might not turn into a swamp" ("que fagan alimpiar loz azarbes mayores de la huerta porque no se faga armarial [sic]); Diaz Cassou, *La huerta de Murcia,* pp. 158–160, privilege of Apr. 28, 1272.

90. ARV, Gobernación, 2236, 2nd hand, fol. 21r (Feb. 20, 1426): "[The miller, Anthoni de Veses, asked the governor] que manets al dit cequier e veedors qui nos entrametre en les dites barbacanals com no sia de son hofici ne jurediccio ans . . . que manets al dit cequier que faça scurar la cequia com sia tota engranada e reblida, e aço es hofici de cequier."

91. ARV, Gobernación, 2292, 14th hand, fols. 1r, 2r (Aug. 29, 1458): "Costuma . . . es en la present ciutat de Valencia e entre los cequiers de totes les cequies de aquella que cascun any en lo mes de abril . . . tollen les aygues e fan fer scurament de totes les cequies e dins lany no es costum scurar les cequies maestres."

92. Benacher–Faitanar Ordinances, 1435: "Item que lo dit cequier per cascuna vegada que scurara les dites cequies ans que torn laygua sia tengut mostrar lo dit scurament als dits sindich diputats e vehedor e tornar la dita aygua dins VII dies e si contrafara que no pusca rebre res del dit cequiatge en aquell any."

93. AMC, Libres de Consell, 20, Apr. 7, 1415: "Item a la proposicio de la scura de la cequia maior de la dita vila feta per los . . . jurats feta acorda lo honorable consell que dimarts primer vinent en viii dies fos feta e que tallasen laygua per scurar aquella."

94. ARV, Gobernación, 2247, 20th hand, fol. 25v (Aug. 13, 1432). Johan Mercader, owner of a mill on the Mislata Canal, testified that a feast day had fallen during the previous April's cleaning but that he had cleaned on the holiday anyway because of the great need.

95. AMC, Libres de Consell, 35, July 12, 1436. The man who had contracted to clean the cequia maior of Castellón had not done so, either in April or in May, and it was growing so late in the season as to cause harm to the town.

96. Ibarra y Ruiz, *Riego de Elche,* p. 189.

97. AMV, Libros de Sotsobreria de Muros, 2, fol. 26r (Feb. 14, 1390): "A erbejar ab dalls e corbelles la cequia den fluvia." For a drawing of a *dall,* see Alcover, *Diccionari,* IV, 10; of a *corbella, ibid.,* III, 520.

98. AMV, Libros de Sotsobreria de Muros, 2, fol. 25v (Feb. 9, 1390). Two

men were paid 2s. each to "throw fire around the canal" ("gitar foch entorn la cequia"); *ibid.,* fol. 31v (Mar. 22, 1390); 5 men were paid to "burn and scythe stubble on the banks of the canal" ("cremar e esdallar romagueres en les ores de la dita cequia").

99. *Ibid.,* fol. 27r (Mar. 5, 1390): "loguer iii dalls de ix dies per al erbejar . . . e ii corbelles les dites dies per avinença."

100. For a drawing of a *pala,* an instrument with an iron head and a wooden handle, see Alcover, *Diccionari,* VIII, 120, esp. no. 4 *(pala de cavar).* For reference to *pales de ferre,* see AMC, Libres de Consell, 35, June 17, 1436. A variety of hoes was also used, notably the *lego (lligó, llegó)* and the *exada (aixada)*; see AMV, Libros de Sotsobreria de Muros, 2, fol. 26r (Feb. 11, 1390): "dos exades stretes e dos legons." For drawings, see Alcover, *Diccionari,* VII, 21 *(lego)* and I, 353–354 *(exada).* The diggers also used a kind of fork *(forca de paler)*; one is listed in an inventory of 1348 (AMV, Notals de Domingo Juan, sig. L, 1. June 30); see Alcover, *Diccionari,* V, 972.

101. Archivo Municipal de Elche, Actas de Consell, May 22, 1444; the *clavari* was to purchase *cabaçets* and other items needed for the cleaning of the cequia maior. See also, AMV, Libros de Sotsobreria de Muros, 2, fol. 25v (Feb. 8, 1390): "una dotzena de corbaços de espart."

102. AMV, Libros de Sotsobreria de Muros, 2, fol. 40r (Apr. 15, 1390): "tira calç reble e arena a la cequia nova den fluvia per tancar los forats que son fort grans en la dita cequia."

103. AMV, Manuals de Consell, 34, fol. 123r (Aug. 30, 1448). The cequia maior of the Marjals was to be cleared of vegetation, scoured, and staked in a certain way ("erbejar scurar e stacar en certa forma"). See *ibid.,* fol. 139r (Oct. 19, 1448): Miquel Assensi was paid for the "old and new stakes ('staques noves e velles') which he had placed in the New Canal."

104. AMV, Libros de Sotsobreria de Muros, 2, fol. 36r (Apr. 7, 1390): "abaxar lo tarquim vell de la obra de la cequia per ço quel altre tarquim si adonas." Folio 36v: "abaxar lo tarquim vell de la ora de la cequia nova." Folio 49v: "aplanar lo tarquim." The silt was used to repair nearby roads: ". . . adobar lo cami, ço es a tirar lo tarquim de la cequia" (*ibid.,* fol. 42r).

105. *Ibid.,* fol. 31v (Mar. 22, 1390): "miga dotzena de cabaços despart e fils despart per a tenir senyals a la amplia de la cequia." *Ibid.,* fol. 33r: "iiii cordes despart per asenyalar la cequia."

106. *Ibid.,* fol. 83v (Aug. 18, 1390): "segadors qui stant en la aygua."

107. AMV, Manuals de Consell, 34, fol. 170r (Mar. 6, 1449). On Dec. 4, 1458, the jurates offered to pay only 6d. per ell for the cleaning of the Cequia dels Codonyers, also in the Marjals (*ibid.,* 36, fol. 128v).

108. AMV, Libros de Sotsobreria de Muros, 2, fols. 73r–74v.

109. Earl J. Hamilton, *Money, Prices and Wages in Valencia, Aragon and Navarre, 1351–1500* (Cambridge, Mass., 1936), pp. 64–78.

110. AMV, Libros de Sotsobreria de Muros, 2, fol. 25v (Feb. 8, 1390): "Item aquest dia mateix doni an Artus trompeta de la ciutat per si e sos companyons ço es trompadors e taballers per una crida que feu per la ciutat xvi dies ja passats que tot hom que affrontats ab la cequia appellada den Fluvia dins xv dies hagues escurat son front de la dita cequia certifficant aquells que no osarien que pagarien la dobla."

111. ARV, Gobernación, 2271, 6th hand, fols. 14r–21v, and 10th hand, fols. 18r–21v (Mar. 7, 1444): "e axi ho ha vist pratiquar en alguns molins del terme e orta de Valencia que scuraven tansolament devall moli fins a la primera parada o fins al primer partidor." Compare ARV, Gobernación, 2211, 8th hand, fol. 48r.

112. AMV, Manuals de Consell, (20, fol. 134v (Nov. 7, 1393). There were innumerable legal tangles over whether the city or the hereters were responsible for cleaning certain sections. On one occasion a commission of jurates and lawyers had to go to the *traversa* of the En Fluvia Canal to ascertain who was responsible (*ibid.,* 23, fol. 231v, Aug. 13, 1407).

113. ARV, Gobernación, 2208, 12th hand, fol. 35 (June 5, 1414). The distinction was that up to the two divisors the water "was not divided by the commons" ("no sia partida per germania") and therefore was still part of the main canal, no matter what the physical contour may have been.

CHAPTER III. INTRACOMMUNITY CONFLICT AND ITS RESOLUTION

1. Jaume Roig, *Spill o libre de consells,* ed. R. Miquel y Planas, 2 vols. (Barcelona, 1929–1950), ll. 14785–14792. Fina Querol Faus sees in this passage "an implicit comparison between the *acequiero* and God, distributor of eternal justice"; *La vida valenciana en el sglo XV: un eco de Jaume Roig* (Valencia, 1963), p. 20. Roig made a detailed description of the irrigation system of Valencia (ll. 14760–14813) based on his own experience as an owner of irrigated land. He died while on a trip to inspect the "diversion dams of the canals" (*Spill,* I, xviii: "Dimecres, primer dia de abril [1478], mestre Jaume Roig, metge, e mossen Lois Masco, ab altres que tenen heretats, anaven a veure los açuts de les cequies, e passat Benimàmet lo mestre Jaume se atura e volch descavalcar per horinar; e al descavalcar . . . ell caigue de la mula, e dona tan gran colp que nos poch levar").

2. Benacher-Faitanar Ordinances, 1435: "Item lo dit cequier o lochtinent de aquell sia tengut de pasegar les dites cequies almenys una vegada cascun mes ab un o mes dels vehedors e ab lo sindich del dit comu."

3. ARV, Gobernación, 2201, 8th hand, fol. 3v (Mar. 30, 1413): "En Gil Roiç de Castellblanch . . . dix . . . que ell com a lochtinent de cequier

de la dita cequia anant per la dita cequia dimecres passat hac huyt jorns en la nit ab sis homens que menava en sa companyhia per evitar los fraus ques poguesen fer en laygua de la dita cequia."

4. It followed that both parties had to be present when sentence was passed. Pere Vives, cequier of Favara, testified that "while he was cequier . . . when a question was before him . . . he had [never] given sentence without the parties being present." ("En pere vives laurador vehi del loch de patrax [sic] . . . dix quell ts. mentres es stat cequier de la dita cequia de favara que questio fos devant ell com a cequier no ha donat sentencia sens que les parts noy fossen presents"; (ARV, Gobernación, 2278, fol. 48r, Dec. 1, 1449.)

5. The cequier's judicial process was identical to that of the mustasaf (see Chapter X). The oral process in this case was an Islamic inheritance, as James II inferred in a charter of 1316: "We understand on behalf of the masters of the city of Valencia that anciently, before the time of the conquest, it was the custom of this city that the mustasaf, summarily and openly, not having received any petition or opinion in writing, but by hearing the statements of the disputants concerning [various urban problems] decides by himself or, should doubts occur to him, with the advice of the jurates and of other masters of the city, and of others who were [formerly] mustasafs"; Francisco Sevillano Colom, *Valencia urbana medieval a través del oficio del mustasaf* (Valencia, 1957), p. 201. See also Arcadio Garcia, "La actuación procesal del mustaçaf," *Ausa* (Vich), II (1955–1956), 301–310. Blasco Ibañez commented, with regard to the Tribunal of Waters, that "the absence of certified documents and of the frightful scribe was the thing which most pleased these folk who were used to looking up to the art of writing with a superstitious fear, inasmuch as they were ignorant of it. There were no secretaries, no pens, no days of anguish awaiting the sentence, no terrifying guards, nor anything more than words"; *La Barraca,* p. 82 (Editorial Planeta ed.).

6. ARV, Protocolos de Juan Domingo, 2974 (Mar. 30, 1419): "E dient lo dit cequier que li plahia que al dit n'Armengol qui present era el hauia request diverses veus que aço reduit fos en scrits e li fos feta carta publica." This document is published in Thomas F. Glick, "Dos documentos medievales referentes al Tribunal de las Aguas," *BSCC,* 53 (1967), 83–84. Compare another sentence of a cequier (a fine for an illegal canal check) recorded by a notary who had been present at the sentencing (ARV, Gobernación, 2278, 1st hand, fol. 41r, Mar. 23, 1450).

7. (a) ARV, Real Patrimonio, Maestre Racional, leg. 482, no. 9852: "Libre del honrat en Miquel Tripo, cequier de la cequia maior de la vila de Castello del any M CCCC XXXXIII [1443]"; (b) ARV, Varia, libro 402: "Libre de la administracio del offici del cequier de la vila de Castello del any M CCCC LXXXVI [1486], cequier, lo honorable en Joan Marques." These two books are hereafter referred to as Castellón Fine-Book (1443)

and (1486). Three similar books from the sixteenth century are preserved in the Municipal Archives of Castellón: (c) "Judiciari del magnifich cequier de la present vila de Castelló del any M D LXXVII [1577], cequier, lo magnifich en Pere Bosch"; (d) "Libre del cequier de la vila de Castelló del any M D LXXVIII [1578], cequier, en Pere Navarro"; and (e) "Libre del cequier de la vila de Castelló del any M D LXXVIIII [1579], cequier, en Jaume Bernat." The sixteenth-century books are not as helpful as the earlier ones inasmuch as only the charges, but not the fines, are specified. Recording of fines, it should be noted, was an administrative act of the cequier and did not alter the oral character of the judicial act.

8. In 1473 Galent and Johan Pelega, Castellón guards, insisted that a resident of Almassora pay this sum, which had been fixed in a prior agreement between the two towns (AMC, Libres de Consell, Feb. 9, 1473); Benacher-Faitanar Ordinances, 1435. Compare ARV, Gobernación, 2278, 1st hand, fol. 41r (Mar. 23, 1450): Andreu Ferrer had twice made an illegal check in the canal of the mill of La Gabia (Favara system) and was fined 6 pounds by the cequier.

9. Castellón Fine-Book (1443): fines of 1d.: Jaume Segarra v. Na Lorexa the Widow (Apr. 27), "tenia parador [sic] en la cequia"; Promens de Canet v. Pere Vilar (Apr. 27), "parador en lo partidor." Fines of 2d.: Johan Guillen v. en (Bernat) Tallant (Jan. 10), "li ha feta la parada en lo seu enfront [?]"; Promens de Canet v. Farrag (Apr. 27), "tenie parada en la cequia"; Promens de Canet v. Ciballa (Aug. 7), "tenie parada en la cequia"; Promens de Canet v. En Sampol (Oct. 10), "tenie parada en la cequia mayor." Fines of 4d.: Promens de Canet v. Serano (Mar. 21), "feu parada en la cequia"; Guillem Bonet v. Berthomeu Mut [?] (Apr. 3), "feu parada en lo cegol." Fine of 6d.: Promens of Canet v. Serano (Apr. 27), "parada en la cequia."

10. Hamilton, *Money, Prices, and Wages,* pp. 52, 72.

11. *The Sonjo of Tanganyika,* pp. 60–61. Although the generalization is suggestive, it does not seem to follow from the evidence. In the village studied by Gray the usual fine for stealing water is one goat, which is said to be "a relatively mild penalty compared to those for other forms of theft." It is hard to understand how one goat can be a small penalty in a society where the average owner has a herd of only fifty-four goats. The fine certainly seems stiffer by far than the penny or two paid by the medieval irrigators of Castellón.

12. *Ibid.,* p. 141.

13. See Table 4, no. 10. The offense was expressed "li levave laygua forcivolment."

14. AMC, Libre de Values, 1462. Asmet had a house (*alberch*) bordering those of Farrag and Dezbatls the shoemaker. In addition, he had 3 fanecates of huerta on the road of St. Mary del Lido, adjoining the field of Anthoni Meria, 5 fanecates of huerta "al moli primer atinent de johan e de mafomat

de xivert," and 3 fanecates of *mallol,* unirrigated, "al fosar dels juheus atinent den marquo e fardo." The estates of Culeyme [Sulaimān] Bocaxo are also listed. Carting, Azmet's profession, was typical of Moriscos; Tulio Halperin Donghi, "Les morisques du royaume de Valence," *Annales,* 11 (1956), 164.

15. On the "Libres de values," see Vicente Traver Tomás, *Antigüedades de Castellón de la Plana* (Castellón, 1958), pp. 81–108. The series extends from 1371 to 1798.

16. According to the Libre de values for 1468, Pere Johan the Younger had a parcel of marjal (probably in the sense of "pasture") in the parish of St. Augustine, bordering land of Johan Valenti.

17. Among the many works which discuss the Tribunal, see Vicente Giner Boira, *El Tribunal de las Aguas de la vega de Valencia* (Valencia, 1960); R. Ros Marín, "El asesor jurídico del Tribunal de las Aguas" (an interview with Giner), *Las Provincias* (Valencia), Nov. 12, 1960; Francisco Xavier Borrull y Vilanova, *Tratado de la distribución de las aguas del río Turia* (Valencia, 1831), chap. vii; Carpentier, *Réglements et tribunaux des eaux,* pp. 91–109; Moncrieff, *Irrigation in Southern Europe,* pp. 180–184; Jaubert de Passa, *Canales de riego,* I, 469–488; Vicente Boix, "El síndico del Tribunal de las Aguas," in *Los valencianos pintados por si mismos* (Valencia, 1859), pp. 101–107; and Blasco Ibañez's incomparable portrait in *La Barraca* (chap. iv).

18. Although there are eight seats, Moncada is not included; but both Quart and Benacher–Faitanar are represented.

19. See Chap. VIII, n. 15.

20. Qadis held court in the mosque, those of Egypt on Mondays and Thursdays (Carpentier, *Réglements et tribunaux des eaux,* pp. 97–99). In Cairo, the *muḥtasib* held court every other day in each of the two principal mosques; Emile Tyan, *Histoire de l'organisation judiciaire en pays d'Islam,* 2nd ed. (Leiden, 1960), pp. 279–283, 631.

21. Giner Boira, *El Tribunal de las Aguas,* p. 9. Giner adds that in former times the president of the Tribunal had indicated the right to speak by a sign of the foot, not the hand, which is an Islamic custom.

22. Julián Ribera, "El sistema de riegos en la huerta valenciana no es obra de los árabes," in *Disertaciones y opúsculos* 2 vols. (Madrid, 1928), II, 312. See Tyan, *Organisation judiciaire,* pp. 212–213 ("La règle de l'unicité"): two qadis can be appointed in the same town, but they cannot act collegiately. Their jurisdiction must be distinct, either territorially or *ratione materiae.*

23. Ros Marín, "El asesor jurídico," and oral communication from Vicente Giner Boira, May 29, 1965.

24. ARV, Real Acuerdo, 70, fol. 121 (Sept. 22, 1775). The 7 syndics or *zequieros* petitioned the King for enforcement of the privileges of James II concerning the tanda with the Pueblos Castillos: "Este dia visita en el

una carta orden de la A. Camara su fecha diez y nueve de este mes firmada por el Sr. Don Thomas del Mello . . . la acompañan, presentada a V.M. por los siete sindicos, o Zequieros de la Vega de esta Ciudad, en que pide se les conforme el Privilegio que les concedió el Rey don Jaime Segundo de Aragón sobre reparto de Aguas en dicha Vega."

25. Francisco Almela y Vives, *Valencia y su reino* (Valencia, 1965), p. 388.

26. The "Actuación sobre el tandeo" addressed to the minister was dated July 8, 1943, and was the work of Vicente Giner Guillot, *asesor jurídico* of the Tribunal and father of Vicente Giner Boira, who succeeded him in that office. The text is reproduced in *Exposición de distintas actuaciones del Tribunal de las Aguas de la vega de Valencia* (Valencia, 1944), pp. 7–18.

27. "Atento el Tribunal de las Aguas a mantener, en todo momento los derechos de las acequias de la Vega de Valencia . . . para que conserven toda su eficacia a través de los tiempos, en beneficio de las tierras que riegan" (*ibid.*, p. 3).

28. The city's leadership in irrigation affairs is the subject of Chaps. IV–VII.

29. AMV, Manuals de Consell, 5, fol. 62v (Aug. 2, 1345): "Oiat lo dit consell ordena e tenc per be que ii jurats e alcuns prohomens cequiers . . . vayen al açut de benaguazir . . . per a regonexer les aygues e que paçen [sic] en manera que la ciutat e la orta daquella hagen sufficiencia daygues."

30. AMV, Cartas Misivas, 12, fol. 180r (Aug. 23, 1414): "per tal com vostra presencia era molt necessaria en los affers de les cequies e fitament de les goles daquelles. . . . On com nosaltres siam en concordia mijançant lonorable governador que disabte primer vinent per lo mati seran tots los cequiers açi en la sala del consell daquesta ciutat."

31. The lake of Tortajada is northeast of Teruel, and the Alfambra is a tributary of the Guadalaviar; see Jaime Marco Baidal, *El Turia y el hombre ribereño* (Valencia, 1960), pp. 34–35, 133.

32. AMV, Manuals de Consell, 36, 8th bk., fol. 13v (July 1, 1457): "[The jurates] elegiren en regidors e administradors de la aygua de la lacuna de Tortajada ques deu dalli traure lançar e decorrer en lo riu d'Alffambra e dalli en lo riu de Guadalaviar que ve a la present ciutat en certa manera e segons ja es stat divisat de paraula, los honorables en Bernat d'Almenara cequier de la cequia de Moncada, en Guillem Lorenç e en Johan Perez, cequiers de la cequia appellada de Miztalla, e en Pere Marti, cequier de la cequia de Favara, presents e acceptants." For more on this mission, see AMV, Cartas Misivas, 23, fol. 109r (July 6, 1457).

33. Glick, "Dos documentos medievales," pp. 81–84.

34. See n. 6 above.

35. Benacher-Faitanar Ordinances, 1435: "Item que les penyores fahedores

per lo dit cequier contra los delinquents se exhegesquen e se hajen a exegir en la plaça de la Seu e de fer lo juhi de aquelles e no en altre loch."

36. Borrull, *Tratado,* pp. 144, 173–175. Borrull's theories on the history of Valencian irrigation are discussed in Chapter VIII.

37. Even today the Tribunal's jurisdiction is not obligatory but voluntary. A delinquent irrigator can choose to be tried by a civil judge (Costa, *Colectivismo agrario,* p. 543). The forces of tradition and consensus were the Tribunal's prime instruments of compulsion. In *La Barraca* the Tribunal's action epitomizes communal disapproval of the protagonist Batiste and his family.

38. See discussion of the *ṣāḥib al-sāqiya,* below, Chapter X.

39. It is typical of traditional irrigation societies that a council of elders makes decisions and judgments concerning the irrigation affairs of the whole community; see Gray, *The Sonjo of Tanganyika,* p. 58.

40. There are numerous parallels between Valencian irrigation law and law merchant as it developed in medieval Europe generally. Both procedures were summary, swift, almost free from appeals, equitable, not technical, and conducted without lawyers. In originally informal courts, such as the English piepowder court, merchants themselves were the judges of the misdemeanors of fellow merchants, just as irrigators of Valencia were judged by other irrigators. See Frederic Rockwell Sanborn, "Law Merchant," in *Encyclopaedia of the Social Sciences,* 15 vols. (New York: Macmillan, 1930–1935), IX, 270–274. The Tribunal of Waters was not a rural court but an urban one; most of the llauradors were citizens of the town of Valencia.

41. Carpentier, *Réglements et tribunaux des eaux,* p. 92, stresses the *public* function of the Tribunal.

42. The episode can be found in ARV, Gobernación, 2201, 8th hand, fols. 3v–4r (Mar. 30, 1413).

43. Beatriu de Miranot, widow of Bernat de Codinachs, lord of Albalat and Benimàmet. In 1416, represented by her son, Mossen Pere Codinachs, *cavaller,* she charged the lord of Vinalesa, Francesch Blanch, with building a new work in the Moncada Canal prejudicial to the hereters of that canal, of whom she described herself as one (ARV, Gobernación, 2214, 9th hand, fol. 27r, Apr. 11, 1416).

44. There is perhaps a play on words here, unclear to me, which resulted in the misunderstanding.

45. ARV, Gobernación, 2208, 19th hand, fol. 11r (Aug. 4, 1414). Another more violent incident befell the cequier of Alcúdia de Carlet, Johan Milla, who, as he came from "diverting the water of that canal" ("vinent de girar laygua de la dita cequia") one day in 1455 was attacked by Ramon Ferrer and Frances Valles the Younger, of Alcúdia, who inflicted upon him two lance wounds in the head and two in the arm (Archivo Municipal de

Alcira, Llibres de Consells, 45, fol. 126, Feb. 28, 1455). Milla related this to the justiciar of Guadasuár, his place of residence, as a result of which the same Ramon Ferrer attacked the justiciar's lieutenant, wounding him with two blows of the lance (Archivo Municipal de Alcira, Llibres de Consells, 46, fol. 10, June 12, 1455).

46. ARV, Gobernación, 2288, 25th hand, fol. 3v (July 27, 1456): "haian menaçat e costumen menaçar de bastonades als dits guards a guardes . . . de la dita aygua [donant?] los gran terror e menaçant los de mort, per la qual raho aquest part no troba guard e guardor de aygua e li ha convengut e cove en les dies de la dita tanda hauer un porter.

47. ARV, Gobernación, 2202, 11th hand, fol. 1r (May 2, 1413): "[The commons had provided in its regulations that] dos bons homens stiguen guardant la cequia de Beneger e de Quart ço es del loch de Alaquaç tro a la çut e los roylls de Manizes e aquelles stiguen de nit e dia guardant que nenguna persona no prengua de la aygua de la dita cequia."

48. AMC, Libres de Consell, Mar. 17, 1473: "donant los tot lo poder que necessari es a donar laygua a qui mester la haia e fer tots los manaments e acces necessaris."

49. See Appendix 6, no. 54 (drought citation of May 1, 1443).

50. As expressed in the Murcian jingle "él que está denantes bebe antes" (Diaz Cassou, *Ordenanzas y costumbres,* p. 109) that corresponds to the Arabic saying "the top field drinks first."

51. ARV, Gobernación, 2246, 3rd hand, fol. 15r (Apr. 9, 1432):

> [Miquel states before the Governor that he] posseheix una vinya situada en terme de la dita vila de murvedre en la partida appellada de almudafer la qual ha rech de la cequia maior de la dita vila e lo dit en miquel haia request diverses vegades an matheu folquera cequier maior e regidor de la aygua de aquella e an jacme cilia lochtinent del dit en matheu que li donasen la tanda de la dita aygua a ell pertanyent per regar la dita vinya segons era costums . . . semblants en temps de secada e aquells dits cequier e lochtinent de cequier james lo haien volgut donar tanda de la dita aygua per regar la dita vinya ans quel peius es als hereters situats per orde deius la vinya del dit en miquel aquells dits cequie[r]s . . . haien donat laygua per regar les heretats e possessions axi vinyes e mallols com altres que deius la dita vinya son situades e james haien volguda donar la dita aygua al dit en miquel.

52. There were, of course, internal disputes involving large numbers of irrigators based on nonspatial factors — such as those involving interpretation of the regulations — in which the entire community opposed the officials. A dispute between the hereters and the cequier of Rascanya in 1414 regarding responsibility for cleaning the canal fronting on certain properties was submitted to the judgment of the whole commons, convoked

in an extraordinary meeting (ARV, Gobernación, 2208, 12th hand, fol. 35r, June 5, 1414).

53. See Markham, *Report*, p. 81, and drawing on p. 82.

54. ARV, Gobernación, 2211, 6th hand, fol. 2r (Apr. 3, 1415):

> On com lo dit cequier . . . no vulla complir alcuna de les dites coses al dit loch de Puçol . . . ans persevera en la dita sua provessio feta, ço es de tallar la aygua de la dita cequia dilluns primer vinent lo qual tallament pogera be fer en la fi del present mes de abril segons es acostumat fer cascun any o en altre temps e los splets serien mes creguts que a present no son e pus forts en tal manera que serien restaurats del perill en lo qual a present son maxime puix a aquella los donat llur dret de la dita aygua segons la prachmatica e ordenacio feta per lo molt alt senyor lo rey en Pere de bona memoria la qual lo dit honrat cequier no vol seruar.

(Fol. 2v:) On Tuesday, Apr. 4, Antoni Sanchez, the court's porter, related that the governor's order had been forwarded to the cequier.

55. The text is given in ARV, Gobernación, 2262, 7th hand, fols. 5v–6r.

56. Privilege of July 5, 1321; see below, Chapter VI.

57. The priority of the four lower canals was asserted in a decision of the Valencia council in 1313 (AMV, Manuals de Consell, 1, fols. 91–93).

58. In the words of the privilege, Moncada keeps 3 *canals* of water and gives the rest to the city; apparently this alludes to the 4 control gates of the canal.

59. Precisely *which* almenares became a point of contention. The *Exposición de distintas actuaciones*, p. 9, says that there has been discussion concerning whether this aid ought to be given through the Almenara Real of the diversion dam (see drawing in Markham, *Report*, p. 82), or through the Almenara Tandera, located 1 or 2 km. from the dam. The *Exposición* says that the Almenara Real is indicated, because it has control gates and the other does not. But Lop (*Murs e valls*, p. 356) stated that aid for the 4 lower canals is to be given "not from the Almenara Real, as some have wished it, but from the Almenara Tandera, which took this name from the *tandas* in times of need." Lop says that the subvention must come through the Almenara Tandera so that it cannot be taken by Quart, Tormos, or Mislata (whose points of diversion are higher than the Almenara Tandera). See Jaubert de Passa, *Canales de riego*, I, 495.

60. ARV, Gobernación, 2262, 7th hand, fol. 6r:

> E que hun dia dels dits quatre dies de la setmana tota la dita aygua de la dita cequia de muncada discorrega e vaja e sia donada als hereters e habitadors dels dits lochs del puig e de puçol per moldre e per regar lurs splets possessions e plantes de aquells a arbitre de cequier de la dita cequia e que en lo dit dia totes les files o rolls corribles de la dita

cequia tro als dits termens del puig e de puçol stiguen tanquades e tanquats e que algu no gos pendre la dita aygua o toldre aquella als dits hereters o habitadors dels dits lochs del puig e de puçol sots pena de LX sous.

The royal letter had been obtained on behalf of Puig and Puçol by the Queen, as lady of Puig. The city, acting on complaints from some hereters of Moncada — evidently *sobirans* — dispatched messengers in an attempt to head off the royal action (AMV, Manuals de Consell, 16, fol. 207v, July 14, 1374). The jurates were angry about the king's intervention and wrote to the city's delegates at court that the 2 [sic] letters which the king had written at behest of some hereters of Moncada (i.e., the downstream irrigators) "better had remained in his pen" ("foren mils romases en la ploma"). "If the water is to be well administered and with equality, the direction ought to be at the disposition of the cequier of Moncada, to do as he wishes" ("car a la veritat si laygua es ben regada e ab egualtat, a tot bastara a gran compliment, mas quel comportiment haia esser a disposicio del cequier de muncada segons vol ho f a r a"); AMV, Cartas Misivas, 3, July 18, 1374, n.p.

61. The bishop was the owner of Puçol (Burns, *Crusader Kingdom*, pp. 137, 139).

62. ARV, Gobernación, 2211, 6th hand, fol. 40r (Apr. 24, 1415): "[The cequier's refusal was] total destruccio dels dits lochs e dels habitadors en aquells qui per no haure la dita aygua a ells pertanyent son fets de sequa e son prests venir per la dita raho en depopulacio e ruhina."

63. Valencia, May 9, 1268 (Branchat, *Tratado de derechos*, III, 300–302).

64. The function of the *ferma de dret* was to stop the action in question and make sure that the defendant appeared before the court; Bienvenido Oliver, *Historia del derecho en Cataluña, Mallorca y Valencia*, vol. III (Madrid, 1879), 499–502.

65. ARV, Gobernación, 2215, 13th hand, fol. 45; (letter) 20th hand, fol. 19v.

66. ARV, Gobernación, 2216, 23rd hand, fol. 3r: "de aquells capitols no es stat usat o acostumat usar sino en temps de habundancia daygues e no en temps de fretura de aquelles almenys quant se deduhia en juhi ans en lo dit temps de fretura de aygues ses acostumada servar la dessus mencionada provisio del senyor rey en pere."

67. ARV, Gobernación, 2233, 12th hand, fol. 39r (Nov. 20, 1424). The plaintiffs are Johan Mason, procurator of the Bishop of Valencia, lord of Puig ("located within the district of Valencia" — "situat dins terme de Valencia"); Francesch Scola, procurator of the justiciar, jurates, and people of Valencia city, "en nom de senyors per la dita universitat del loch de Puig e de la baronia vulgarment appellada de Cebolla" (that is, the barony of Puig; Puig means "hill" and was so called by the Muslims, *jabal*, then corrupted to *cebolla*, "onion," Burns, *Crusader Kingdom*, p. 492 n. 94);

Johan Borrell, syndic of the abbess of the Franciscan convent of Játiva ("abadessa de les menoretes de la ciutat de Xativa, senyora en la ciuilitat e en les rendes de dit loch del puig"); and finally Bernat Steve, representing the residents of Puig.

68. ARV, Gobernación, 2233, 18th hand, fols, 2v–3r: "lo dit en jacme dalmenara cequier que desus ha juredicio ordenaria e contensiosa en la dita cequia de muncada e ayga de aquella e en los heretes e regants de la ayga . . . en e per fer pagar lo cequiatge e destreny[er] forçar e complir aquelles heretes e regants . . . que no volen . . . pagar lo dit cequiatge."

69. There is some evidence that compromise was achieved by electing men of the middle villages as cequier. Bernat Campmar, cequier in 1415, was from Meliana. Other fifteenth-century cequiers were Ferrer Sarnes from Albalat and Roiç d'Espejo from Farnals (ARV, Gobernación, 2262, 7th hand, fols. 14r, 15r). The same was true of the Favara system, where cequiers were often from the centrally located village of Patraix. See ARV, Gobernación, 2288, 25th hand, fols. 19r, 25r: "En Pere Vives de Patraix . . . es stat cequier de la dita cequia; en Pere Marti, de Patraix, cequier de la dita cequia de Favara [in 1457]."

70. ARV, Gobernación, 2233, 18th hand, fol. 4 (Mar. 31, 1425):

> Et . . . vehents los dits vehins e habitadors dels dits lochs de puçol e del puig que los heretes sobirants de la dita cequia de muncada indirectament los volen levar la dita ayga fahent molts statuts e ordenacions en la dita cequia per los quals los dits vehins e habitadors dels dits lochs de puçol e del puig no poden obtenir justicia alguna de frau o tolliment quels fos feyt de la dita aygua . . . [because] los dits heretes sobirants en la dita cequia han statuit e ordenat que de alguna questio o debat que fos per raho de la dita ayga no pogues esser haud recos a algun altre oficial o jutge si no als jutges per ells constituits en la dita cequia los quals son en tot o en la maior part los dits heretes sobirants de la dita cequia, de ques seguex que com als vehins . . . dels dits lochs de puçol e del puig era tolta la dita ayga en sa tanda o jornada de regar per algun o alguns dels dits heretes sobirans los dits jusants de puçol e del puig hauien anar e demanar justicia devant los dits heretes sobirans en la dita cequia. [The sobirans] eren participants en lo dit tolliment dayga e axi eren fets part e jutge de per tant los dits jusans nunqua podien obtenir justicia.

71. ARV, Gobernación, 2262, 2nd hand, fol. 37r (June 26, 1438): "Dels sindichs del puig e de puçol contra lo cequier de la cequia de muncada. [Ponç testified on July 2] que en e de la dita cequia de muncada han molts alters . . . los quals no poden regar sino quant tota laygua de la dita cequia discorre en una. Item, diu . . . que quant tota laygua de la dita cequia de muncada va als dits termens del puig e de puçol los dits alters han e poden haure rech de aquella."

72. *Lo roll de maçamagrell.* In the inventory of 1771 (Jaubert de Passa, *Canales de riego,* I, 277) the roll of this name was not an alter but was located on the floor of the canal. Several other intakes in this district were alter, however, e.g. "un caño o rollet altero llamado de na Carbonella, a dos palmos de suelo de acequia, riega altera." Also mentioned in this document are the roll of Farnals and another, unnamed, in Rafelbunyol. According to the 1771 survey the roll called La Cruz or Puebla de Farnals was on the floor, but the other 4 gates in Farnals were alters (*ibid.,* 278–279). In Rafelbunyol most of the turn-outs were alters in 1771 (*ibid.,* 279–281).

73. ARV, Gobernación, 2262, 2nd hand, fol. 37v.

74. *Ibid.,* fol. 47r.

75. *Ibid.,* 7th hand, fol. 13r: "E dix que el ts. en los anys xxxv, xxxvi e xxxvii es stat cequier de la cequia de muncada e en los dits anys se recorde que ell com a cequier prega als de maçamagrell per ço quels de pug e de puçol haguessen compliment de aygua que lo jorn de la lur tanda los plagues de tanquar lo roll que de la dita aygua discorre en lo terme del dit loch de maçamagrell e james ho pogue obtenir ans tots temps stava ubert."

76. From the Arabic *al-rīf,* "countryside," particularly the countryside bordering the sea, as in the Moroccan Rif; R. Dozy, *Supplément aux dictionnaires arabes* 2 vols., 2nd ed. (Leiden, 1927), I, 575. This district is a good example of a marshy area connecting two huertas and capable of being irrigated from either. There was a larger, analogous marjal area to the south of the city watered in parts by the Royal Canal of the Júcar and in other places by the tails of the Favara and Quart systems. A lower branch of the Quart system (Faitanar) crosses the Gully of Torrent, irrigates part of the districts of Torrent and Picanya, then joins the Júcar system in Beniparrell (Facundo Burriel y Garcia-Polavieja, "La huerta de Valencia (geografía geogenia, historia)," unpub. ms. p. 104).

77. By a privilege of Barcelona, June 5, 1260; Antonio Chabret, *Sagunto: su historia y sus monumentos,* 2 vols. (Barcelona, 1888), II, 371. The diversion would have been between Rafelbunyol and Puig.

78. Valencia, Jan. 23, 1301 (text in Chabret, *Sagunto,* II, 488–489). Chabret's title misleadingly indicates that the privilege refers to the condition of the Arrif Canal; in reality the king was acting on petition from the irrigators of Arrif and ordering those of Moncada to get their canal in shape and thus prevent loss of valuable water:

> Pro parte hereditariorum del Arrif termini Muriueteris fuit nobis humili supplicatione monstratum quod cum ipsi consueuerint hactenus rigare hereditates in dicto termino sitas ex aqua cequie de Monchada. Et nunc vicini dicte Cequie illi videlicet qui sunt termini Ciuitatis Valencie intendentes eis maliciose subtrahere et auferre aquam Cequie supradicte nolint dictam Cequiam emundare et in tantum ampliauerint los Roylls ols ullars ipsius Cequie quod aqua que deberet conuerti

in hereditatum comodum predictarum inutiliter diffunditur versus Lalmargil totaliter discurrendo.

79. ARV, Gobernación, 2277, 2nd hand, fol. 46v (May 6, 1449): "los splets o fruyts que son a present en lo rech de la dita cequia de la riff qui totalment se sequen es perden com los forments e altres splets tro a huy no sian regats nos sien poguts regar per no haure pogut haure laygua de la dita cequia e aço per gran culpa del cequier e hereters sobirans de la dita cequia los quals han permes e huy en dia permeten que laygua de la dita cequia vaja a perdres per los marjals."

80. *Ibid.,* 7th hand, fol. 36r (May 17, 1449). There follows an explanation of administrative procedures in time of drought, namely that in addition to the cequier and çabacequiers 3 or 4 other "good men" — or more — were elected to supervise the division of water and to decide which "fruits of the earth" had greatest need of water.

81. *Ibid.,* fol. 37r (May 26, 1449): "ell ts. ha vist e veu huy en dia que los de puçol se clamen que no poden hauer aygua e veu clamar a molts altres que son primers en lo rech quels de puçol clamant se que nols basta laygua de la dita cequia. E creu el ts. que sino per lo bon orde e regiment que es en la dita cequia no bastaria a la mitat del rech que es en aquella."

82. ARV, Gobernación, 2288, 25th hand, fol. 1r (1457): a description of the course of Favara from its açut above the mill of Mossen Jacme Gil to Albal. The document also contains an inventory of the divisors of the canal, dated Dec. 1, 1362 (fol. 8).

83. *Ibid.,* fol. 1r. This is the position of the upstream irrigators.

84. *Ibid.,* fol. 3v (July 27, 1456); see n. 46, above.

85. The aqueduct is described by Markham, *Report,* p. 86.

86. ARV, Gobernación, 2226, 3rd hand, fol. 26r (Mar. 29, 1421): "lo partidor de la cequia del dit loch de Quart alli hon se parteix entre Quart e Benejer la aygua que ve e discorre per la cequia dels archs." If repairs to the arches were needed, the cequier of Benacher and Faitanar had to expend up to 50s. yearly to make them (Benacher–Faitanar Ordinances, 1435).

87. Marco Baidal, *El Turia,* p. 321.

88. ARV, Gobernación, 2263, 6th hand, fol. 12v (Apr. 3, 1435). The Governor first heard the dispute on Aug. 27, 1434; records are not extant.

89. See Thomas F. Glick, "Levels and Levelers: Surveying Irrigation Canals in Medieval Valencia," *Technology and Culture,* 9 (1968), 178.

90. ARV, Gobernación, 2260, 8th hand, fol. 16r (June 22, 1437): "Mossen Jacme Gil fou elet cequier de la cequia de Rascanya."

91. See n. 119 below.

92. They were present at the general meeting of the commons of Benacher and Faitanar on May 29, 1435, at which the *Capitols* were adopted.

93. ARV, Gobernación, 2253, 6th hand, fol. 39r (Apr. 3, 1435): "cascun dels cequiers que huy son en les dites cequies sien tenguts de tornar e tornen qualseuol penyores que fetes haien e en llur poder sien fins a huy franquament e sens paga alguna de colonia o pena a aquells de qui son com lo hu haia usat en part de la iurisdiccio pertanyent al altre."

94. ARV, Protocolos de Juan Peris, sig. 2864 (*Capitols* of 1421); see Glick, "Dos documentos medievales," p. 83 n. 5. Faitanar's need for Benacher's support in resisting a challenge from the Chirivella Canal was undoubtedly a spur to the administrative reunification of the 2 branches (see n. 98 below).

95. ARV, Gobernación, 2257, 4th hand, fol. 11r (May 9, 1436): "lo entrevall de una cequia e altra deia ser molt distant . . . en tal manera que per la continuacio dels descorriment de les aygues e derruyment de les ribes e caixers de les cequies la una no destruescha laltra; e la dita obra nova e o obriment de novella cequia se faça de feyt en molta propinquitat de la dita cequia de fauara e a perill de derruir e destruir aquella."

96. Benacher-Faitanar Ordinances, 1435: "Item que lo dit cequier . . . no gos ne presumesca ni meta ni metre faça ell ni altri per ell palesament ni amagada aygua de les dites cequies en la cequia de Mizlata. . . ." Faitanar, on the other hand, was permitted some water from Mislata, which was forbidden to withhold it (ARV, Gobernación, 2378, 2nd hand, fol. 16v, Apr. 15, 1486).

97. ARV, Gobernación, 2271, 5th hand, fol. 8r (Oct. 6, 1444):

> [A letter from the governor to the residents of Burjassot and other hereters of the Tormos Canal stating that Moncada has the right] de tenir tanquat continualment hun roll qui es en la dita cequia prop lo abeurador de burjaçot sino solament en temps que iii o iiii possessions qui han rech per lo dit roll han necessitat. E que ara vosaltres en dan e prejuhi de la dita cequia de muncada e comu de aquells . . . vos fforçats tenir lo dit roll continuament ubert e per una cequia quey fforçats feu portar la dita aygua a la dita cequia de tormos.

The Moncada and Tormos systems were so intertwined in the area of Burjassot and Godella especially that the nearness itself was a constant source of friction. See Moncrieff, *Irrigation in Southern Europe,* p. 173: "It is not so strange that they [the cequiers of Moncada] must positively not have lands on any of the other canals, and especially, it is no wonder that there should be manifested a good deal of reserve towards the Tormos Canal, which irrigates close to that of the Moncada, and being on a lower level must have many opportunities of profiting by its waters."

98. ARV, Gobernación, 2255, 25th hand, fols. 36r–37r (July 5, 1435).

99. Ven Te Chow, ed., *Handbook of Applied Hydrology* (New York: McGraw-Hill, 1964), p. 15–35.

100. For a general description of the modus operandi of medieval levelers, see Glick, "Levels and Levelers."

101. See Chap. XII, n. 35.

102. See ARV, Gobernación, 2271, 6th hand, fol. 14r (1444). Evidence was submitted regarding the responsibility of millers to maintain the canal up to the first diversion downstream to show that, in this case, the check in question was not implanted for the purpose of diversion but rather to slow down the velocity of the water ("lo dit partidor . . . cessa esser partidor com no si fara parada per a regar sino que es limitador daygua").

103. From the point of view of the diverter, the backwater was often the effect desired. Compare ARV, Gobernación, 2215, 15th hand, fol. 23r (Mar. 26, 1416); the *alqueria* of Sagres had the right to divert water from the Rahana Canal (Játiva) Monday through Thursday by placing checks in the canal and thus "making the water flow backwards during the said four days" ("de fer parades en la dita cequia e de fer regolfar la dita aygua en los dits quatre dies").

104. AMC, Libres de Consell, 16, n.p. (Sept. 8, 1408): "Item lo dit honrat consell a suplicacio de alguns vehins de la dita vila heretats en lo regadiu del cequiol prouehi que fos vista la pedra que era stada mesa per lo honrat en pere de reus en lo dit cequiol per girar aygues al abeurador e que si la dita pedra ere sobresalta e fahia gran regolfa en lo dit cequiol que sie baxada en tal forma que en lo dit cequiol nos faça regolfa e lo abeurador haie la aygua que ha necessaria." The Cequiol irrigated the highest fields near the town (Traver, *Antigüedades de Castellón*, p. 144).

105. AMC, Libres de Consell, 17, n.p. (Mar. 15, 1410): "Item fou exposat per alcuns homens regants de la aygua del cequioll que per la parada ques fa al pont della lo moli de oli den berthomeu sales per ço que vingue aygua per al seruei de la obra de nostra dona santa maria la aygua del dit cequiol esta envassada per la regolfa que fa de temps applegue molt tarquim entant que los escurados no volen escurar." The matter was referred to again in the council meeting of Apr. 9: "per la parada ques fa en lo dit cequiol . . . la aygua . . . faça regolfa per la qual sich applegue mes tarquim que no faria si la dita parada no sich fahia."

106. Most of the mills constructed on irrigation canals were overshot-wheel mills. An undershot wheel needs a constant, fast current to operate efficiently, whereas an overshot wheel needs a well-directed and well-regulated water supply and is used to put variable or sluggish flow to best advantage. The miller built up head with canal checks and then diverted the water through a chute (*canal*) to the wheel, set below the level of the supply channel; see Alcover, *Diccionari*, VIII, 507; R. J. Forbes, *Studies in Ancient Technology*, 2nd ed., 9 vols. (Leiden, 1965), II, 91; and Louis C. Hunter, "Water-Mills in Southern Europe," *Technology and Culture*, 8 (1967), 460–464. Mills had 2–10 stones, those nearest the head of the canal with more than those downstream (Marco Baidal, *El Turia*, p. 435).

107. ARV, Gobernación, 2236, 2nd hand, fol. 21r (Feb. 20, 1426). This was a flour mill in Alfafar and the complainants were residents of Massanassa. Veses was accused of rebuilding the intakes (*barbacanals*) at a higher level than that of the original ones, but he claimed that the construction had taken place only after he had consulted the men of Massanassa as well as "bons maestres e liviladors daygua."

108. AMC, Libres de Consell, 17, n.p. (Apr. 9, 1410). The mill had caused harm to the town "per lo feyt de la regolffa que fa la aygua per tant quant mol lo dit moli en la dita cequia derroque los quexers e destrengue los quexers de la dita cequia."

109. ARV, Gobernación, 2273, 13th hand, fol. 46v (Aug. 25, 1446). The argument that the mill should be allowed to run with only one stone was rebutted by alluding to the observed "gentleness" with which the same mill ran with two: "lo testimoni de la dita na ferrada consta quel dit moli draper no pot obrar ab una roda com aquell no puixa obrar sino ab molta aygua [but the court's assessors, with their own eyes] ha vist obrar lo dit moli ab dues rodes gentilment."

110. ARV, Gobernación, 2272, 3rd hand, fol. 7r (Aug. 20, 1446); "Jatsia molt amplament e bastant axi per los testimonis ex aduerso produhits . . . que lo moli draper no pot obrar ab una roda anant per ço com volent ho fer per lo regolf que fa lança laygua de la cequia en la cequia den bonuehi appellada." Bonvehi was the canal of a mill on the Rambla branch. By 1452 Perfeta had stopped using the other wheel as well (see n. 121, below).

111. AMV, Manuals de Consell, 6, fol. 60r (Sept. 27, 1346): "Item fou ordenat en lo consell . . . que los jurats vayen ab alcunes bones persones . . . en los quals haye maestres dobra de vila, ço es a la almenara del vall de la ciutat per la qual es girada laygua va al moli den caner, per regonexer e veure en qual manera pora ser profitos a la ciutat o que torn la dita almenara en lestament que era o si paria ser profitos a la ciutat ab lo dit en caner e que la dita almenara o parada sie de tot en tot remoguda per la qual almenara o parada laygua del dit vall ragolpsa [sic] . . . e sen pot enseguir gran dampnatge com la vall no pot exaguar segons que deuria e esta a scurar unes vegades que no faria." The matter is continued on fol. 75r (Oct. 17, 1346). Nicolau Primitiu located the site of the almenara of this mill; "D'arqueologia," *ACCV*, 2 (1929), p. 86.

112. ARV, Gobernación, 2208, 19th hand, fol. 42r (Aug. 23, 1414). One of the aggrieved irrigators was Pere de Reus (see n. 103, above).

113. ARV, Gobernación, 2288, 25th hand, fol. 1r: "los partidors o braços appellats de na scodella e de na piquabaralla."

114. See the inventory of the partidors of Benacher–Faitanar, ARV, Gobernación, 2253, 9th hand, fol. 27 (June 20, 1435). An irrigation improvement attributed to Pere Bou, lord of Callosa, is the rebuilding of all the canals of that place and the remaking of the divisors in stone (ARV, Gobernación, 2377, 46th hand, fol. 13r, June 1, 1486).

115. ARV, Gobernación, 2288, 25th hand, fol. 8 (Ap. 17, 1457, including a copy of document dated Dec. 1, 1362). The war allusion is to the expected invasion of Peter the Cruel of Castile (see Chap. IV, n. 33).

116. In fact, one divisor of the Na Rovella Canal was actually called The Level (*lo livell*); ARV, Gobernación, 2257, 4th hand, fol. 35r, May 19, 1436.

117. The most typical divisor was one to one (as in Figure 3). That most divisors were 1:1 is indicated by the fact that no ratio at all is given in most documental references, whereas *unequal* divisors, which must have been relatively infrequent, are specifically mentioned (ARV, Gobernación, 2285, 14th hand, fol. 2r, Mar. 1454: regarding division of the Mislata Canal into two unequal parts [*dues ineguals parts*]: when one side was irrigating, the other had to be closed "with a slat, or with silt, earth or brush" — "ab post o ab tarquim terra o broça").

118. ARV, Gobernación, 2260, 8th hand, fols. 40r–42r (July 18, 1437). The sentence was given on July 29 (fol. 42r) and has survived in the *Litium* series (rather than Executions) because Quart appealed, on Aug. 3 (fol. 42v). The sentence provided

> que los solars dels dits partidors sien alçats mig palm per egual livell. E a la part damunt dels dits partidors sia feta al costat dels dits solars una filada de pedra o solar. E semblantment a la part davall dels dits solars e partidors sia feta altra fila de pedra, o solar, on aygua dels dits partidors o solar, la dita aygua a la dita cequia a cascuna part. E de aqui avant sia fet a cascuna de les dites cequies per egual uns xv o xx palms ab pedra o morter e tant com tendran les dites filles o solars, e solar de argamassa sia feta a cascuna part de les dites cequies caxers de pedra per ço que laygua sia presa de la dita cequia per cascun dels dits partidors.

Raising the level of this divisor had been an issue 16 years before. See ARV, Gobernación, 2226, 3rd hand, fol. 26r (Mar. 29, 1421). Some irrigators of Benacher wanted to raise the level of Quart's side, reducing the amount of water flowing into the Quart Canal but leaving the level on Benacher's side unchanged.

119. ARV, Gobernación, 2281, 2nd hand, fol. 28r (Apr. 21, 1452): "com la dita cequia dalgiroç e la cequia de rambla partesquen la llur aygua en hun partidor constituhit dessus los molins del dit mossen jacme perfecta lo qual partidor livella e fa particio de les dites aygues e del curs de aquelles . . . e en cascuna de les dites cequies lla hon es la lengua del dit partidor ha sos caxers e sos lindas de pedra fets a lliuell en los quals nos pot fer novitat alguna com se deia seruar lo antich e acostumat curs de les aygues."

120. Mercader owned this mill as early as 1430, when he was involved in a conflict with V. Ros who, in making a drinking trough for his cattle, had broken the bank of the Mestalla Canal below Mercader's mill (ARV, Gobernación, 2241, 6th hand, fol. 26r, May 19, 1430).

121. The same mill which Perfeta was operating in 1446 with only 1 of its 2 wheels.

122. ARV, Gobernación, 2282, 12th hand, fol. 3v: "[The divisor had been built] ab certa mesura e ab certs lindars e ab cert livell. Empero per lo dit mossen jacme perfeta en gran emulacio prejuhi e detraccio dels dits hereters de la dita cequia de rambla es procurada e donada causa que molta mes aygua descorre e reb la dita cequia de algiroç al dit partidor que ne acostumava . . . e per conseguent molta menys aygua discorre en la dita cequia de rambla."

123. It is not clear whether the fulling mill was an overshot or an undershot mill. It seems to have been undershot, built across the canal; if the wheel were removed, an impetuous flow certainly would have resulted. The fact that it had bleeder channels, however, would indicate an overshot mill.

124. Some others were Bernat Steve, who represented Puig and Puçol in half a dozen cases between 1415 and 1438 in which these tail irrigators on the Moncada Canal sought to defend their rights against the upstream men, and Dionis Dolit, who represented Burriana in a number of irrigation cases in the governor's court between 1417 and 1438. See T. F. Glick, "Abogados de aguas en la edad media valenciana," *Las Provincias* (Valencia), June 22, 1967.

125. Regarding the widening, see ARV, Gobernación, 2223, 34th hand, fol. 9r (1419).

126. See Chap. XIII, nn. 52–53.

127. ARV, Gobernación. 2282, 12th hand, fol. 13r:

> Car diu . . . que [si] en lo temps ques partiren les aygues fos pensat del dit moli pilater no fora prudent provisio donar avantatge cert en lo partidor de la cequia de algiroç per lo impediment incert ques poguera pensar que degues dar lo dit moli obrant ans es persumidor e versemblant que les dites aygues sens haure sguart algu al dit moli pilater foren diuidides ab egualtat entre les dites cequies e si avantatge algu reb de aygua la cequia de algiroç, ço que nos creu, aquell tal avantatge seria stat donat a la dita cequia de algiroç per haure mes regants segons dit es haud sguard que no rega sino ab la terça part de laygua.

128. *Ibid.*, fol. 6v: "nos podia pensar mesurar ne arbitrar per home desta vida lo exercici fahedor de aquell que havia mester."

129. *Ibid.*, fol. 6v: "Item, per ço . . . car per sentencies es stat guanyat quel dit moli draper no pot fer regolf per lo qual se perda aygua de la dita cequia poqua o molta ans ha a dar salt al partidor en manera que laygua una vegada saltada no puixa tornar. E axi per les dites sentencies appar de que resulta que lo dit moli per no fer lo dit regolf ha [de] tenir les dites canals ubertes e dar auiament a tota laygua que cau del partidor e no pot fer empaix que perda aygua." There would have to be a drop before the mill in any case in order to provide it enough head.

130. *Ibid.*, fol. 7v:

Item diu . . . que los lindars de les dites cequies de rambla e dalgiroç son constrohits en tal liuell e mesura que per molt auiament que qualsevol de les dits cequies done davall lo partidor no pot rebre mes aygua de son dret si donchs laltra de les dites cequies no stava enrunada es dava causa en aquella fer tal regolf per lo qual no pogues rebre son dret, car posible es que en tal cas que laygua que aquella axi enrunada o ab gran regolf no poria rebre per son empaig diverteria en laltra per haure lo dit discurs deliure e veuria sens vici e causa de laltra que rebria lo augment e per ço no sen poria formar accio contra aquella mas contra aquell qui tal causa o empaig procuraria.

131. *Ibid.*, fol. 13r:

Item diu . . . que quant laygua de alguna cequia mare se parteix per egual entre dues cequies en la construccio dels lindors e dels partidors de les dites cequies se ha e acostuma haure sguard qual de aquelles cequies a les quals la dita aygua se deu partir ve pus dreta al discurs de la cequia mare. E per ço com trobant se pus dreta alguna de les tals cequies per haure e rebre laygua pus delliure directa via ne pendria mes que no laltra que no la rebria axi dreta via per tal raho se sol e sacostuma dar avantatge de qualque pocha latitut o amplaria al partidor de la cequia qui no ve tan dreta e se fa pus stret lo partidor de la cequia que ve dreta, por ço que per aquell poch avantatge la cequia que no ve dreta no reba mes aygua que laltra que no ve tan dreta ans cascuna de aquelles reben laygua ab egualtat."

This is completely logical: a curved channel offers more resistance to flow because more energy is needed to push the water around a curve. Making one side of the divisor wider would reduce the friction loss and allow more flow to take place.

CHAPTER IV. THE CITY'S ROLE IN IRRIGATION DEVELOPMENT

1. ARV, Varia, 126, fols. 80r–82v. See Appendix 4 for translation of the document. The manuscript is not dated, but Garcia Mezquita was working on other irrigation projects in the Valencian region in the second and third decades of the fifteenth century (see Glick, "Levels and Levelers," p. 179).

2. ARV, Gobernación, 2206, fol. 38r (Oct. 22, 1414): "Item dien que si en quant es a los adobar los açuts dessus dits allo es a carech e despeses de les terres regants e dels senyors dels molins poblats en les dites cequies mares."

3. ARV, Gobernación, 2262, 1st hand, fol. 40 (Mar. 8, 1438): "Den Luis

Eximenez contra lo sindich de la cequia de Rascanya." A charter included in the evidence documents the general meeting in 1436 of the "deputies, cequier, inspectors, and hereters of the Rascanya Canal, and the owners of the mills" on the main canal and the Axera branch.

4. AMC, Libres de Consell, 17, Mar. 15 and Apr. 9, 1410. On Sept. 30 the council paid 230s. in order to extend the Cequiol to the town square to facilitate work on the church. See also *ibid.*, 22, Mar. 13, 1417, regarding apportioning the cequiatge.

5. ARV, Protocolos, 3253 (Sept. 26, 1351):

> On com lo dit en Ramon del Mas per vigor de la dita procuracio e sindicat haia pres e reebut diverses quantitats de diners axi de juheus usurers comsimplament de xpians per a obs e necessari de la dita cequia per vigor de la qual manleuta fou feta collecta per tots los hereters de la dita cequia la qual dita collecta colliren e reeberen en Pere Moragues e en Pere Ffrancolli los quals devien de liurar la dita collecta e moneda la qual ells colliren de aquella an Simon Çestrens lo qual dit en Simon degues pagar tots los deutes los quals eren feyts per lo dit en Ramon.

6. AMV, Manuals de Consell, 22, fol. 288v (Apr. 24, 1404):

> Del ensaig e cost de la cequia del canyo . . . lo concell attes que la cosa si pot haure acabament es molt profitosa e deu esser ensajat ab poca messio per no enbolcar la ciutat a gran despeses e infructiferes si la cosa no havia manera, provei concordantment e hac per bo que ab cequia poca fos fet en saut si la dita aygua podia esser tirada e engravar tro al loch livellat e que de la peccunia de la ciutat fos pagada la meytat del cost del dit ensaig es vist la dita cequia poder e deure esser feta e per aquells la dita aygua poder esser tirada, en tal cas la meytat de la aygua de la dita faedora cequia haia e sia de la dita ciutat e tanta o tal part com muntara segons layrata de ço que la ciutat pagara en la faccio de la dita faedora cequia. See also AMV, Manuals de Consell, 22, fol. 268v (Nov. 24, 1403).

7. AMV, Cartas Misivas, 8, n.p., Oct. 14, 1404.

8. This is the origin of the Cequia del Canó discussed in Chap. II, n. 62.

9. Antonio Joseph Cavanilles, *Observaciones sobre la historia natural, geografía, agricultura, población y frutos del Reyno de Valencia*, 2 vols. (Madrid, 1795–1797), I, 171. Cavanilles discusses the area surrounding the Albufera where the water table was so high that in "infinite fields" one had to dig only a foot or less to find water. The reason for the spread of swamps from their original bounds, he thought, was the diversion of water from its original course without previous preparation of canals solid enough to contain it. He did not doubt that the Royal Canal of Alcira was in itself responsible for most of the water in the Rio de los Ojos (the Riu dels Ulls of the medieval documents), owing to filtration. Those who built

these old canals were interested only in making the channel capable of carrying water to certain places, but they never examined the nature of the land with regard to its water-holding capacities.

10. The Punta d'En Silvestre, now called simply La Punta, is located at the turn in the river below the city where today the Alicante highway, lined by *barracas,* leads southward. See description of the vineyard of Anthoni Benahull, llaurador, located at the Punta d'En Silvestre, whose bounds were the field of Pasqualet, the En Fluvia Canal, the La Conqua road, and "the road which is in the midst of the canebrakes" ("la dita vinya affronta ab terra de Pasqualet, ab la cequia d'en Fluvia, e ab lo cami de la Conqua, e ab lo cami d'en mig dels canyars"); ACV, Actos Capitulares 1400–1411, sig. 3579, 4th bk., fol. 11r, July 15, 1405. The end of the En Fluvia Canal seems to have been especially unstable; some documents indicate that it ran only as far as Massanassa, others that it had an extension leading into the Albufera. See AMV, Manuals de Consell, 22, fol. 229r (Jan. 12, 1403), "De la cequia de l'Albufera": the city provides that a canal be dug at its expense from the Canal of En Fluvia and "taken up to the Albufera, through which the water carried by the En Fluvia Canal might run and drain into the Albufera without harm to the Marjals" ("sia feta e treta una cequia de la cequia appellada den fluvia e menada tro a la dita albufera per la qual laygua ques pendra de la dita cequia den fluvia descorrega e tir tro a la dita albufera e entre en aquella e aço sens dampnatge de les marjals"). In 1406, in reference to the above provision of 1403, the council was asked to clean and finish the New Canal, which the city had built in the marshland (*almarjal*) to carry water from the En Fluvia Canal to the Albufera (AMV, Manuals de Consell, 23, fol. 51, July 27, 1406). In 1448 the En Fluvia Canal is described as follows: "la qual cequia comença a masanasa e termena en lo riu de godolaviar e en la punta denominada den siluestre, la qual cequia decorre de la marjal de la orta." In any case, the water flowing in En Fluvia was to have drained in the Albufera. Because of conditions of the Marjals, however, there were long periods when the exit to the Albufera was blocked. The Acequia del Oro, in the nineteenth century, was lined with rubble (*mamposteria*) to prevent filtration (Martínez Aloy, *Provincia de Valencia,* p. 289).

11. ACV, Pergamino 5597 (July 4, 1342): "Guillermo de Fluvia . . . novellament ha feytes e no ces continuament de fer cequies per menar algunes aygues [in the Marjals]." The expression for improving land was *panificar* — to bring bread from the land.

12. These canals were functioning early in the fourteenth century; see ACV, Pergamino 7429 (May 31, 1316), a dispute about irrigation on the canals of Castelló and Cavaller, in Russafa. This was a typical upstream-downstream dispute about contributions for cleaning, the men of the tail claiming that the *sobirans* received more water and hence should pay more.

13. See Jaubert de Passa, *Reglamentos y ordenanzas,* p. 344.

14. A canal crossing En Fluvia ("la traversa de la cequia den fluvia") is mentioned in 1407 (AMV, Manuals de Consell, 23, fol. 231v, Aug. 13). In 1444 the *almarjals* below Russafa and Alfafar were said to be in great danger of infection because canals were "broken down, were not flowing, and the water was stagnating" ("les cequies son riblertes e no corren e les aygues stan embassades"). The peril could be overcome, it was suggested, if the water of the Cequia de Castelló could be made to flow into that of En Fluvia; AMV, Manuals de Consell, 32, 2nd bk., fols. 197r–198r (Apr. 22, 1444), and 201v (May 8, 1444). The canal was built the following year in the Marjals et Castelló d'En Arrufat; it started in the Cequia del Castelló, crossed the Cequias del Cavaller and del Bisbe and joined En Fluvia (AMV, Manuals de Consell, 33, fols. 92r–93r [Apr. 30, 1445]); this canal was evidently the same as the "cequia de la traversa de la cequia de Castelló" (AMV, Manuals de Consell, 34, fol. 102v [May 30, 1449]).

15. ACV, Notals de Bonanat Monar, libro 3518, fol. 167^{2}r: "Contracte de la ferma dels capitols sobre lo fet dels marjals." The agreement was embodied in a royal privilege of Peter IV of Aragón dated Barcelona, Aug. 16, 1386. A Castilian translation was printed in pamphlet form in Valencia in 1766 and is found in Lop, *Murs e valls,* pp. 318–328.

16. ACV, Notals de Bonanat Monar, libro 3518, fols. 167^{2}v–168v.

17. AMV, Libros de Sotsobreria de Muros, 2, fol. 340v.

18. AMV, Manuals de Consell, 20, fol. 134v (Nov. 7, 1393).

19. *Ibid.,* fol. 296 (Apr. 1, 1396).

20. AMV, Manuals de Consell, 23, fol. 51 (July 27, 1406); 32, segon libre, fol. 18r (Sept. 18, 1441); 32, fols. 92r–93r, 96 (Apr. 30, May 14, 1445); 34, fol. 102v (May 30, 1448).

21. AMV, Manuals de Consell, 33, fol. 148r (Oct. 4, 1445): "Primo fou pres de la terra den berthomeu guerau una fanecada e hun quarto e onze braçes franch a raho de vint lliures caffiçada fon tatxat iiii lliures."

22. *Ibid.,* fol. 149r: "Item en bernat gomiç al alqueria den seguer havia y quatrecentes vint quatre braces axi per la terra com per lo destrich de la canal que no ha pogut sembrar, v lliures."

23. *Ibid.* To cite some further examples of expropriation, when the Count of Cocentaina built a new canal to bring water from the Riu dels Ulls to Pardines, some landowners complained that the count's factors and agents "had broken up diverse lands and possessions where the canal passes and have done much harm to the neighbors and residents of the said town, which harm up till now has not been satisfied . . . but, what is worse yet, they changed the course of the canal in another direction and again cause more damage, breaking up the fields and cutting down trees and grain (*blat*) of the townsmen without paying them" (Archivo Municipal de Alcira, Llibres de Consell, 47, fol. 299v, Apr. 29, 1457). The privilege of John I granting the city of Valencia the right to take water from the Júcar at Tous and conduct it to the Guadalaviar was granted on the condition that

the owners of the land be reimbursed: "sic tamen quod illis quorum fuerint possessiones per quas dicta aqua transitum fecerit, pro damnis eis in dictis possessionibus inferendis debita satisfactio tribuatur" (Branchat, *Tratado,* III, 209, Nov. 15, 1393).

24. For Rovella and sewage problems, see AMV, Manuals de Consell, 4, fols. 457v–459r (Mar. 4, 1344), including a privilege of James II of Aragón, dated Apr. 15, 1311 (fol. 458v). By terms of James' privilege, the city had the right to use water from the Guadalaviar to flush the sewers from Saturday vespers to Sunday vespers. The privilege does not mention the canal of Na Rovella, but the identification is implicit. The council decreed that on the said days the water of that canal would run through the "sewers and main drains of the city . . . removing the filth and waste material" therein ("e lo dit consell . . . accorda . . . que la aygua de la cequia apellada de na Revella [sic] que tro del riu de Godalaviar . . . decorrega per los valls e mares dels albellons de la ciutat, del disapte de hora de vespres a enant tro al seguent dia del diumenge en hora de vespres, a remoure les inmundicies e leges que en lo dit vall e albellons son e paran, per tal que les dites inmundiciones no donen corrupcions ne malauties a les gents"). See also Nicolau Primitiu, "D'arqueologia," *ACCV,* 2 (1929), 162–163 n. 2, for disposition of council of Aug. 20, 1389. Below the city the Rovella Canal irrigated fields in Russafa and the Marjals; yet the irrigators living by the side of the canal seemed to suffer no ill effects, because the earth of the channel and fields acted as a filtration and leaching system. The farther downstream from the city the purer the Rovella water grew, until at the sea it was almost completely purified; Constantino Gomez Reig, "Las aguas de las cloacas empleadas para el riego en la huerta de Valencia," *La crónica médica* (Valencia), 3 (1879–1880), 322–326, 353–357).

25. The Na Rovella Canal was the industrial canal par excellence of the medieval kingdom. Along its banks were the shops of weavers (*texidors*), dyers (*tintorers*), tanners (*blanquers*), matmakers (*estorers*), and cloth finishers (*perayres*), among others. All depended on water and all trod a thin line between legal use and infraction of municipal ordinances against befouling it. See AMV, Manuals de Consell, 1, fol. 189r (Apr. 26, 1322): "no one . . . dare throw . . . into the canal of the Moorish quarter called Na Revella [sic], stones, waste, refuse from houses, nor empty the bellies of sheep, goats or other beasts, nor iron waste, nor leftovers from mats, nor fibres, nor waste material of dye-shops, under pain of five sous." The owners of workshops (dyers, matmakers, and tanners are specified) were obliged, like other water users, to clean their *frontres* on the canal. Weavers claimed to have the right to wash wool in the canal (ARV, Gobernación, 2208, 17th hand, fol. 36r, July 24, 1414).

26. AMV, Manuals de Consell, 27, fols. 100r, 144v (Mar. 9, Aug. 14, 1419). A more economical alternate plan, whereby water would be drawn from the Favara Canal, was also discussed.

27. AMV, Manuals de Consell, 13 (4th fasc.), fols. 41v–42r (Mar. 20, 1360).

28. AMV, Manuals de Consell, 14 (4th fasc.), fols. 7v–9r (June 8, 1363). The mill was to be built in the Plaça dels Preycadors (now the Plaza de Tetuan). Regarding the shortage of mills see fol. 8v: "considerada la gran necessitat e fretura de molins que es en Valencia a temps de present e lo gran poble que habita es recull en la dita ciutat e la honor e fama ques crexeria de haure molins daygua dins los murs de la dita ciutat." The reference to the Castilian army concerns the war (1359–1363) between Peter IV of Aragon and Peter I the Cruel of Castile. In 1363 Peter the Cruel actually reached Valencia and took up residence in the Royal Palace outside the walls until the approach of the Aragonese army forced him to retire to Morvedre; H. J. Chaytor, *A History of Aragon and Catalonia* (London, 1933), p. 185.

29. AMV, Manuals de Consell, 33, fol. 125r (July 24, 1445). The need for exact information concerning how much flour could be expected from the Moncada mills was crucial at this time. On July 6 the jurates had informed the townsmen that a bounty of 2s. per *caffiz* would be paid to anyone bringing flour from a distance of 3 leagues or more from Valencia, so great was the shortage of grain (*ibid.*, fol. 123v: "Crida de la ajuda de les farines").

30. AMV, Manuals de Consell, 34, fol. 140: "Crida per sanitat e pluja a Santa Maria de Misericordia." On this church, a Dominican convent just beyond the southern wall of the city, see Burns, *Crusader Kingdom,* p. 232). In the accounting of the year's expenses, dated May 31, 1449 (Manuals de Consell, 34, fol. 196r), the expense of the procession is entered although no sum is specified: "Item la despesa feta en lo processio qui ana fora la ciutat per que nostre senyor deu nos dona pluja."

31. The corvée was known in Valencia but seems to have been employed only as a final recourse in times of catastrophe. In 1407 the citizenry was grouped into tens and fifties to fight a plague of locusts in the fields of the huerta; Alvaro Santamaria Arandez, *Aportación al estudio de la economia de Valencia durante el siglo XV* (Valencia, 1966), p. 83.

32. Elche also resorted to corvée labor in times of disaster, as in the flood of February 1435 when the main canal was cleaned by six *quadrelles* of ten men each. The *sobrecequier* could also call out the *quadrelles,* in accord with an established turn, to clean the canal when the responsible hereters defaulted (Ibarra y Ruiz, *Riego de Elche,* p. 188).

33. José Sánchez Adell, "Las murallas medievales de Castellón," *BSCC,* 28 (1952), 54–55. *N.b.* doc. 10 (p. 55 n. 2): on July 25, 1409, the council ordered that "the whole town be divided into tens (sie adeenada per deenes) in case they should have to go forth from the town." The city wall was divided into 6 equal parts, with a parish responsible for each.

For work on the wall the men of the parish were grouped into *cinquantenes* and *deenes.*

34. AMC, Libres de Consell, 15, n.p., Apr. 29, 1406: "que paguen tres diners [?] o pus vaien tres homens per deenes."

35. AMC, Libres de Consell, 27, July 22, 1422: "fon prouis que la dita gola fos scurada ab les deenes que obs hic fosen no ab homens a lloguer."

36. *Ibid.,* July 26, 1422: "sil dit en johan vilaroia se volia abstenir de tornar la dita cequia que la vila li fahes ajuda de cent homens de les deenes en ajuda de scurar lo dit enrrunament."

CHAPTER V. THE SEARCH FOR NEW SOURCES OF WATER

1. Martínez Aloy, *Privincia de Valencia,* pp. 57–63 (Turia), 63–74 (Júcar); Marco Baidal, *El Turia,* p. 405; Foster, *Geographic Structure,* pp. 43–44.

2. AMV, Manuals de Consell, 16, fol. 225r (Oct. 2, 1374): "de ques porien regar moltes partides que ara nos reguen e poria dar creximent a laygua daquest riu e daqui present a la ciutat e a la orta, maiorment a temps de seccades segons es lo present temps e alcuns altres ia passats."

3. *Ibid.,* 17, fol. 20 (Oct. 3, 1374): "E jassia que en la dita ciutat hagues de bons maestres de la dita art de geometria e de livell, empero vraisemblant lo dit fet mils e pus madurament pora esser vist . . . per maior nombre de maestres que per menor; per aquesta rao concordantment proveuren e tenguen per be quels dits jurats haien aquells mes e mellors maestres de la dita art axi re la dita ciutat com daltres persones que haure puxen, e per ço tremeren missatger o missatgers . . . en altres parts per cercar e haure e menar alts dits maestres."

4. "Livells axi d'entreguart com de pas." The *entreguart* was an alidade; the *livell de pas* (Castilian, *nivel de tranco*) was a larger version of the masons' common A-shaped level. See Glick, "Levels and Levelers," p. 172.

5. AMV, Manuals de Consell, 17, fols. 38v–39r (Feb. 11, 1376):

> [The masters and levelers] havien regonegut e fet regonexer e livellar e mesurar tot aquest fet a lost lochs del açut e de la cequia faedor e faedora començant prop e sobre lo dit castell de Thous en lo sol o pla del dit riu, e continuan per la vayll de Thous tro al açut de la cequia del rey ques pren del dit riu e rega en los termes de la ciutat de Xativa e de la vila d'Algezira. E dally anant continuan per muntanyes e per barranches e per plans tota vegada ab livells axi dentreguart com de pas per hoc encara ab la via e mesura de la dita cequia del rey aytant com durava. E trobaren segons los dits livells e mesures que la dita faedora cequia prenent aquella prop e sobre lo dit castell de Thous venia o venir poria e traure e menar aygua del dit riu de Xuquar, passan sobre los lochs d'Antella, de Raçalany, de la Alcudia,

de Benimodol [sic] de Carlet, de Alginet, d'Espioca, de Picasen, d'Alcacer, de Torrent, e engravar en lo dit riu de Godalaviar sobre lo loch de Manises, empero dejus laçut de Moncada. E que tota aquesta obra segons taxacio dels dits maestres e altres persones expertes a lor arbitracio ampliament segons dehien feta poria costar de trent e cinch en quaranta mila libras.

6. *Ibid.*, fols. 40v–41r (Feb. 12, 1376).

7. AMV, Manuals de Consell, 20, fol. 71v (Apr. 2, 1393): "Item quels honrats jurats de la dita ciutat tracten ab lo senyor rey que sia haud del privilegi que la ciutat del riu de Xuquer puxa traure daygua una gran cequia capdal per menar e metre aquella a la orta e al riu daquella."

8. The text of the privilege dated Tortosa, Nov. 15, 1393, is reproduced in Branchat, *Tratado de derechos,* III, 209–210, and in A. Guillén, *El Tribunal de Aguas de Valencia* (Valencia, 1920), pp. 109–110.

9. See Appendix 6, Table A, nos. 23–27.

10. AMV, Cartas Misivas, 6, fols. 361v–362r (May 26, 1400). This request recalls the arrangements made between city and bishop regarding improvement of the Marjals in 1386.

11. AMV, Manuals de Consell, 22, fol. 120v (July 24, 1400). This text is printed in Guillén, *El Tribunal de Aguas,* pp. 112–113.

12. AMV, Cartas Misivas, 2, n.p. (July 23, 1401).

13. AMV, Cartas Misivas, 7, n.p. (July 26, 1401). The text, and complete list of magnates, can be found in Guillén, *El Tribunal de Aguas,* pp. 110–111. Gaspar Escolano states that the reason for the meeting in Torres Torres was that King Martin was there at the time (*Décadas de la historia de la insigne e coronada ciudad y reino de Valencia,* 2 vols. [Valencia, 1878], I, 434).

14. Part of Ramoneta's interest might have been that such a project would have reduced the city's pressure on the Pueblos Castillos, of which Ribarroja was one.

15. AMV, Cartas Misivas, 7, n.p. (Aug. 18, 1401).

16. AMV, Cartas Misivas, 8, n.p. (Oct. 14, 1404). See Chap. IV, n. 8. Another plan to "bring water of the Royal Canal in the district of Alcira to the Guadalaviar" was discussed, and another investigatory commission appointed, in October 1406; AMV, Manuals de Consell, 23, fol. 92v (Oct. 25, 1406).

17. AMV, Manuals de Consell, 26, fol. 70v (Sept. 23, 1415).

18. AMV, Manuals de Consell, 50, fol. 200r (Sept. 3, 1500): "cent sous per una carta en la qual sera pintada laygua que sera a portar axi de Castello e encara de lo part del riu de Xuquer."

19. AMV, Manuals de Consell, 63, n.p. (Apr. 16, 1529): "[The officials of the city] prometen Andreu Puig, obrer de vila, que mostrat lo cami per lo qual diu antiguament se prenia una sequia de aygua del riu de

Juquer que venia fins al pla de quart ac ab tot effecte y essent effectuada la sequia a consell de aquell a regantse lo dit pla de quart li prometen de ferlo franch de sixanta cafficades de reg de la dita sequia e noresmenys li prometen donar cinquanta ducats per los treballs e endustria que haura sostengut en la dita sequia."

20. *Ibid.* (Dec. 11, 1529). I have relied on the text as reproduced in Guillén, *El Tribunal de Aguas,* pp. 113–115. The project is also referred to in AMV, Manuals de Consell, 63, for Nov. 19 and Dec. 9, 1529.

21. Guillén, *El Tribunal de Aguas,* p. 113: "Memorial del livell per hon pot venir laigua de Xuquer fins a Valencia sens assut y sens mina."

22. See Rafael Tasso Yzquierdo, *Algunos datos sobre la historia, descripción y actuación de la Acequia Real del Júcar,* 2nd ed. (Valencia, 1964), p. 23. This is the same route suggested in the privilege of King Martin, Jan. 16, 1404 (Branchat, *Tratado,* III, 201).

23. Guillén, *El Tribunal de Aguas,* p. 114. How this fila was calculated is not revealed; probably it was based on the passage of water through a fixed orifice, but at any rate it was only to give a rough idea.

24. He may have had in mind the ruins later described by Jaldero and Valls David (see Chap. IX).

25. Guillén, *El Tribunal de Aguas,* pp. 114–115: "Lo pla de quart se pot tot regar de aquesta manera, prenent laygua de pedralba per hon se acostumaua de regar lo pla de quart antigament y laygua ab la qual se ha de regar lo dit pla de quart se ha de pendre del riu de Valencia de aquesta manera, ço es que les quatre cequies, bennager fauara mizlata e rouella, se poden girar al dit pla de quart."

26. AMV, Manuals de Consell, 131, n.p. Text in Guillén, *El Tribunal de Aguas,* pp. 115–121. On Negret see Borrull, *Tratado,* p. 10.

27. Escolano, *Décadas,* I, 434.

28. *Ibid.*: "En nuestros dias, por el celo de un su hijo llamado Fonte, ha vuelto a rebullir, y ha estado tan adelante con su magestad del Rey nuestro señor, que podria ser que el nacimiento de tan deseadas aguas, se debiese a este Fonte, o Fuente." I have been unable to trace the reference to Benedict XIII.

29. Joan Font, *Discurso breve con que se prueva la possibilidad de sacar agua del Rio Xucar, para los llanos de Quart, Liria, Morviedro, y otros* (Valencia, 1628). Both 1627 and 1628 were drought years. See D. Alvaro y D. Diego de Vich, *Dietario valenciano (1619–1632)* (Valencia, 1921), pp. 78–79, 246, and Mosén Juan Porcar, *Coses evengudes en la ciutat y regne de Valencia (1589–1629)* 2 vols. (Madrid, 1934), II, 266.

30. Font, *Discurso breve,* p. 6.

31. See, for example, Miguel Angel Gilbao y Castro, *Impulsos a la execución del bien común de España en discurso breve sobre el riego que se ha de sacar del rio Xucar* (Valencia, 1652), and (Domingo de Usenda), *Relación de la visura hecha por el capitan de cavallos . . . Domingo de*

Usenda y Mansfield para sacar agua de los rios Júcar y Gabriel y regar con ella los llanos de Cuarte (Valencia, 1658).

32. Archivo Municipal de Elche, Actas de Consell, May 27, July 14, July 30, 1420.

33. *Ibid.*, Aug. 14, 1420: a letter to Chinchilla seeking a safe-conduct for levelers and another letter to the leveler of Játiva.

34. *Ibid.*, Sept. 1, 1420:

> vos respondemos . . . que vuestros vezinos e los lliuilladores de la agua entren e vayan e passen e lliuellen nuestro termino seguramente en lo que toque a nos que todos seran bien agogidos et los sera dado fauor e ajuda segun se pertanesce; pero en razon de la costa sabed que de presente nos non podemos contribuir ni dar cosa alguna por muchas dubdas que tenemos que la dicha agua non podra venir aca pero si la dicha agua sallir podia e seyendo puesca en tales terminos que la espirencia de los omes conosciesen que podrian venir a effecto a la sajon faremos e pagaremos sueldo por libra lo que razonable fuere de faser.

35. *Ibid.*, Apr. 23, 1528. On Feb. 14, 1531, the council agreed to pay the expenses of Juliá Rosell, "master of bringing water," who was to divert two and one-half filas more of water from the springs of Crevillente than the amount already coming from them.

36. Ibarra y Ruiz, *Riego de Elche*, pp. 21–24. Verde was also responsible for works on the main canal of Elche and was the author of a short treatise on its physical characteristics (*ibid.*, pp. 163–177).

37. Rafael Viravens y Pastor, *Crónica de la muy ilustre y siempre fiel ciudad de Alicante* (Alicante, 1876), p. 232; Vicente Bendicho, *Crónica de la muy ilustre, noble y leal ciudad de Alicante* [1640] (Alicante, 1960), p. 45: "En 1628 reanudáronse trabajos preparatorios iniciados en 1558, para traer un brazal del Júcar a estas comarcas y las de Elche, pero el proyecto no pasó adelante por haberse opuesto el Síndico de Valencia, originándose un pleito, actualmente en *litis pendencia*." This is doubtless the matter referred to in a letter from the King to the jurates and syndic of Valencia in 1632 with regard to a dispute with Alicante, "en querer sacar agua de la acequia" (AMV, Cartas Reales, 9, fol. 144r, Aug. 7, 1632).

38. Font, *Discurso breve*, p. 1. Alicante was still interested in the Júcar as late as the nineteenth century, and Valencians still felt obliged to oppose such a move. See *Documentos para conocerse y juzgarse las inconveniencias de distraer aguas del Rio Júcar hacia la Provincia de Alicante según el proyecto de D. Juan Bautista Peyronet: memoria sobre el proyecto del Canal del Rio Júcar* (Valencia, 1860); Antonio Sancho and Sebastian Monleon, *Dictamen facultativo sobre el proyecto formado per D. Juan Bautista Peyronet para canalizar el Rio Júcar y poder fertilizar la Provincia de Alicante* (Valencia, 1860); *Apuntes breves y sencillos en impugnación de*

la larga memoria de Don Juan Bautista Peyronet sobre su proyectado canal del Júcar para la Provincia de Alicante (Valencia, 1860).

39. The Cabriel rises in the province of Teruel, flows south to Requena, and joins the Júcar at Cofrentes.

40. AMV, Manuals de Consell, 16, fols. 62v–63r (Feb. 20, 1372). The privilege was said to be "en lo archiu de la sala de la dita ciutat."

41. *Ibid.,* fol. 81r (May 15, 1372).

42. See Chap. III, nn. 31–32.

43. AMV, Manuals de Consell, 17, fol. 47v (Mar. 26, 1376). The Cabriel is also mentioned.

44. AMV, Manuals de Consell, 25, fol. 237 (June 23, 1413): "Encara comana lo dit concell als dits honorables jurats que facen veure a regonexer la sima de la aygua de Santa Cruç ques diu ques pot girar al riu de Godalaviar e fer hi ço que sia obs. E aximateix sobre los açuts ques pertenen esser fets en lo dit riu contra forma de privilegi e ques torne a rao; hoc encara sobre los cent caffis de forment que demana la vila de Moya a subvencio sua per sos diners, e ells daran loch a la dita aygua de la dita sima."

45. AMV, Manuals de Consell, 26, fol. 13r (May 8, 1415). The matter is also discussed, together with the Júcar project, in *ibid.,* fol. 27v (May 28, 1415).

46. On the physical discontinuity of the cultivated zones, see Brunhes, *L'Irrigation,* p. 55, and Julián San Valero Aparisi, "El campesino valenciano y su obra," in *Notas sobre la antigüedad de la agricultura y el regadío en tierras valencianas* (Valencia, 1964), pp. 76–77. San Valero, commenting on Wittfogel's theory, asserts that the fragmented nature of the Valencian countryside inhibited the development of an institutional revolution (i.e., one leading to a true hydraulic society) completing the agricultural revolution which produced the intensively irrigated huertas. The huertas are the work of towns, not of states.

47. The mutual compatability of feudal service tenure and the ecological requirements of irrigation agriculture is suggested by E. R. Leach, "Hydraulic Society in Ceylon," *Past and Present,* no. 15 (April 1959), 17.

CHAPTER VI. REGIONAL ARRANGEMENTS

1. AMV, Manuals de Consell, 1, fols. 91r–93r (Aug. 21, 1313).

2. See Nicolau Primitiu Gomeç, "Contribució al estudi de la molinería valenciana mijeval," *III Congreso de Historia de la Corona de Aragón* 2 vols. (Valencia, 1923), II, 700n.

3. AMV, Manuals de Consell, 1, fol. 92r: "Encara ordenaron que per la mirva de la aygua que avien per la dita raho o particio los lochs del Putg a de Puçol que son derrers a aver aygua de la cequia de Muncada que tots los rolls e files hixens de les cequies de Muncada, de Quart, de

Tormos, e de Miçlata sien tancats totes nits, ço es a saber del sol post tro al sol exir."

4. *Ibid.,* fol. 91v: "fora gitat amor hodi parentesch corrupcio e subur-nacio." Five *veedors* are identified by residence; they were apparently from the Moncada Canal (Massarrojos, Moncada, Albalat, Puçol, and Puig).

5. Branchat, *Tratado de derechos,* III, 306–308. The three privileges are discussed by José Latour Brotons, "La distribución de las aguas en la huerta de Valencia," *Astrea* (Valladolid), 2nd semester (1957), 5, and in the *Exposición de distintas actuaciones,* pp. 8–10.

6. Branchat, *Tratado de derechos,* III, 204–205.

7. *Ibid.,* 206–207. The terms of the privilege specify that the city's 4 days follow those of the Pueblos Castillos ("et habitores Civitatis Valentiae aliis quatuor diebus et quatuor noctibus subsequentibus").

8. AMV, Manuals de Consell, 12, fol. 52r (Jan. 2, 1355). During a drought Moncada was measured, under the direction of a jurate; as it was found to have 11 palms of water it was decided that it could spare ½ palm for the lower canals; on Jan. 15 (fol. 55v) it was reported that the cequier of Moncada had closed the almenara, and the jurate, Pere Mascho, was ordered by the council to open it again so the city would not lose its possession. Compare AMV, Cartas Misivas, 7, n.p. (July 21, 1401), and 13, fols. 11r, 13r (July 17 and Aug. 9, 1415), letter from the jurates to the syndics and cequier of Moncada.

9. AMV, Cartas Misivas, 7, n.p., July 15, 1401: "aquesta ciutat que segons sabets es mare de tot lo regne e la restauracio de la qual es conservacio de tots sos menbres passa distret daygues e per ço hauen fet ab lo cequier de la cequia de moncada que ell tinga laygua tres dies e valencia quatre dies los quals començen lo dissapte a vespre ora acostumada e valrias poch aquesta ordinacio si no hauien tota laygua per tal vos pregam affectuosament que aquests iiii dies hauam laygua daqui tenir la vosaltres iii dies e la ciutat quatre dies segons muncada. Et dur aquesta ordinacio per unes tres o quatre tandes." Similar letters were sent to Ramoneta of Riusech, lady of Ribarroja, and Hugh of Moncada, lord of Vilamarxant.

10. AMV, Cartas Misivas, 13, fol. 11r (July 15, 1415):

> Al honrat en Berenguer Rochafort, cequier de les aygues de Valencia . . . Nosaltres [i.e., the jurates] a pregaries dels noble e honorable mossen pere de moncada e mossen johan de sent feliu hauem acordat que de la aygua de la tanda que ara han les nostres cequies ne lexets passar par les cequies de Beniguazir e Vilamarxant quatre canals, ço es, per cascuna, dues canals, car ells nos han consentit graciosament esta tanda que dels quatre dies de la lur tanda nos lexen per a les cequies de aci hun dia axi que aquesta tanda ells hauran tres dies e les cequies jusanes v dies.

See letter to cequier of Moncada, July 17, 1415.

11. See Chap. III, n. 24.

12. AMV, Manuals de Consell, 15, fol. 8 (July 31, 1368). In 1413 the city had to carry out its threat of arms (see Chap. VII). The modern modus operandi, whereby cequiers of the huerta implemented the turn by proceeding upstream and opening the gates of the Pueblo canals, had evolved by the beginning of the sixteenth century. See AMV, Cartas Misivas, 33, n.p. (May 29, 1501): a letter informs the officials of Benaguasil, La Pobla de Vallbona, Vilamarxant, and Ribarroja that because Valencia needed water the cequiers of Mislata, Mestalla, Favara, Rascanya, Rovella, Quart, and Tormos were being sent to "derrocar almenares, pendre tandes e posar les aygues en la present ciutat en la forma acostumada."

13. AMV, Cartas Misivas, 1, fol. 41r (Aug. 28, 1335): "les cequies de la orta son mirves daygues en tant que les esplets de la orta son en perill de perdre per si deus e bons amichs no hi acorren."

14. *Ibid.*: "La qual provisio a bona perfeccio venir no poria sens ajuda e secos de les aygues dels dits lochs sobirans, es assaber de pedralba, de villamarxant, de beniguazir e de riba roya, los senyors e alcayts dels quals esguardan lo ben deure de vehinatge."

15. AMV, Cartas Misivas, 7, n.p. (May 18, 1401), a letter to the Lady of Ribarroja asking for "two additional *canals* of water from now until the forthcoming feast of Quinquagesima" (that is, that she open two or more of her control gates from the present until the fifth Sunday before Easter of the following year). Also, AMV, Cartas Misivas, 13, n.p. (May 6, 1415), letter to officials of Benaguasil asking for one day of water to alleviate the shortage. Latour Brotons ("Distribución de las aguas," p. 6) describes this *agua de gracia* as a step taken just short of putting the general river turn (*tandeo*) into effect.

16. See Chap. VIII, nn. 26, 28.

17. AMV, Manuals de Consell, 36, 7th book, n.p. (May 2, 1516): report by Johan de Burgos, mason, and Johan Corber, stonecutter, on their trip to the diversion dam of Vilamarxant to estimate damages caused by Jaume de Pertusa, wood merchant (*cabanyer de la fusta*). The cequier and *alcayt* of Vilamarxant stated that the damage was caused by poor maneuvering by the hookmen (*ganxers*) who were guiding the wood past. Burgos and Corber had also inspected the dam and canal of Bugarra.

18. AMV, Cartas Misivas, 11, fol. 101r (Oct. 10, 1412): the city needed the "water of the Knights" (i.e., Pueblos Castillos, under the suzerainty of various lords) for two days in order to float the wood, just as it had already secured that of Moncada. See *ibid.*, 12, fols. 163r (July 14, 1414), 168r (July 26), and 170r (Aug. 1). By a privilege of James I (Alcira, Jan. 30, 1267) wood cut anywhere in the kingdom could be floated to the city via the Júcar and Guadalaviar (Branchat, *Tratado de derechos,* III, 195–196; see also privilege of James II, June 1, 1314, *ibid.*, pp. 200–201).

19. AMV, Cartas Misivas, 19, fols. 5v–6r (June 20, 1437).

20. AMV, Cartas Misivas, 22, n.p. (May 17, 1454): "lo debat que es entre los de la pobla de benaguatzir sobre lo regar de les ortalices de la cequia del alguatzir." Compare another dispute between these two places, Cartas Misivas, 30, n.p., May 9, 1483.

21. AMV, Cartas Misivas, 23, fol. 185v (Aug. 1, 1458): letter to Pere Ramon de Muncada. The jurates of Benaguasil and La Pobla had complained to those of Valencia that they had had to appoint *sobrecequiers* to guard the water of their canal that passes through the district of Pere's village of Vilamarxant. Benaguasil's cequier sought damages from Pere because its rice was without water and could not be irrigated from any other canal.

22. AMV, Cartas Misivas, 23, fol. 94r (Mar. 24, 1457), to the Baron Viscount of Gayano. The jurates had received his letter about poor repair of the canal of Benaguasil, which had caused damage to his canal of Ribarroja. ("Molt noble baro: Una letra vostra hauem rebut feta en ribarroia a xxii del present sobre la cequia de benaguazir la qual diem que per no tenirla en condret . . . es cayguda sobre la vostra cequia en gran dan de aquella.")

23. AMV, Cartas Misivas, 5, fol. 215r (Aug. 12, 1393): a letter to the Lord of Chestalgar requesting that he let the water flow to the city for 4 continuous days.

24. AMV, Cartas Reales, 7, fols. 27r, 28v–29r (Apr. 24, 1606). The archbishop, chapter, and jurates of Valencia wrote to the king about a canal that the jurates of Teruel were planning to build to conduct water from the Guadalaviar. The King wrote to Teruel that taking the water would divert it from greater needs, particularly those of the city of Valencia in the summer.

25. AMV, Cartas Misivas, 7 n.p. (July 20, 1402). The jurates wrote to Jacme de Valldenoç, who was guarding the water, that little upstream water (*aygua del riu alt*) was reaching the city.

26. AMV, Manuals de Consell, 5, fol. 62v (Aug. 2, 1345). The council ordered 2 jurates and 2 cequiers to proceed to the diversion dam of Benaguasil to inspect (*regonexer*) the water and to see that the city and the huerta were receiving their due.

27. ARV, Gobernación, 2202, 16th hand, fol. 28r (July 4, 1413). The subsyndic of the city asked that the Governor or his Lieutenant go upstream to investigate the new diversion dams and canals being made on the banks of the Guadalaviar, and others which had been enlarged "contrary to ancient form," resulting in great harm to the city.

28. *Ibid.*, 12th hand, fol. 30r (May 25, 1413) and the countersuit, 2203, 28th hand, fol. 14r (May 30, 1413).

29. AMV, Cartas Misivas, 8, n.p. (Aug. 14, 1405).

30. See Appendix 3.

31. AMV, Manuals de Consell, 16, fol. 170v (Aug. 23, 1373):

> Item fou proposat en lo dit consell que alcuns cequiers dalcunes de les cequies de la ciutat e de la orta de poch temps aença contre furs e privilegis us e costum tots temps observats efforçaven vedar a les gents de no lavar e haure altres aempriu en o de les aygues de les dites cequies ans los dits cequiers per avaricia o per malicia assaiaven penyorar o fer rescatar les gents lavants o aemprants les dites cequies. [Thus the council ordered] que cascu puxa e li sia legut en qualsevol de les dites cequies e en les aygues daquelles . . . lavar e fer lavar lana, draps, bugades e totes altres coses.

In fact, most of the canals had regulations against such industrial use, but here the city was acting in what it considered a higher economic interest.

32. AMV, Cartas Misivas, 12, fol. 180r (Aug. 23, 1414). The cequiers were convened in the council's chambers (la Sala del Consell) to discuss "los affers de les cequies e fitament de les goles daquelles."

33. For the duties of the office of bailiff, see Danvila, *La germania de Valencia,* pp. 431–432.

34. ARV, Fondos en depósito, Pergaminos, caja 6, no. 2 (also, *Aureum opus,* fol. 33, chap. xxii): "Item statuimus et ordinamus et prohibemus sub pena lx sol. quod cequiarii de Muncada et de Fauara et alii non mitant tantam aquam per cequias suas in prejudicium hereditariorum per ultra unam mulnariam de qualibet cequia possit inde perdi uel amiti im amargalibus."

35. *Fori regni valentiae,* "In extravaganti," fol. 10:

> Item suppliquen los richs homens, e cavallers de la ciutat, e daltres viles del regne que com los cequiers en molts casos facen molts e diuersos greuges a ells, e als homens lurs, e donen lo dret del aygua a altres, e reguen aquella hajen gran iuredictio axi dels juhis de aquells no puixa hom a alcun recorrer. Que sia vostre merce ordenar que a cascun cequier cascun any sia feta inquisicio per aquells qui per nos seran ordenats segons que per fur nou en los officials es acostumat de fer. Majorment senyor com ells sien officials, e han poder vostre, e ells fan penyores, e leuen calonies segons que ells se volen. Concordant richi homens, et generosi, et persone Ecclesiastice. Contrasten hi los jurats de la ciutat. E semblantment hi contrasten los sindichs de les viles reals.

36. AMV, Cartas Misivas, 3, n.p. (July 18, 1374): "per part dels alcuns hereters de la cequia de Moncada son estades imperades dues letras del senyor Rey les quals parlant ab la sua reverencia e de qui signades les ha foren mils romases en la ploma . . . car a la veritat si laygua es ben regada e ab egualtat a tot bastara a gran compliment, mas quel com-

portiment haia esser a disposicio del cequier de Muncada segons vol ho fara . . . hon vos pregam affectuosament e decontinent haiats revocats de les dites letres."

37. ARV, Gobernación, 2247, 12th hand, fol. 41r; 17th hand, fol. 25r (Nov. 15, 1432). Rascanya and Rovella specifically asked that they receive the same amount of water as Favara and Mestalla, since they irrigated the same amount of land. (All 4 canals received 14 filas.)

38. Branchat, *Tratado de derechos,* III, 155–156. See also *ibid.,* p. 139, privilege of James II, Lérida, July 24, 1317, giving the bailiff jurisdiction over concessions made by the king for the construction of mills, ovens, and baths. Emphyteusis was a civil law contract by which a grant was made of a perpetual right to possession of property at a fixed annual rent. Such property was heritable and alienable.

39. ACV, Pergamino 3924 (Sept. 14, 1310): "El dit batle respon que . . . com per lonch us de temps per moros e de crestians sia acostumat de levar la dita almenara e dar aygua als valls e a les cequies de la ciutat. . . . E com al dit batle fos request per los dits hereters quels fer complir la forma del dit privilegi lo dit batle ana a la damunt dita cequia e feu levar la dita almenara per dar la dita aygua segons la forma damunt atorgada."

40. Branchat, *Tratado de derechos,* III, 308–309 (Valencia, February 1358).

41. *Ibid.,* p. 308: "cequiae superiores dicti rivi et specialiter illa quae vocatur de Muncada, in tantum redundant desuper, et cum tanta implentur affluentia, quod magna pars ejusdem aquae in mari et alias amittitur quasi continue et frequenter." Compare *Fori regni valentiae,* fol. 91r (1510), wherein it is asked that the governor and bailiff be forbidden to intervene in irrigation matters: "Ad supplicationem brachis regis. Item senyor com per furs e privilegis del dit vostre regne de Valencia, la conexença de les cequies e aygues del riu de Guadalaviar pertanga als jurats de la dita ciutat, excepto la cequia real appellada de Moncada."

42. ARV, Gobernación, 2241, 6th hand, fol. 32r (May 24, 1430).

43. For the duties of the office, see Danvila, *La germania de Valencia,* pp. 429–431.

44. On the Royal Canal of Alcira the opposite was the case: the Governor was forbidden to interfere in the affairs of this canal, whose jurisdiction was that of the Cequier Real. But appeal was to the bailiff; Branchat, *Tratado de derechos,* III, 316–321, privilege of John II, Valencia (July 11, 1467).

45. *Ibid.,* 202–203. In similar dispositions regarding the Moncada Canal, however, the bailiff was charged with destroying any new works and with preventing further alterations of the canal's physical dimensions (*ibid.,* 302–306, privilege of Peter IV, Barcelona, Oct. 5, 1368, reaffirming that of James II, Trayguera, Apr. 20, 1298).

46. See, e.g., ARV, Gobernación, 2236, 2nd hand, fol. 21r (Feb. 20, 1426) regarding Anthoni de Veses' mill in Alfafar (Favara Canal); *ibid.*, 4th hand, fol. 43r (June 1, 1426) regarding Luis Bonet's mill (Mislata Canal); *ibid.*, 2278, 1st hand, fol. 41r (Mar. 23, 1450) regarding the mill of La Gabia (Favara Canal).

47. ARV, Gobernación, 2272, 3rd hand, fol. 7r (Aug. 20, 1446): "Et apres ab lo dit en jacme perfeta tantsolament segons forma de fur ha e deu esser tractada devant vos e cort vostra qui sois jutge ordinari de les causes de e sobre ques litiga entre les dites parts axi per ço com si tracta de molins e de cequies e de aygues e del liure descorriment de aquelles com a e segons apar per les actes del dit."

48. ARV, Gobernación, 2213, 36th hand, fol. 31r (1415): "E per conseguent appar clar e indubitat que del greuge del dit cequier non pot esser jutge lo dit cequier e qui vos Mossen lo governador ne sots jutge e no altri." In this case, between the cequier and a miller of the Mislata Canal, the cequier was accused (fol. 31v) of favoring his relative ("lo dit cequier per favorejar son parent ha fet son poder que la aygua del dit royll, del qual es la present questio, sia tolta als hereters jussans"). By not returning the water to the main canal the cequier had contravened the disposition of the fuero, contrary to his office, and for this reason "reasonably and justly he ought to be corrected by you Mossenyor [Governor], to whom it pertains to correct the grievances caused by whatever officials of the realm" ("rahonablement e justa deu esser corregut per vos Mossenyor al qual se pertany corregir los greuges fets per qualsevol officials del dit regne").

49. For the office of the civil justiciar, see Danvila, *La germania de Valencia*, pp. 443–446.

50. *Aureum opus*, fol. 77, chap. clvi ("Que justicia in civili de causis cequiarum et aquarum non se intromittat: scilicet cequiarii cum provisoribus"), Barcelona, Aug. 18, 1326. Also, privilege of Tortosa, Apr. 6, 1318 (*ibid.*, fol. 60, chap. lxxxix).

51. On its course see Carlos Sarthou Carreres, *Provincia de Castellón* (Barcelona, n.d.), pp. 52–61.

52. For summaries of the distribution of Mijares water in La Plana, see *ibid.*, pp. 65–71; Vicente Gimeno Michavila, "Los riegos de la comarca de La Plana," *BSCC*, 19 (1944), 139–145, and 20 (1944), 17–32, 205–222; Benito Traver, *Historia de Villarreal* (Villarreal, 1909), pp. 471–484; J. M. Oliver Moros, "Las aguas del Mijares: distribución de las mismas entre las comunidades que componen la Junta de Aguas de la Plana; cesión posterior de Burriana a los regantes de Nules," *Buris-ana*, vol. X, no. 103 (1967), 18–19; and José Dolz Moros, "Antecedentes, peculiaridades y datos curiosos sobre la Junta de Aguas de la Plana," *ibid.*, pp. 16–17.

53. The area of the Valencian huerta is 297 km^2, that of the huerta of La Plana approximately 90 km^2. In 1418 Valencia had 8,000 hearths; the

figures for La Plana were Burriana 592, Castellón 1,110, Villarreal 598, a total of 2,300 (Francisco A. Roca Traver, "Cuestiones de demografía medieval," *Hispania*, 13 [1953], 18).

54. After the cequier's election, the council would resolve that this information should be communicated to Almassora (see AMC, Libres de Consell, Jan. 18, 1414). The letter actually sent informed Almassora of the naming of the cequier and guards, and in some cases (see Libres de Consell, Jan. 8, 1392) reiterated certain regulations, such as those forbidding the placing of checks in the canal or not returning water to the main channel.

55. AMC, Libres de Consell, Feb. 9, 1473: a letter entitled *Acces del cequier* was sent from Johan Marti, cequier of the cequia maior of Castellón, to the justiciar of Almassora requesting that fines levied by guards J. Pelega and Nadal Galent upon a citizen of Almassora be carried out. The fine of 20s. was for having put a check in the canal contrary to regulations agreed upon by both towns. Similarly (*ibid.*, Apr. 26, 1473), Marti wrote to Almassora asking that 3 specified men appear before him and his guards to hear certain claims made against them by the guards of Castellón.

56. Archivo de la Corona de Aragón, reg. 13, fol. 287 (cited by Santamaria Arandez, *Aportación al estudio de la economia de Valencia*, p. 91 n. 39).

57. The number 60 was not fortuitous; see Chap. XI.

58. These proportions remained in force until 1789, when, as a result of separation of the waters of Castellón and Almassora made possible by rebuilding the cequia maior and construction of a house to guard the divisor (*la Casa de las Rejas*), the proportion was changed to 17⅕ for Castellón and 9⅘ for Almassora (Dolz Moros, "Antecedentes, peculiaridades y datos curiosos," p. 16).

59. The text of the *Sentencia arbitral* is widely reproduced, e.g., in Traver, *Historia de Villarreal*, pp. 472–477, and Sarthou Carreres, *Provincia de Castellón*, pp. 66–67.

60. AMC, Libres de Consell, 20 n.p. (Mar. 9, 1415).

> Als molts honrats los jurats de les viles de castello de vilareal e de almaçora . . . de nos los jurats de la vila de burriana, saluts e honor. Com segons la sentencia en la particio de la aygua del riu de millars entre vos e nos promulgada cascuna universitat haie liure certa part de la aygua del dit riu. Et nos ne haiam nostra part de aquella segons la dita sentencia. Per tal la amistat de cascun de vos requirim e pregam vos placie tratmetre vostres partidors dilluns primer vinent de mayti al loch acostumat per donar son dret a cascuna part e partesque aquella segons la dita sentencia deu volent lo nostre partidor sera. Com nos siam appellats fer per cascun de vos coses semblants a maiors justicia

migançan. Dat. Burriana ix die martii anno a nat. domini M CCCC XV.

The council of Castellón indicated its willingness to comply in a reply of the same date.

61. AMC, Libres de Consell (1472–1473), n.p. (Mar. 31, 1473):

Dels trenchs fets en los açuts. Fonch proposat per los honorables jurats com en la present vila eren venguts misatgers de almaçora e de vilareal e hauien parlat ab ells com los açuts de castelló e de almaçora e encara la . . . almenara de vilareal eren stats sportellats de molts portells e le dita almenara era stada leuada be que tota la dita aygua se trobava en la cequia de borriana e se hauien alguns sentiments que homens de borriana ho hauien fet. E aximateix dels dits trenchs ne hauia feta relacio a ells lonorable en guillem castell qui es partidor e axi que lo honorable consell prouehis en lo que fos fahedor. Lo honorable consell prouehi que hun home de almaçora per les tres viles ço es per castello vilareal e almaçora vaga a Valencia e sia stat als aduocats de les viles e que haia consell que es lo ques pot fer de justicia e que ço que consellaren que porten la scriptura ordenada e ques pose es persegueixca lo dit fet.

62. ARV, Gobernación, 2214, 7th hand, fol. 13r (Mar. 11, 1416).

63. Some of the pertinent court cases are ARV, Gobernación, 2219, 3rd hand, fol. 35r (Feb. 13, 1417); 2219, 7th hand, fol. 3r (Mar. 26, 1417); 2242, 11th hand, fol. 45r (Oct. 4, 1430); and 2272, 2nd hand, fol. 48r (Aug. 14, 1446).

64. The first three sentences governing Castellón and Almassora were those of July 11, 1275, Oct. 31, 1290, and June 13, 1297. The second is preserved on parchment in the Municipal Archives of Castellón, as are further sentences of Aug. 14, 1338, May 7, 1347, and Mar. 28, 1355. Many of the dispositions concerning Almassora in the Libres de Consell have to do with canal checks made by men of Almassora in the cequia maior contrary to the form of the sentence (council minutes of Aug. 12, 1386; Feb. 9, 1473; Feb. 17, 1473). Almassora sued in defense of its rights according to the sentences, in the governor's court, on Oct. 25, 1407 (ARV, Gobernación, 2195, 15th hand, fol. 43r). According to the sentences all disputes had to be arbitrated by a panel of 2 men each from Castellón and Almassora and 1 from Burriana.

65. See ARV, Gobernación, 2258, 15th hand, fol. 9r (Nov. 16, 1436); 2262, 1st hand, fol. 35r (Feb. 7, 1438); 2272, lst hand, fol. 11r (Mar. 27, 1446); 2315, 15th hand, fol. 26r (Nov. 28, 1465); Branchat, *Tratado de derechos,* III, 343–348 (Mar. 12, 1484); and a postmedieval document, *Traducción al castellano de la escritura de concordia en latín y lemosín otorgadas entre las villas de Burriana y Nules sobre aguas de riego, con fecha 28 de Julio de 1662* (Castellón, 1963).

66. AMC, Libres de Consell, 32, n.p. (Feb. 12, 1430): "el senyor rey los donaue un cequier de Nules."

67. AMC, Libres de Consell, 17, n.p. (Apr. 9, 1410): "la conexença de aquella [i.e., the cequia] ordinariament es pertany e es del dit honrat cequier e no del dit honrat batle . . . com ell dit cequier es e fos jutge ordinari de la dita cequia." On the same day the council rebuked the bailiff again for interfering in a question regarding "el escorrer les aygues de la orta en les marjals." See also Libres de Consell, 35, n.p. (Nov. 25, 1436): the bailiff of Castellón forbade the cequier to order a miller to clean his share of the main canal, and the council ordered the bailiff not to interfere, asserting that because the canal belonged to the town only the cequier had jurisdiction.

68. José M. Cueco Adrian, *La Font de la Vall de Segó* (Valencia, 1965), p. 75.

69. See a letter to *El Mercantil Valenciano* (Apr. 26, 1878), p. 2: "Siete siglos . . . que los árabes fueron expulsados de esta feráz comarca enriquecida bajo los reinados de los Abderrahmanes, y siete siglos que la Vega de Valencia se retuerce bajo la tiránica presión de los señores de los pueblos denominados Castillos."

CHAPTER VII. THE HUERTA IN CRISIS

1. For Mediterranean weather, and that of Valencia in particular, see Houston, *Western Mediterranean World*, pp. 10–36; Paul Kunow, *El clima de Valencia y Baleares* (Valencia, 1967); Casas Torres, *Huerta de Valencia*, pp. 26–34; Fontavella, *Huerta de Gandia*, pp. 38–41.

2. Houston, *Western Mediterranean World*, pp. 23–24.

3. *Ibid.*, p. 710.

4. *Ibid.*, pp. 27–30. Houston points out that June rainfall and temperature data must be interpreted carefully because the dry season usually begins after the middle of the month.

5. Foster, *Geographic Structure*, pp. 38–41; see also Gustaf Utterström, "Climate Fluctuations and Population Problems in Early Modern Economic History," *Scandinavian Economic History Review*, 3 (1955), 40–41.

6. Vicente Giner Boira, "La sang de les Santes Escudelleres," *Las Provincias* (Valencia), July 20, 1965. For the process of rubefaction see Bunting, *Geography of Soil*, p. 180.

7. Foster, *Geographic Structure*, pp. 41–42.

8. *Ibid.*, pp. 43–44; Houston, *Western Mediterranean World*, pp. 33–35.

9. Ellsworth Huntington and S. S. Visher, *Climatic Changes* (New Haven, 1922), chap. vi.

10. Emmanuel LeRoy Ladurie ("Histoire et climat," *Annales*, 14 [1959], 33) rejects the anecdotal method in historical climatology in favor of more

accurate data such as dates of harvests, growth periods, and similar evidence. Nevertheless, prayers for rain can at least give a rough graphic indication of concern, useful when other data are lacking. For this method see Emilio Giralt, "En torno al precio del trigo en Barcelona durante el siglo XVI," *Hispania*, 18 (1958), 48–49. Most years in sixteenth-century Barcelona had 65 days or more when such prayers were recorded.

11. See Appendix 6, Table A, nos. 23–29.

12. See *ibid.*, no. 31.

13. AMV, Cartas Misivas, 11, fol. 207r: "Gran necessitat nos constrem axi de fretura e distret daygues com de minua de farines per proviso de la ciutat maiorment car per lo present adveniment del senyor rey es obs siam provists en dar loch e manera de haure hi gran copia de farines. On vos pregam . . . vos plaria dar o lexar nos laygua de la cequia de muncada dema en tota la nit seguent sots condicio e retencio que sens per ço no us sia feyt prejuhi als privilegis e libertats de la dita cequia."

14. ARV, Gobernación, 2202, 11th hand, fol. 1r.

15. AMV, Cartas Misivas, 11, fols. 210–211r: "Honrat senyor: Per habundar aquesta ciutat de farines de les quals hauem per la poquea de laygua del riu trametem los portadors de la present als molins daquexes parts per molre forment propi daquesta ciutat per que us pregam affectuosament que continent farats donar loch e manera de molre en los vostres molins lo forment quey sera portat per tal que sens triga puxam haure habundament de farines e non façam enuig als hereters de laygua del riu."

16. ARV, Gobernación, 2202, 12th hand, fol. 30r.

17. ARV, Gobernación, 2203, 28th hand, fol. 17r.

18. AMV, Manuals de Consell, 25, fol. 196r.

19. AMV, Cartas Misivas, 11, fol. 233r: "Sobre lo fet de les aygues trametem a vosaltres en Pere Colomines veguer nostre be informat de nostra intencio; pregam vos que lo donets fe e creença sobre tot ço que us dira toquant lo dit fet de les aygues e complirho per obra ab acabament e sera cosa de quens farets gran plaer."

20. AMV, Manuals de Consell, 25, fol. 202r.

21. The jurates' letter to Segorbe is not extant; its contents are reiterated in Segorbe's reply on June 16 (AMV, Protocolos de Jaime Dezpla, 25, loose sheet): "Rebuda havem vostra letra entre altres moltes coses continent que novellament vos seria stat a entendre que nosaltres donaven obra a algunes novitats sobre la particio de les aygues no seruants la forma antigada e usitada en temps passat, en tant que donavem obra a la despoblacio e destruccio dels lochs de Altura e de les Alcubles e encara del monastir de Vall de Crist qui son vostres ciutadans e vassalls en certs casos."

22. *Ibid.* The letter goes on to discuss relations between the cequiers of Altura and Segorbe and whether Valencia has any jurisdiction in the matter.

23. AMV, Manuals de Consell, 25, fol. 237r.

24. AMV, Cartas Misivas, 12, fol. 9v: "Mossen lo prior: de gran necessitat e cuyta hauem obs aquest vostre frare qui senten de livellar e traure aygues, on vos pregam affectuosament quel nos trametats de fet en tal manera que dema ans de dinar sia açi ab nosaltres."

25. ARV, Gobernación, 2202, 16th hand, fol. 28r: "alguns açuts e cequies noves sien estades començades a fer e fetes en les ribes del Riu de Godalaviar e . . . ampliats los açuts e cequies contra forma antiga e en gran dampnatge e prejuhi de la dita ciutat . . . per tal cas laygua del dit riu per alla no puxa venir librement e integra a la dita ciutat e orta daquella." (This is contrary to fueros and privileges, notably that of James II, Aug. 1, 1318.)

26. AMV, Cartas Misivas, 12, fol. 14v: "Nosaltres hauem ordenat que tres deenes dels flaquers daquesta ciutat, de les quals es cap en Guillem Feltrer, flaquer, vaien a molre a vostre moli per ço que la dita ciutat sia de farines abondada de les quals per lo gran destret que es daygues en lo nostre riu hauem fretura."

27. *Ibid.*, fol. 15v.

28. *Ibid.*, fol. 16r: "Per la gran fretura de aygues que segons sab vostra noblea hauem en nostre riu some estats en gran necessitat de farines, maiorment es aquesta necessitat en los flaquers daquesta ciutat del pa o pastor dels quals viu gran part del poble de la ciutat dessus dita, per que affectuosament pregam vostra savia noblea que placia a aquella donar a tots nostres flaquers los quals divertiran a vostre moli una mola qui don prest espatxament de farines als dits flaquers."

29. The population of Valencia in 1418 was 8,000 hearths; see Chap. VI, n. 53.

30. AMV, Cartas Misivas, 12, fol. 16r:

> Sabut hauem com per ordinacio vostra son estades exemplades les goles dels açuts del vostre riu de que e per la pocha aygua de aquell los molins no poden molre ne lavorar axi com seria obs. E com segons creem sabets nosaltres siam constituits en gran penuria de aygues, loat ne sia ihu x̄st, en tant que no podem hauer farina don se puxa proveir tanta multitut de poble lo qual per gracia divinal habita en la ciutat ne puxam bastar a proveir los qui ostegan per lo senyor rey a Bunyol per fretura de farina. Per ço vostra bona amistat pregam affectuosament . . . que us placia sens tot vostre prejuhi almenys defferint al temps present e a nosaltres per la justant necessitat proveir que les dites goles se puxen en tal manera artar . . . que si puxa molre molt forment del qual . . . entenem trametre a coll de besties en aquexa part.

31. *Ibid.*, fol. 23v: "Si us ne venits som encapats car açi es molta la pressura que passam per fretura de farines per penuria daygua. E com no

siats aqui nos faran rescatarla, molra o no molran [sic], per que us pregam que almenys la present setmana acurets aqui provehint, ordenant e disponent, axi com aquell qui havets comunio o participi de nostra congoxa e pressura en manera ques puxam deffendre de brogit o dar qualque remey e mitigacio a les gents."

32. AMV, Manuals de Consell, 25, fol. 248v.

33. AMV, Cartas Misivas, 12, fol. 24v: "Al molt honrat en ffrancesch çavall, batle de la baronia de oliva e de rebollet."

34. *Ibid.,* fols. 24v–25r.

35. AMV, Manuals de Consell, 25, fols. 252r–253r.

36. *Ibid.,* fol. 258v.

37. *Ibid.,* fols. 258v–260r.

38. *Ibid.,* fol. 258v: "per esguard de la gran fretura e steleritat [sic] e destret que de present eren daygues en lo riu de guadalaviar per ço car no podien haure aygues per obs de molre lurs blats de que los vehins e habitadors de la dita ciutat per molre ab rescats e extorcions indebites hoc encara los havents terres vinyes e possessions no podents haure aygues perdessen llurs esplets en notori dan llur e de la cosa publica de la dita ciutat e orta daquella don havia gran rumor querela e brogit en la dita ciutat e orta."

39. *Ibid.,* fols. 260r–261r.

40. The text of the letters is in *ibid.,* fol. 261r–261v. The Count of Luna's is dated July 12, 1413, Pere de Moncada's May 26, 1413. Both are in Latin and both set fines for incurring the royal wrath by disturbing rights of the 3 villages to water from the river. Obviously the Pueblos Castillos had planned their resistance carefully for several months.

41. *Ibid.,* fol. 261v: "E com segons la tenor e narracio de aquelles [the letters], aquelles sien vistes surrepticies e tacite veritate impetrades e acaptades donants los acaptants entendre al senyor rey que son en possessio de ço que allys conte ço que cessa esser ver parlant ab humil subieccio del dit senyor, per tal que haud acord farien ço que fossen tenguts fer."

42. *Ibid.,* fols. 262r–266r; Jacme Serra is listed as present at this meeting.

43. The King had besieged the rebellious Count of Urgel, who had invaded Aragón with promises of help from the English (Chaytor, *A History of Aragon,* p. 208).

44. AMV, Cartas Misivas, 12, fols. 35v–37r.

45. AMV, Manuals de Consell, 25, fols. 266v–267r.

46. *Ibid.,* fols. 267r–268r.

47. *Ibid.,* fols. 268r–269r. The same practice is followed now, except that the decision to implement the tanda is taken by the *acequieros* rather than the jurates. The *acequieros* decide on Thursday whether or not a turn is necessary; if so, they inform the Pueblos Castillos that the turn is in effect, and the city's first day therefore falls on Monday.

48. Some flooding had occurred in October 1413, paradoxically having

the same effect on the mills as the drought: the rush of water in the river broke the diversion dams and no water was reaching the mills (AMV, Cartas Misivas, 12, fol. 62v, Oct. 14, 1413). By the end of November "penury of water" had returned (*ibid.*, fol. 83r, Nov. 22, 1413).

49. See Chap. VI, n. 10.

PART TWO: CULTURAL CONTINUITY IN IRRIGATION

CHAPTER VIII. ARABS, ROMANS, OR CHRISTIANS? THE HISTORIOGRAPHY OF SPANISH IRRIGATION IN THE NINETEENTH CENTURY

1. The *Claridad de la acequia de la villa de Elche,* partly a recapitulation of customs and ordinances and partly a description of the physical characteristics of the irrigation system of Elche, is reprinted in Ibarra y Ruiz, *Riego de Elche,* pp. 102–145.

2. It is peculiar that Ortiz derived Palombar (from the Latin *palumba,* dove) from the Arabic when he correctly established the etymologies of the other 3 names: Atufá (Arabic *tufāḥa,* apple), Alingasa (Arabic *injāṣ,* pear), and Alausa (Arabic *lauz,* almond). Some 3 centuries later Markham and Badger (see nn. 34–39 below) fared considerably less well, associating Atufá with *ṭufūḥ,* fullness, and Alausa with *ladhdha,* delight, enjoyment. Paradoxically Ortiz, who was closer by far to the true etymologies than was Markham, rejected Islamic influence categorically, while Markham accepted it uncritically.

3. Ibarra y Ruiz, *Riego de Elche,* pp. 105–106.

4. *Ibid.,* p. 108. Confusion of the Arabic article *al-* with Romance *al* (to the [one]) is a common component of popular etymologies. Compare the derivation of the sales tax, *alcabala,* from *al que vala.*

5. *Teatro crítico universal,* 9 vols. (Madrid, 1726–1740), VIII, disc. 12.

6. Francisco Javier Borrull y Vilanova (1745–1838), secretary of the Holy Office, professor of law at the University of Valencia, named to Royal Judgeship of Tithes by Charles III, elected deputy to the Cortes of Cádiz (1810–1813) where, on July 31, 1812, he petitioned for the survival of the Tribunal of Waters under the terms of art. 278 of the Constitution of 1812, whereby the Cortes was permitted to authorize the preservation of certain special tribunals even though art. 148 had outlawed private jurisdictions. On July 31, 1813, he spoke again, giving a short history of the Tribunal (*Discurso sobre la distribución de las aguas del Turia y deber conservarse el Tribunal de los Acequieros de Valencia* [3rd ed., Valencia, 1828]. This speech was the basis of the *Tratado* of 1831 (see above, Chap. III, n. 17). Fernando VII renamed him to his old judgeship, in which office he served

both before and after his exile to San Sebastian during the Trienio Liberal (see Almela y Vives, *Valencia y su reino,* pp. 385–386).

7. Borrull, *Tratado,* p. vii.

8. *Ibid.,* chap. iii.

9. *Ibid.,* chap. iv. On p. 106 Borrull cites Cavanilles' description of the aqueduct of Chelva.

10. *Ibid.,* chap. v.

11. *Ibid.,* chap. vi, p. 113: "It would be easy to find out which princes ordered this admirable distribution of the water of the Turia, the minister in charge of it, the architect who conceived the plan and who carried it into effect, if during the lifetime of Cardinal Giménez more than 80,000 Arabic volumes had not been burned, and thereafter several more in the fire of the Royal Monastery of the Escorial." The facility with which Borrull projected back into time the contemporary modus operandi of public works projects is remarkable.

12. *Ibid.,* pp. 114–133.

13. *Ibid.,* pp. 133–134.

14. *Ibid.,* p. 136; but see pp. 137–139, where Borrull discusses intelligently some of Conde's errors. A much cited passage in Conde deals with agricultural improvement during the reign of al-Ḥakam II: "During the long peace which King Alhakem [sic] maintained, agriculture was encouraged in all the provinces of Spain. Irrigation canals were built in the *vegas* of Granada, Murcia, Valencia, and Aragon. Lakes or *albuheras* [sic] for irrigation were constructed, and crops of all kinds were grown according to the nature and climate of the provinces" (*Historia de la dominación de los árabes en España,* 3 vols. [Madrid, 1820], I, 487). No indication of the sources is given. Conde's methodology of careless synthesis was well known to nineteenth-century scholars. Markham, in the bibliography to his *Report,* called Conde's work "a confused mass of translations from Moorish chronicles" (p. 101; see also p. 75).

15. The millennial celebration was held in November 1960. The featured speaker at the commemorative *acto* on November 13 was Minister of Public Works Jorge Vigón, who spoke on the recent history of water confederations. In an interview in *Las Provincias* (Valencia), Nov. 12, 1960, Vicente Giner Boira, who had organized the celebration, was asked how it was known that the Tribunal had been founded 1,000 years ago. He replied: "Scholars believe that it was founded during the reigns of ʿAbd al-Raḥman III and al-Ḥakam II around the year 960. Based on this theory, we celebrate it today." See also *Las Provincias* and *Levante* (Valencia) for Nov. 11–13, 1960, and *Hoja del Lunes* (Barcelona), Nov. 14, 1960.

16. For Jaubert de Passa's biography, see Chap. I, n. 2.

17. Jaubert translated Borrull's *Discurso* and inserted it in his *Voyage en Espagne . . .* (Spanish trans., *Canales de riego:* I, 481–487). Borrull wrote (*Tratado,* p. v) that Jaubert in his book "makes clear the epistolary

correspondence which he had with me, the different notices which I gave him with which to illustrate his writings and [he shows himself] to conform to my hypothesis concerning the ordering of both establishments [i.e., the canals and the Tribunal] by the Saracens."

18. *Voyage en Espagne dans les années 1816, 1817, 1818, 1819, ou recherches sur les arrosages,* 2 vols. (Paris, 1823).

19. *Canales de riego* (see above, Chap. II, n. 1). The Sociedad Económica de Amigos del País of Valencia, recognizing "the great merit of the French work, which extolled the state of our agriculture so highly, particularly in regard to irrigation," offered a prize for the best translation. The prize was won in 1835 by the magistrate Juan Fiol, but funds were lacking for immediate publication. When finally published in 1844, Jaubert's original text constituted vol. I; vol. II consisted of two distinct subvolumes, separately numbered. The first contains several monographs, including one by Juan Roca de Togores y Albuquerque, "Memoria sobre los riegos de la huerta de Orihuela" (1st fasc., pp. 5–150), which won the prize of the Sociedad Económica of Valencia in 1831. On p. 37 he cited Borrull's theory on the origin of the Turia system during the reigns of ʿAbd al-Raḥman and al-Ḥakam. Other monographs are those of Pedro de Lara y Meliá, "Memoria sobre la huerta de Gandia, sus riegos y productos" (1st fasc., pp. 151–205), which won a gold medal in the Sociedad's competition for 1831, and F. de P. A., "Memoria y plan sinóptico de las acequias del río Turia" (1st fasc., pp. 207–228). The second part contains the *Reglamentos y ordenanzas* of the huerta canals.

20. *Canales de riego,* I, 4, 8. Ibn al-ʿAwwām was the most widely read of the Andalusí agricultural writers because of the translation into Spanish by J. A. Banqueri (*Libro de agricultura* [Madrid, 1802]).

21. *Canales de riego,* I, 16.

22. *Ibid.,* p. 390.

23. *Ibid.,* p. 391.

24. *Ibid.,* pp. 470–471 n. 1.

25. Richard Ford, *A Hand-book for Travellers in Spain,* 2 vols. (London, 1845), I, ix.

26. *Ibid.,* pp. 430–431.

27. See *ibid.,* p. 360.

28. Compare E. W. Lane, *An Account of the Manners and Customs of the Modern Egyptians,* 2 vols. (London, 1846), I, 31. Ford used the first edition of Lane (1836) and followed his spellings of *shadoof* and *sackiyeh.*

29. For Markham's biography see above, Introduction, n. 1.

30. Vicente Boix y Ricarte (1813–1880). Boix's views of irrigation history were Borrull's exactly. See *Apuntes históricos sobre los fueros del antiguo reino de Valencia* (Valencia, 1855), pp. 102–103: "El Rey D. Jaime I halló ya concluídas en 1238 las grandes obras de esta hermosa canalización, que deben sin duda atribuirse al gobierno de los árabes Abderrahman–Anasir–

Ledinala [sic] y Alhaken Almostansir Bilah, su hijo, por los años 911 al 976." The "sin duda" is indicative of his unquestioning acceptance of the Borrull thesis.

31. Markham, *Report,* pp. 93–94. Pedro Salvá y Mallen (d. 1860), a bibliophile, was the son of the philologist and book collector Vicente Salvá y Pérez, whose collection he catalogued (*Catálogo de la biblioteca de Salvá,* 2 vols. [Valencia, 1872]). Doubtless under Salva's aegis — perhaps in his home — Markham perused the tomes listed in the bibliography of his *Report,* pp. 101–104. The list includes Conde's *Historia de la dominación de los árabes en España,* Banqueri's translation of ibn al-ʿAwwām, Rafael de Mancha's *Memoria sobre los riegos de la huerta de Murcia* (Murcia, 1836), Borrull's *Tratado,* Jaubert de Passa, and Cirilo Franquet y Bertrán's *Ensayo sobre el origen, espíritu, y progresos de la legislación de las aguas,* 2 vols. (Madrid, 1864).

32. *Report,* pp. 6–7 and n.

33. *Ibid.,* p. 49.

34. Sir Albert H. Markham, *The Life of Sir Clements R. Markham* (London, 1917), p. 168.

35. George Percy Badger (1815–1888), a scholar of Arabic and author of *An English-Arabic Lexicon* (London, 1881).

36. *Report,* pp. 49, 51.

37. *Ibid.,* p. 50. Markham was not a careful translator, even from Spanish. See Harry Bernstein and Bailey W. Diffie, "Sir Clements R. Markham as a Translator," *Hispanic American Historical Review,* 17 (1937), 546–557.

38. A similar line of argument was used by the Marqués de Molins, Mariano Roca de Togores y Carrasco (1812–1889) in his discourse upon reception into the Academy of History, June 29, 1869, entitled "Antigüedades de Elche," reprinted in his *Discursos académicos,* vol. I (Madrid, 1890), pp. 261–395. "Of the twenty-four *partidores* or channels through which the water runs," he argued, "few lack an Arabic name" (p. 334); one name so identified is Forat. He listed the names (pp. 335–336) but made no attempt to establish etymologies for any but Alingasa (Alintjasa), Atufá, and Alausa, which he must have known from the *Claridad* of Ortiz and which, strangely, Badger had missed (see n. 2, above).

39. *Report,* p. 90 (italics mine).

40. Aymard's informants in Valencia were mostly technicians, including José Ortega (chief engineer of the province), Jaime Zacares (secretary of the Tribunal of Waters), and Juan Vallejo (president of the General Sindicate of the Turia).

41. Aymard, *Irrigations du midi de l'Espagne* (see above, Intro., n. 2), p. 11.

42. Loaysa made a survey of the irrigation canals of Granada in the 1570's.

43. *Irrigations,* pp. 269–271. Translation adopted from that of W. H. Hall, *Irrigation Development* (see above, Intro., n. 3), p. 489.

44. Aymard's attitude toward the irrigation system of Granada — or lack of system, as it were — is similar to René Tresse's final judgment of the Ghūṭa of Damascus: "une situation aussi anormale," resulting from the operations of cleaning and repair having been left for many centuries to the direction of peasants, without any qualified administration. To Aymard and Tresse, who reflected French attitudes toward administration, a "normal" situation was a controlled one, especially control involving a central authority (Tresse, "L'Irrigation dans la Ghouta de Damas," p. 551). Aymard was aware of similarities between Spanish and North African water measures (*Irrigations,* p. 24 n. 1).

45. *Huerta de Murcia,* pp. 157–158.

46. *Ordenanzas y costumbres,* pp. 54–55.

47. Fontavella, *Huerta de Gandia,* fig. 12.

48. *Influencia que ejerció la dominación de los árabes en la agricultura, industria y comercio de la provincia de Castellón de la Plana* (Castellón, 1889).

49. *Ibid.,* pp. 40, 63; see also p. 44: the poverty of agricultural lore in the *Etymologies* of Isidore of Seville and the *Forum Iudicum* is adduced to support the proposition that huerta cultivation was rare in Roman and Visigothic Spain.

50. *Ibid.,* pp. 45–54, 72–92. Ford also mentioned plants introduced by the Arabs; see *Hand-book,* I, 431 (rice), 432 (carob).

51. *Influencia,* p. 92.

52. In an appendix (pp. 109–113), selections from the *Repartimiento* and other donations of James I are reproduced to substantiate the impression of an intensive irrigation regime at the moment of conquest: proof is in the presence of key words such as *ortum* (the assumption being that an *ortum,* orchard, must be irrigated), *vineam* (vineyards were customarily irrigated in medieval Valencia), *regadiyo,* mentions of canals, etc.

53. Their references (*Influencia,* p. 104 n. 2) are to historians of Islamic Spain like Conde and Viardot and to Antonio Garcia Maceira's pamphlet *Apuntes y noticias sobre la agricultura de los árabes españoles* (Madrid, 1876).

54. *Influencia,* pp. 105–106.

55. *Ibid.,* pp. 107–108.

56. *El archivo* (Denia), 4 (1890), 190.

57. *Ibid.,* p. 191. Chabás's appraisal of the changeover after the reconquest is curious in view of the fact that several years later he himself published a document proving that the Christian knights actually summoned the Muslim irrigation officials and interrogated them about the custom: *Distribución de las aguas en 1244 y donaciones del término de Gandia por D. Jaime I* (Valencia, 1898).

58. "Mons parturiens," *El archivo,* 2 (1887–1888), 232–235.

59. *Orígenes del justicia de Aragón* (Zaragoza, 1898).

60. The theory is expounded in *Origenes,* pp. 250–263. For a discussion, see Thomas F. Glick and Oriol Pi-Sunyer, "Acculturation as an Explanatory Concept in Spanish History," *Comparative Studies in Society and History,* 11 (1969), 145.

61. *Origenes,* pp. 300–301.

62. *Ibid.,* pp. 37–38.

63. In 1906 Ribera wrote: "At the present I still consider definitive that which I then discovered concerning the origins of the justiciar; I still believe that the principle of imitation pervades human actions; but I have completely abandoned the vain notion that history might be converted into science by such a method" (*Lo científico en la historia* [Madrid, 1906], p. 62).

64. "Sistema de riegos" (see Chap. III, n. 22). The article originally appeared in *Almanaque "Las Provincias"* for 1908.

65. Ribera, "Sistema de riegos," p. 310.

66. *Ibid.,* pp. 311–312; see Chap. III, n. 22.

67. *Ibid.,* pp. 312–313. Bellver and Cacho reversed the proposition: intensification of cultivation was more the cause of increased population than the effect.

68. *De historia arábigo-valenciana* (Valencia, 1925), p. 16.

69. *Pallantia, (vulgo) Valencia la vieja* (Vinaroz, 1902).

70. Valls may have had in mind either Jaldero (*Pallantia,* p. 166) or Augusto Danvila, who had described walking toward Ribarroja and observing "los negruzcos machones del acueducto que, allá por el siglo III antes de Jesucristo, proveía de agua el *castrum* romano . . ." ("Las ruinas de Pallantia," *Almanaque "Las Provincias"* [1890], p. 212).

71. *Pallantia,* pp. 117–118. Even some historians have accepted uncritically the oral tradition as fact. For example, the Marqués de Molins, "Antigüedades de Elche," p. 333, says that "in order to prove [the Arab origins of the irrigation system] it is not necessary to make profound linguistic studies, dust off documents, or decipher inscriptions. The rudest laborer, when asked the name of the estate he works, the district where he lives, and of the canal from which he irrigates, will give you testimony of our assertion." Compare attitudes in Syria (Chap. IX, n. 73, below) and in Latin America, Fernando Horcasitas, "La mentalidad difusionista popular," *Anales de antropología* (Mexico), 6 (1969), 153–167.

72. *Pallantia,* p. 123.

73. *Ibid.,* pp. 26, 180.

74. *Oriental Despotism,* p. 215.

75. See *ibid.,* p. 29, and Forbes, *Studies in Ancient Technology,* II, 6.

76. Administratively the Reconquest was more Oriental than Wittfogel imagines. The massive land inventories known as the Libros de Repartimiento were a typical agromanagerial instrument, with more clearly Islamic antecedents than those Wittfogel posits for the Domesday Book (*Oriental*

Despotism, pp. 213–214; the early date of the Domesday Book makes any connection with Islam, via Sicily, quite improbable, whereas the Repartimiento of Mallorca was redacted both in Romance and Arabic).

77. *Oriental Despotism,* pp. 216–219: Wittfogel attributes the later strength of Spanish absolutism more to responses to the frontier situation than to submissiveness to authority instilled in the Oriental style. But the more usual interpretation of the Spanish frontier was that it had a leveling effect and was conducive to democratic or decentralized institutions rather than absolutist ones.

78. "Hydraulic Society in Ceylon," p. 5.

79. *Ibid.,* p. 13. Even when archeological evidence shows that such works are developed over a long period of time, local myth "invariably attributes their construction to the initiative of a single outstanding monarch." A great part of the literary evidence regarding ancient irrigation (e.g., Hammurabi's direction of irrigation projects) must therefore be taken with more than just a grain of salt.

80. *Ibid.,* p. 14.

CHAPTER IX. THE CLASSICAL INHERITANCE

1. In the spread of technology from the ancient Near East both eastward and westward, Islamic culture has a focal character. See Joseph Needham, *Science and Civilisation in China,* vol. I (Cambridge: Cambridge University Press, 1954), pp. 220–223. In the spread of technologies westward North Africa had a particularly focal role. First, the southern coast of the Mediterranean was the route by which Islam spread from the East to Morocco and Spain in the West; second, northern Africa acted as a great storehouse of pre-Islamic agricultural techniques upon which the Muslims later drew, using them to maximum advantage (see Houston, *Western Mediterranean World,* p. 118).

2. For the history of the sinking bowl, a device used for measuring irrigation time, see Thomas F. Glick, "Medieval Irrigation Clocks," *Technology and Culture,* 10 (1969), p. 425.

3. Henri Goblot, "Dans l'ancien Iran: les techniques de l'eau et la grande histoire," *Annales,* 18 (1963), 514–516. Goblot, however, is more interested in the antecedents of storage dams.

4. On the *sudds* of South Arabia, see Adam Mez, *The Renaissance of Islam* (Patna, 1937), p. 451: "in S. Arabia, where the object was to collect running water for use, they had pools (*masani'*) bordered with pebbles; but farther up the mountains (as in San'a) they had dams (*sadd*), with openings below; the water was distributed by means of canals. This method was such a specialty of S. Arabia that ibn Rustah finds it necessary to explain the word." Ibn Rustah's definition of *sudd* can be found in *Les atours*

précieux, trans. Gaston Weit (Cairo, 1955), p. 126: "Cette digue consiste en un barrage établi au débouché des montagnes en tourant le périmètre de ces domaines, et en bas duquel on a ménagé des orifices." See also al-Hamdanī, *The Antiquities of South Arabia* (*al-Iklīl*), trans. Nabih Amin Faris (Princeton, Princeton University Press, 1938), pp. 67–69: "Of the dams of al-Yaman which are of Ḥimyarite origin." For modern diversion dams in South Arabia see Etore Rossi, "Note sull'irrigazione, l'agricoltura e le stagioni nel Yemen," *Oriente moderno,* 33 (1953), 349; and R. B. Serjeant, "Some Irrigation Systems in Ḥaḍramawt," *Bulletin of the School of Oriental and African Studies* (London), 27 (1964), 54.

5. Goblot, "Techniques de l'eau," p. 515. Concerning the Marib dam see Adolf Grohmann, "Ma'rib," in the *Encyclopaedia of Islam,* 1st ed., 4 vols. (Leiden: E. J. Brill, 1913–1934), III, 286–291.

6. Richard LeBaron Bowen takes great pains to show that Marib was a purely diversionary dam which could not have been used for storage: "Irrigation in ancient Qatabān (Beiḥān)," in R. L. Bowen, Jr., and Frank P. Albright, *Archaeological Discoveries in South Arabia* (Baltimore: The Johns Hopkins Press, 1958), p. 72. A typical Islamic way of storing water was in cisterns, which supplied *qanats,* wells, or *sudd*-canal systems; see Marcel Solignac, "Recherches sur les installations hydrauliques de Kairouan et des steppes tunisiennes du VIIe au XIe siècle," *Annales de l'Institut des Etudes Orientales* (Algiers), 10 (1952), 168.

7. Solignac, "Recherches," 11 (1953), 61.

8. The two tributaries of the Guadaletin — the Luchena and the Velez — run parallel for about 8 km.

9. *Le péninsule ibérique au moyen-âge d'après le Kitāb ar-Rawḍ al-mi'ṭār fī habar al-akṭār d'Ibn ᶜAbd al-Munᶜim al-Ḥimyarī,* ed. and trans. E. Lévi-Provençal (Leiden, 1938), pp. 172 (Arabic), 207 (trans.); hereafter cited *Rawḍ al-miᶜṭār.* The Hispano-Arab writer al-ᶜUdhrī also mentions the raising function of diversion dams; *Fragmentos geográfico-históricos de al-Masālik alā gamīᶜ al-Mamālik* (Madrid, 1965), p. 2.

10. John Q. Ressler, "Indian and Spanish Water-Control on New Spain's Northwest Frontier," *Journal of the West,* 7 (1968), 13–14; Wells A. Hutchins, "The Community Acequia: Its Origin and Development," *Southwestern Historical Quarterly,* 31 (1927–1928), 275; N. J. Schnitter, "A Short History of Dam Engineering," reprinted from *Water Power,* April 1967, p. 6.

11. B. Laufer, "The Noria or Persian Wheel," in *Oriental Studies in Honour of Cursetji Erachji Pavry* (London, 1933), p. 250. Laufer thought the original engineer an Iranian, "probably a Sogdian. The designation 'Persian wheel' is valid. It cannot be fortuitous that this name appears in Europe, in Bukhara, as well as in India." But Joseph Needham prefers an Indian origin; *Science and Civilisation,* vol. IV, pt. 2 (Cambridge, 1965), p. 361

12. Needham, *Science and Civilisation,* IV, pt. 2, 360; Julio Caro Baroja, "Norias, azudas, aceñas," *RDTP,* 10 (1954), 35–36 (citing Strabo and Pliny).

13. Caro Baroja, "Norias, azudas, aceñas," p. 154.

14. For a typology of lifting devices, see the helpful table in Forbes, *Studies in Ancient Technology,* II, 38–39. Strictly speaking, a *noria* is moved "by the force of the river alone." The wheel powered by animals is the *sāniya,* but in Spanish usage (both medieval and modern) the two are confused; in the Catalan zone *sinia* became a general term, referring to all hydraulic wheels used for irrigation.

15. Caro Baroja, "Sobre la historia de la noria de tiro," *RDTP,* 11 (1955), 53.

16. Caro Baroja, "Sobre cigüeñales y otros ingenios para elevar agua," *Revista de Guimarães,* 65 (1955), 162. This primitive device was doubtless known in pre-Roman Spain; see Jorge Dias and Fernando Galhano, *Aparelhos de elevar a água de rega* (Porto, 1953), p. 231.

17. R. Capot-Rey, *Le Sahara Français* (Paris, 1953), pp. 323–324 and map 7; G. S. Colin, "La noria marocaine," *Hespéris,* 14 (1932), 23. The desert environment, lacking in wood and other materials necessary to the construction of water wheels, was an effective barrier limiting the *noria* to north Saharan and pre-Saharan regions.

18. Caro Baroja, "Norias, azudas, aceñas," p. 119; Leopoldo Torres Balbás, "Las norias fluviales en España," *Al-Andalus,* 5 (1940), 206.

19. Caro Baroja, "Noria de tiro," p. 179; some of Caro's arguments were anticipated by Bellver and Cacho (*Influencia,* p. 53) and Jaubert de Passa (*Canales de riego,* chap. iii, "Norias de la alta Cataluña," I, 28–31).

20. See Dan Stanislawski, *Portugal's Other Kingdom: The Algarve* (Austin: The University of Texas Press, 1963), pp. 59–61; Dias and Galhano, *Aparelhos,* pp. 168–169. Compare map 5 (*ibid.,* p. 168, desenho 70) with that showing the more northerly concentration of the *shadūf* (Portuguese *picota, cegonho*) in Portugal (*ibid.,* p. 173, fig. 71).

21. Juan Torres Fontes, ed., *Repartimiento de Murcia* (Madrid, 1960), p. 202. Some grants cite the land as having "riego de acenia" (pp. 189, 202). Others indicate irrigation with a *shadūf* ("se riega de çegunnal," p. 197) and with a *dāliya,* a scoop balanced by the human body ("la açequia que se riega de alhataras," p. 202). For definitions of these devices, see Forbes, *Studies in Ancient Technology,* II, 39.

22. Martínez Morellá, *Cartas de los reyes de Castilla a Orihuela,* p. 50.

23. "Repartimiento de Valencia," p. 326: "Predicatores: I rotam tantum in illa cequia que vadit ad molendina B. de Turolio et possint eam tantum elevare ita ut possint abstrahere unum bonum riguarium de aqua de die et nocte continue et semper et [. . .] aliquid in aptanda vel mudanda cequia."

24. See Casas Torres, *Huerta de Valencia,* pp. 95–98 and figs. 8–9; Fontavella, *Huerta de Gandia,* p. 92 and photo 24.

25. Georges S. Colin, "L'origine des norias de Fès," *Hespéris*, 16 (1933), 156–157. Concerning the great *noria* of Toledo see Leopoldo Torres Balbás, "La Albolafia de Córdoba y la gran noria toledana," *Al-Andalus*, 7 (1942), 464–465.

26. Only a few Ibero-American *norias* have been described, but these are found in such widely separated places as the lower part of the Sao Francisco River (Pernambuco, Brazil) and the Mexican state of Zacatecas; Caro Baroja, "Norias, azudas, aceñas," p. 149 n. 295; George M. Foster, *Culture and Conquest* (Chicago: Quadrangle, 1960), pp. 63, 69 n. 9. There is a reconstructed *noria* at Mission San Antonio de Padua, Jolon, California (Ressler, "Indian and Spanish Water-Control," p. 17 n. 17).

The distribution of *noria* place names (La Noria, Las Norias, Noria) in Latin America makes clear the diffusion of the technique in Spanish colonial times. The U.S. Board on Geographic Names gazeteers yield 5 such names in Chile, 1 in El Salvador, 2 in the Dominican Republic, and 143 in Mexico (as compared with 16 in Spain and 8 in Portugal). Whether all these names actually correspond to *noria* sites is doubtful. Further study is needed and is no easy task, inasmuch as electrification has put an end to many *norias* (in many parts of Latin America the term has passed into local speech as a synonym of "well"). From La Noria Street in Acapulco to Norias in Kenedy County, Texas, and Noria, New Mexico, however, the place names bear witness to the device's wide diffusion in colonial Mexico. Names such as La Noria del Burro (24° 02′ north, 103° 28′ west) leave no doubt as to their origin, and Professor Angel Palerm has told me that he has observed wells in El Bajío which could only have served for *norias* (personal communication, Apr. 30, 1968). Humboldt and other travelers commented on their frequent use.

27. The *qanat* literature is extensive; for works that primarily discuss diffusion, see Paul Ward English, "Qanats in the Old World," *Proceedings of the American Philosophical Society*, 112 (1968), 170–181; Goblot, "Techniques de l'eau;" J. Humlum, "Underjordiske vandingskanaler: kareze, qanat, foggara," *Kulturgeografi*, 90 (1965), 81–131, and an earlier version, "Karezes: su construcción, funcionamiento y extensión en el viejo mundo," 20th International Geological Congress, Mexico, 1956, *Publicaciones*, 4 (Mexico, 1957), pp. 155–171; Carl Troll, "Qanat-Bewässerung in der alten und neuen Welt," *Mitteilungen der osterreichischen geographischen Gesellschaft*, 105 (1963), 314–330.

28. English, "Qanats in the Old World," p. 175; Goblot, "Techniques de l'eau," p. 501; Forbes, *Studies in Ancient Technology*, I, 154; II, 11–12.

29. Solignac, "Recherches," 10 (1952), 60; C. Germain de Montauzan, *Les aqueducs antiques de Lyon* (Paris, 1908), pp. 284–290.

30. English, "Qanats in the Old World," p. 177; Goblot, "Techniques de l'eau," p. 513; Solignac, "Travaux hydrauliques hafsides de Tunis," *Revue africaine*, 79 (1936), 556, 561; Jaime Oliver Asín, *Historia del nombre*

"Madrid" (Madrid, 1959), pp. 85–86 (*foggaras* in Algeria), 86–89 (*hattaras* in the Tafilalet and in Marrakesh), 359–367 (summarizing Solignac plus additional data on Ifni); Colin, "La noria marocaine," pp. 37–40.

31. There were other Iberian *qanat* systems in the Middle Ages (see n. 33, below). Al-Ḥimyarī mentions two subterranean canals to the north of Murcia provided with breathing wells; *Rawḍ al-Miʿṭār,* p. 220. Oliver Asín (*Historia del nombre "Madrid,"* pp. 140–141 n. 2) found vague references to probable *qanat* systems in the valley of Sidonia (Cádiz) and in the province of Seville. For similar works in Catalonia and the Canary Islands, see Humlum, "Kareze, qanat, foggara," pp. 166–168.

32. Oliver Asín, *Historia del nombre "Madrid," passim.*

33. The geographer Chris Field made a survey of Spanish irrigation sites in 1968–1969. He found the Crevillente system locked up, but the guard's description was similar to that of Cavanilles (personal communication, Apr. 22, 1969). Field also found 4 other filtration systems in the vicinity of Crevillente, and others at Rodalquilar, Puerto Lumbreras, and Huercal Overa.

34. Cavanilles, *Historia natural,* II, 275–276. Sir Clements Markham (*Report,* pp. 38–43) also described the *qanats* of Crevillente, comparing them to those of Peru.

35. Oliver Asín, *Historia del nombre "Madrid,"* pp. 137–140, emending Colin, "La noria marocaine," pp. 38–39.

36. All the New World *qanat* sites are problematical, however. The filtration galleries of Los Angeles are of Spanish origin, but, lacking wells, are not true *qanats*; George B. Cressey, "Qanats, Karez, and Foggaras," *Geographical Review,* 48 (1958), 44. The Tehuacan *qanats* of Mexico are almost certainly of Spanish origin, but Humlum believes that those of Parras (Coahuila) and Pica (Chile) may be pre-Columbian (Humlum, "Kareze, qanat, foggara," pp. 108–113). Troll, who discovered the Tehuacan system in 1954, believes that the purely Spanish vocabulary describing features of the water system indicates its Old-World origins; comment on Iwao Kobori, "Human Geography of Methods of Irrigation in the Central Andes," in *The Natural Environment of the Central Andes* (Paris: UNESCO, 1962), p. 137.

37. The story was originated by al-Masʿūdī; see S. Tolkowsky, *Hesperides: A History of the Culture and Use of Citrus Fruits* (London: J. Bale, 1938), pp. 123–124. For a Spanish echo, see Bellver and Cacho, *Influencia,* pp. 80–81.

38. Tolkowsky, *Hesperides,* p. 101.

39. *The Muqaddimah: An Introduction to History,* trans. Franz Rosenthal, 3 vols. (New York: Pantheon, 1958), II, 295.

40. Quoted by Tolkowsky, *Hesperides,* p. 114.

41. See above, Chap. I, nn. 58–63.

42. Tolkowsky, *Hesperides,* pp. 260–263.

43. "Only two percent of [today's] Spanish crop are Valencias, and perhaps only half of that comes from the groves of Valencia itself." Blood oranges and navels have now replaced the original Valencia in its native land; John McPhee, *Oranges* (New York: Farrar, Straus and Giroux, 1967), p. 13.

44. *El Mediterraneo y el mundo mediterraneo,* I, 216.

45. C.H.W. Johns, *The Oldest Code of Laws in the World* (Edinburgh, 1903), p. 14. See also G. R. Driver and John C. Miles, *The Babylonian Laws,* 2 vols. (Oxford: Clarendon Press, 1952–1955), I, 151 n. 3.

46. Driver and Miles, *Laws,* I, 152. Wittfogel (*Oriental Despotism,* pp. 268–269) treats such gatherings of elders as a civil jury, dealing with legal matters of a local nature, operating under the king's control. But insofar as implementing local irrigation procedures is concerned, the king's control would have been very indirect and indeed inconsequential regarding development of local institutions and customs of distribution and regulation.

47. The Babylonian corvée had some singularly undespotic elements: for example, 2 wage scales for canal excavation, one for the easier work near the surface and another for the harder labor of removing earth from deeper levels; O. Neugebauer and A. Sachs, *Mathematical Cuneiform Texts* (New Haven, 1945), p. 88.

48. Forbes, *Studies in Ancient Technology,* II, 43.

49. In Greece the practice of irrigation never attained universality, although drainage waters were commonly diverted into channels and made available for cultivation and small-scale techniques were widely appreciated, as Homer's description of a peasant clearing his ditch of obstructions and watering his orchard attests (*Iliad* 31.257–262). A. Jardé contrasts the necessity for cooperation that irrigation imposes upon the members of a society with the spirit of independence and self-sufficiency that characterized the ancient Greeks; *The Formation of the Greek People* (New York: Alfred A. Knopf, 1926), p. 42.

50. Ellen Churchill Semple, *The Geography of the Mediterranean Region: Its Relation to Ancient History* (New York: Henry Holt, 1931), pp. 434–436. Forbes (*Studies in Ancient Technology,* I, 162), in stressing the hygienic public and household uses of water in the classical world and, similarly, the dominance of dry-farming methods, may under-rate the role of irrigation.

51. On Roman irrigation in North Africa see Jean Baradez, *Fossatum africae* (Paris, 1949); J. Despois, "Development of Land Use in Northern Africa,' in L. Dudley Stamp, ed., *A History of Land Use in Arid Regions* (Paris: UNESCO, 1961), pp. 222–224; Semple, *Geography of the Mediterranean,* pp. 456–457; and M.-R. de la Blanchère, "L'aménagement de l'eau et l'installation rurale dans l'Afrique ancienne," *Nouvelles archives des missions scientifiques,* 7 (1897), 1–109. La Blanchère believed that North Africa retained more traces of Roman irrigation than any other place in

the Empire. The best summary of classical irrigation practices is his article "Fossa," in Charles Daremberg and Edmund Saglio, *Dictionnaire des antiquités grecques et romaines*, 5 vols (Paris, 1877–1919), II, pt. 2, 1321–1333.

52. See Semple, *Geography of the Mediterranean*, pp. 463–464; Adolf Schulten, *Geografía y etnografía antiguas de la península ibérica*, vol. II (Madrid, 1963), pp. 118–120.

53. Cavanilles, *Historia natural*, II, 81–83 (chap. xc) and author's drawing opp. p. 82.

54. Francisco de Paula Jaldero, *Monografía y planos de los antiquísimos canales existentes en el término de Ribarroja que vienen de Villamarchante y se pierden en Manises* (Valencia, 1853); and Domingo Fletcher Valls, "Acueductos romanos en término de Ribarroja del Turia (Valencia)," in *Notas sobre la antigüedad* . . . pp. 93–102.

55. Nicolau Primitiu, "D'arqueología," *ACCV*, 4 (1931), 156.

56. Solignac, "Recherches," 10 (1952), 133, and 11 (1953), 134. See 10 (1952), 145, on the different placement of archstones in Roman and Islamic aqueducts.

57. In discussing the ends for which Romans built canals in Gaul, Etienne Carpentier asserts that such canals were usually the work of one man and usually for milling rather than community irrigation (*Réglements et tribunaux des eaux*, pp. 7, 11, 14).

58. Tarradell, in *Història del País Valencià*, I (Barcelona, 1965), pp. 147–156. These pages on irrigation are an excellent commentary, bringing to bear a deep knowledge of Roman civilization upon a largely undocumentable problem. I disagree only with the statement that a collective complex of canals is more proper to the Roman mentality than to the Arab. Given the tribal structure of Arab society, I should think the opposite would more nearly be true.

59. José Llorca-Rodriguez, "Romanidad de los riegos de la huerta valenciana," in *Notas sobre la antigüedad* . . . pp. 105–114; NB photos pp. 107–108

60. Houston, "Urban geography of Valencia," p. 21.

61. *Digest* 43.20.1.22, "cottidie, alternis diebus . . . quinto die."

62. See also *Digest* 8.6.7, 43.20.5, and 43.20.5.1 (whereby 2 parties conveying water in the same channel separately at certain hours — "certis horis separatim" — agreed to swap hours).

63. *Digest* 8.3.17.

64. Diocletian and Maximian ordered, in January 294, that "it is not acreage, but the use to which water is put, that measures the right to water" (*Corpus* 3.34.12). Water for household consumption has first priority, followed by use for livestock and vegetable gardens, then field crops, and finally industrial uses; Eugene F. Ware, *Roman Water Law* (St. Paul, Minn., 1905), p. 94n.

65. "Ea namque est quam omnes uillae tractus eius per uicem in dies modulosque certos dispendatum accipiunt," Frontinus, *De aquis urbis Romae*, I, 9, in Clemens Herschel, *The Two Books on the Water Supply of the City of Rome of Sextus Julius Frontinus* (Boston, 1899), pp. 12–13. The water rights were attached to the land, for Cicero wrote: "In my Tusculum villa I must pay a tax for the Crabran water, because I received my estate subject to this liability," *De lege agraria*, 3.2, Herschel, p. 168.

66. Emphasis mine. *Historia naturalis*, 18.188: "felix super omne miraculum riguo . . . sed certis horarum spatiis dispendatur inter incolas"; trans H. Rackham (Cambridge, Mass.: Harvard University Press, 1950), V, 307.

67. *Corpus inscriptionum latinarum* (C.I.L.), VIII, doc. 18587; also Hermann Dessau, *Inscriptiones latinae selectae*, doc. 5793. For comments on this inscription see Semple, *Geography of the Mediterranean*, p. 457, and W. E. Heitland, *Agricola: A Study of Agriculture and Rustic Life in the Greco-Roman World from the Point of View of Labour* (Cambridge: Cambridge University Press, 1921), p.293.

68. For the fall equinox as the traditional time of sowing see Columella, *De re rustica*, II.8.2. The equinox also divided the periods of summer water (*aqua aestiva*) and daily water (*aqua cottidiana*), *Digest* 43.20.1.32.

69. Dessau reads 408, but this is not in accordance with the number of hours stated.

70. C.I.L., VIII, 1781.

71. *Voyage en Espagne d'un Ambassadeur Marocain (1690–91)* trans. H. Sauvaire (Paris, 1884), p. 9.

72. *Rawḍ al-miʿṭār*, p. 220. Medieval Muslims also thought that the agricultral development of Orihuela was a Roman legacy; E. Lévi-Provençal and E. García Gómez, *Una crónica anónima de ʿAbd al- Raḥman III al-Nāṣir*, (Madrid-Granada: CSIC, 1950), Arabic p. 53, trans. 121–122.

73. The tradition of a Roman past — curiously enough — is as current among present-day Muslim irrigators as the Islamic past is among Spanish irrigators. Whenever one asks a resident of the Ghūṭa of Damascus about the origin of some hydraulic work attribution is to the Romans (Tresse, "L'Irrigation," p. 472). One of the canals crossing the plain of Enfida (Algeria) was still in the last century called by the natives Saguiat-er-Roumi (i.e., *sāqiyat al-rūmī*, irrigation canal of the Romans, or Christians; La Blanchère, "Fossa," p. 1326 n. 9).

74. See summaries based on this material by E. Lévi-Provençal, *L'Espagne musulmane au Xème siècle* (Paris, 1932), pp. 166–167; Manuel Sanchis Guarner, in *Història del país valencià*, I, 340–45; S. M. Imamuddin, *Some Aspects of the Socioeconomic and Cultural History of Muslim Spain, 711–1492* A.D. (Leiden, 1965), pp. 74–79; and esp. Leopoldo Torres Balbás, "Los contornos de las ciudades hispanomusulmanas," *Al-Andalus*, 15 (1950), 437–486.

75. For example, Ibn Baṣṣāl, *Libro de agricultura* (*Kitāb al-filāḥa*), ed. J. M. Millás Vallicrosa (Tetuan, 1955), chap. i.

76. The most interesting surviving document is ARV, Procesos de Madrid, Letra S, no. 429, a legal instrument recording the settling of a dispute over division of water between Torres Torres and the lost village of Carcer, dating from the decade before the Reconquest (1223); the document is described by J. M. Cueco Adrian in "Un documento árabe inédito," *Levante* (Valencia), Aug. 1, 1965. This is the only known surviving Arabic irrigation document from the Valencian area. All the other Arabic material has been published by Angel González Palencia: these documents are 1) a record of a sale of water (in "eighths") from the region of Zaragoza, A.D. 1245 ("Notas sobre el régimen de riegos en la région de Veruela en los siglos XII y XIII," *Al-Andalus*, 10 [1945], 79–80); 2) an agreement between 2 villages near Guadix concerning division of irrigation water between them, A.D. 1187 ("Documentos árabes del Cenete," *Al-Andalus*, 5 [1940], 321–323); 3) rental of water belonging to Jérez by the residents of a neighboring village for 35 years, A.D. 1479 (*ibid.*, pp. 375–376); and 4) various notarial records from the twelfth and thirteenth centuries (*Los mozárabes de Toledo en los siglos XII y XIII* — see summary of water material in the *Volumen preliminar* [Madrid, 1930], pp. 289–290). Sarthou Carreres (*Provincia de Castellón*, p. 165) states that a grant of the next to last Muslim king of Valencia, Sīd Abū Saʿīd, concerning the waters of the Mijares, was recorded on an Arabic parchment that "perhaps is still preserved in the Archives of Villahermosa."

77. "Contornos de las ciudades hispanomusulmanas," p. 438.

78. "The bridge of Ashkābo," which Lévi-Provençal identified as Alcantarilla (*Rawḍ al-miʿṭār*, p. 220 n. 3).

79. Al-ʿUdhrī, *Fragmentos geográfico-históricos*, p. 1.

80. Ibn Khāqān, *Qala'id al-ʿIqyan* (Marseilles-Paris, 1860), p. 72: "wa-fī sāqiya al-kubra dawlāb."

81 *Rawḍ al-miʿṭār*, pp. 66–68.

82. Torres Balbás, "Contornos de la ciudades hispanomusulmanas," p. 464; see also Elías Terés, "Textos poéticos árabes sobre Valencia,"*Al-Andalus*, 30 (1965), 291–307.

CHAPTER X. IRRIGATION ADMINISTRATION IN AL-ANDALUS: THE ṢĀḤIB AL-SĀQIYA

1. The story is told in the *Dhaḥīra* of ibn Ḥaiyān and is quoted by ibn ʿIdhāri, *Al-Bayān al-Mughrib*, ed. E. Lévi-Provençal, vol. III (Paris, 1930), p. 158: "kānā waliyā . . . wakālat al-sāqiya bi-balansiya." *Sāqiya* in this context may indeed mean "irrigation," by extension; but the sense might

better be that of one canal, i.e., a municipal one. See Ambrosio Huici Miranda, "Dos acequieros musulmanes, reyes de Valencia," *Las Provincias* (Valencia), May 21, 1961.

2. Manuel Garrido Atienza, *Los alquezares de Santafe* (Granada, 1893), p. 40: "Hamet fijo de Abdalla, el conosçido repartidor del dicho río." The partition was certified by the chief qadi of Granada (p. 42) who naturally did not participate in any of the technical details of irrigation. This document, entitled *Repartimiento de las aguas del Río Genil,* was translated from the original Arabic by the royal scribe Ambrosio Xarafy on Feb. 12, 1502.

3. González Palencia, "Documentos árabes del Cenete," pp. 321–325.

4. ARV, Procesos de Madrid, Libro S, no. 429.

5. A pre-Islamic Arabian official, the *ṣāḥib al-siqāya,* is a possible model; he was not, however, an irrigation officer, but merely one who provided drinking water for pilgrims to the Kaʿba in Mecca.

6. Neuvonen, *Arabismos,* p. 245: "Fueros de Aragón: Quando alguno es acusado sobre furto . . . d'aquel qui guarda el agua o la çequia, qui es clamado cauaçequia. . . ." See Table 22, no. 6, of the present vol.

7. Tyan, *Organisation judiciaire,* p. 559. This is the only reference to the *qāḍi al-miyāh* known to Tyan. Elsewhere (p. 111) he points out that, particularly in Spain, the title of qadi lost its original restricted meaning: there judges with reducd and special jurisdictions were called qadi. Ibn Pascual called the police officer usually known as *ṣāḥib al-shurṭa* by the term *qāḍi al-shurṭa.*

8. *Ordenanzas y privilegios de la muy noble y leal ciudad de Lorca* (Granada, 1713), nos. 105, 294, 298 (pp. 33, 129, 131). The *sobrecequier* of Elche was also called the *juez de aguas,* a translation of *qāḍi al-miyāh* (Ibarra y Ruiz, *Riego de Elche,* p. 98).

9. Tyan, *Organisation judiciaire,* pp. 576, 601, 646.

10. Neuvonen, *Arabismos,* pp. 80, 128. For the office of *zavazouke* or *zavazoure,* see L. G. de Valdeavellano, "El mercado: apuntes para su historia en León y Castilla durante la Edad Media," *Anuario de historia del derecho español,* 8 (1931), 321–322. For the *çalmedinado* of Zaragoza (1391) see Bofarull, *Colección de documentos inéditos,* VIII, 353.

11. For the etymology of *muḥtasib into almotacen, mustasaf,* see Neuvonen, *Arabismos,* pp. 93–94; J. Corominas, *Diccionario crítico etimológico de la lengua castellana,* 4 vols. (Madrid, 1954), I, 159; Arnald Steiger, *Contribución a la fonética del hispanoárabe* (Madrid, 1932), pp. 110 (final *b* into *f*) and 265 (interior *ḥa* dropping out).

12. See Thomas F. Glick, "The Muḥtasib in Islam," unpub. diss. (Columbia University, 1963).

13. Compare the judicial procedure of the medieval mustasaf (Garcia, "Actuación procesal del mustaçaf," pp. 301–310) and that of the medieval cequier, above Chap. III.

14. Claude Cahen, "Amīn" in *Encyclopaedia of Islam,* 2nd ed. (Leiden, Brill, 1960——), I, 437.

15. Corominas, *Diccionario,* I, 77; Dozy, *Supplément,* I, 38.

16. This practice is identical to that of *jaricar* in Lorca. The Alporchón of Lorca, wherein the sale of water takes place, is administered by a *fiel* (*amīn* in Spanish trans.).

17. *Report,* pp. 58–59. Markham's description of Novelda's irrigation system, which he believed was closest to the Arabic model, is still the best.

18. Henri Schirmer, *Le Sahara* (Paris, 1893), p. 286 on Ghadames. On Touat, H. M. P. de la Martinière and N. Lacroix, *Documents pour servir a l'étude du Nord Ouest africain* (Lille, 1897), III, 219–221. The official in Touat was called the *kiel el-ma* (measurer of water, from the root *k-y-l*), a hereditary office. On Touzer, Brunhes, *L'Irrigation,* p. 488. On El Aghouat, *ibid.,* p. 251; also Eugene Fromentin, *Une été dans le Sahara* (Paris, 1857), pp. 157–159.

19. Ibarra y Ruiz, *Riego de Elche,* pp. 241–242: "Yo Obaquer Ismael alami o partidor del aygua."

20. *Ibid.,* p. 238.

21. "Ordenanza de las aguas de la ciudad de Granada (1538)," in Franquet y Bertrán, *Ensayo,* II, 179:

> el dicho azequiero por razón de dicho cargo conforme a la costumbre antigua, tenia de salario de cada molino y batan que muelen con el agua de dichas azequias, tres maravedis cada viernes, y demás de esto tenia cada dia nueve maravedís, los quales le daban y pagaban los arrendadores de la zaquifa [roofed gallery] de los cueros de la Ciudad, y assi parece por cierta información que de ella se huvo, como después que la Ciudad se fundó, assi de tiempo de Moros, como despues de Christianos, se han pagado los dichos derechos al dicho azequiero, y porque el dicho salario es poco, y al presente no se halla persona que sirva el dicho oficio con el; mandamos que al dicho azequiero le den los dichos tres maravedís cada molino y batan que muelen con las dichas azequias cada viernes, y los dichos neuve maravedís cada dia los dichos arrendadores que agora son . . . de la dicha renta de la zaquifa, conforme a la dicha costumbre antigua, y demas de lo susodicho, la Ciudad de Granada le dé de sus propios cuatro mil maravedís de salario en cada un año.

22. Memorandum from J. Lapanne-Joinville to Arthur Maass, typescript, p. 21 ("Seguias communes"); Bruno, *Régime des eaux en droit musulman,* p. 39.

CHAPTER XI. PROPORTIONAL DISTRIBUTION AND THE MEASURING OF WATER

1. *Pul Eliya: A Village in Ceylon* (Cambridge, Eng., 1961), pp. 164–165.

2. *Sentencia arbitral del serenísimo infante D. Pedro (1346)*, in Traver, *Historia de Villarreal*, pp. 472–477.

3. Carlos Sarthou Carreres, *Datos para la historia de Játiva*, 3 vols. (Játiva, 1933–1935), I, 297.

4. J. Musso y Fontes, *Historia de los riegos de Lorca* (Murcia, 1847), p. 34.

5. Fontavella, *Huerta de Gandia*, p. 79.

6. "Antigüedades de Elche," p. 336; cf. Markham, *Report*, pp. 51–52.

7. *Sentencia arbitral (1346)*; see n. 2, above.

8. Ruiz-Funes (*Derecho consuetudinario*, p. 141) states that the turn does not function on Sundays in Murcia. Semitic law generally did not permit irrigation on the Sabbath.

9. *Provincia de Valencia*, p. 294 n. 484.

10. Aymard, *Irrigations du midi de l'Espagne*, p. 22; Jaubert de Passa, *Canales de riego*, II, 217n.

11. Martínez Aloy (*Provincia de Valencia*, pp. 287–288) suggested that the canals of Quart, Tormos, and Mislata were of more recent construction than the other main canals of the huerta because their names are not found in early privileges (particularly that of James II, Apr. 24, 1321). He was unaware that the canals in question are mentioned in the minutes of the city council of Aug. 21, 1313 (AMV, Manuals de Consell, I, fol. 91r). Such reasoning is doubtful in any case because major changes in the system of proportionality would have merited royal attention.

12. The history of these attempts is given by Borrull (*Tratado*, pp. 7–10) and by Manuel María Azofra, *Memoria sobre la exacta medición de agua corriente por medio de muelas, filas y plumas* (Valencia, 1844).

13. Aymard, the first scholar to understand that proportionality is at the heart of the Valencian distribution system, was adamant on the futility of trying to figure out what the fila was in hard measurements (*Irrigations du midi de l'Espagne*, pp. 23–25). Chap. vi of Andrés Llaurado's *Tratado de aguas y riegos*, 2 vols. (Madrid, 1884), I, 83–94 ("Distribución del agua — sistema distributivo fundado en la proporcionalidad") is based on Aymard. One of the few modern writers to appreciate the concept of proportionality was A. Guillén; see *El tribunal de Aguas*, esp. pp. 44–45, following Brunhes. See also *Exposición de distintas actuaciones*, pp. 19–22.

14. Markham's *Report*, pp. 64–65, points out that even when fixed measures are adopted they are not necessarily the same in each place, that of Gandia differing from that of Murcia, for example.

15. Robert A. Fernea, *Shaykh and Effendi* (Cambridge, Mass., 1969), p. 125

16. Rossi, "Irrigazione nel Yemen," p. 352.

17. Tresse, "L'Irrigation," pp. 479–480. In origin the *qīrāṭ* was the counterpart of the *siliquas,* 1/24 of the gold solidus. By extension it came to mean the 24th part of anything.

18. Tresse, "L'Irrigation, pp. 511–512, and Richard Thoumin, *Géographie humaine de la Syrie centrale* (Tours, 1936), p. 55 n. 2.

19. Tresse ("L'Irrigation," p. 475) derives *mezzaz* from *mazza* (to suck). It is more likely derived from *māza,* to separate (see Appendix 8 n. 4), a usage parallel to *almatzem* (Gandia), derived from *qasama,* to divide. For the form of these divisors, compare the pictures in Tresse (pl. 4A) with those of similar divisors in Gandia (Fontavella, *Huerta de Gandia,* photo 20).

20. The Gandia system is extremely complicated and its rationale is not entirely clear. The river has 48 filas at the Casa Fosca dam, just like the Moncada Canal in Valencia. Unlike Valencia and the Ghūṭa, however, the lower diversions of Gandia diminish in filas rather than start each stage anew with 24. Finally, the canals of Gandia and Oliva, containing 14 and 21 filas respectively, are assigned arbitrary quotas of *hilos* in order to make subsequent divisions in a neater fashion. By mathematical juggling, however, the final *hilo* at the lowest diversion has the same value as the original fila (see Fontavella, *Huerta de Gandia,* fig. 11).

21. Compare Ibarra y Ruiz, *Riego de Elche,* p. 201; the *hilo* of the Marchena Canal is always "double" and therefore runs for 6, not 12 hours. In the Marchena system, the *cuarta* is equivalent to 1½ hours.

22. See Chap. xii, n. 29. *Tumen,* the 1435 document form, is closer to the Yemenite pronunciation than the usual castilianization *azumbre.*

23. Markham, *Report,* p. 57; cf. González Palencia, "Riegos de Veruela," Arabic doc. pp. 79–80, trans. 80–83, the sale of "one-eighth of water" ("althaman wāhid min al-mā'"). González Palencia states that the eighth is equivalent to 3 hours of water (p. 81 n. 6).

24. Musso, *Riegos de Lorca,* pp. 34–36; cf. also the very similar system of Librilla (Murcia) in Ruiz-Funes, *Derecho consuetudinario,* pp. 178–179.

25. E. R. Leach, "Primitive Time-reckoning," in *A History of Technology,* ed. Charles Singer, 4 vols. (Oxford: Clarendon Press, 1954), I, 112. The cyclical notion is especially apparent in Arabic expressions for turn (e.g. *daula, nūba, daur*) which connote revolution.

26. The main lines of this hypothesis were worked out in discussion with Professor Arthur Maass.

CHAPTER XII. THE IMPRINT OF ISLAM UPON THE TERMINOLOGY OF IRRIGATION

1. See Caro Baroja, "Norias, azudas, aceñas," pp. 60–61.

2. Neuvonen, *Arabismos,* p. 85. Ramifications of the multiplicity of channel terms are endless: Neuvonen quotes a line of Arabic which mentions a

"canal which is called *sāniya* and which in Romance is called *calicin* [i.e., *cauces*]." He lists *reguera, regadera,* and *aguaducho* as additional synonyms.

3. Rossi, "Irrigazione nel Yemen," p. 356.

4. Jaume Roig, *Spill,* ll. 14779–14782.

5. José Latour Brotons, *Antecedentes de la primitiva Ley de Aguas* (Madrid, 1955), p. 45.

6. *Ibid.,* p. 45; Ruiz-Funes, *Derecho consuetudinario de Murcia,* p. 140; Diaz Cassou, *Ordenanzas y costumbres,* p. 63.

7. The assigning of a different name to each gradation of canal in a hierarchical distributory system has been standard in Mediterranean irrigation systems since Akkadian and Babylonian times. See Jørgen Laessøe, "The irrigation system at Ulhu, 8th century B.C.," *Journal of Cuneiform Studies,* 5 (1951), 25–26, on the variety and hierarchy of canal names: "*Hirītu* seems to be a main canal from which smaller and narrower ditches branch off. *iku, palgu,* and *atappu* all seem to belong in this second category." In Babylonia, the larger canals were called *narū,* i.e., river, cf. Arabic *nahr* (R. J. Forbes, *Studies in Ancient Technology,* II, 21). In Iraq today — to cite but one example drawn from the Islamic world — the main canal is called *jadwal;* the primary and secondary feeders *bada* and *naharān;* still smaller channels *ʿumud;* followed by the equivalent of a Valencian *cequiola,* the *mirriyān,* which runs only the length of a plot; finally there is the *sharūgh,* or irrigation furrow (Fernea, *Shaykh and Effendi,* p. 122). Three thousand years of continuity in irrigation terminology indicated by Babylonian *naru,* Arabic *nahr, naharān,* and Valencian *almenara,* from the same root (see below, nn. 15, 54) is truly impressive.

8. Caro Baroja, "Norias, azudas, aceñas," pp. 52–59.

9. F. W. Maitland described the instability of agrarian terminology in Norman England whereby Anglo-Saxon, French, and Latin terms were "rudely intermixed" during the transitional period following the Conquest and finally emerged as an organic whole as some synonyms perished and others survived; *Domesday Book and Beyond* (London, 1897), pp. 8–9. A similar situation was produced in Valencia after the Reconquest, with the difference that no kingdomwide standardized irrigation terminology ever evolved. Frequently Latin documents included both the arabism and the Romance equivalent in such phrases as "azutum sive resclosa" or "cequia sive aqueductus." But in the majority of Valencian documents the arabism was preferred to the Romance form; see documents of 1270 and 1318, where açut is declined in Latin (*azutorum, azuto cequiae, azuta*) — Branchat, *Tratado de derechos,* III, pp. 196, 202. The use of arabisms to describe institutions or artifacts of long establishment in Christian Spanish culture is perplexing to Neuvonen (*Arabismos,* p. 307). The problem, really, is not one simply of the long existence of a given object but of the relative value ascribed to it by each culture. See Bellver and Cacho's comments on the use of "cequia" for all open water-conducting ditches beside the continued use

of the Latin *canalis* (in the derived forms *canal, canalat*) but only in the restricted sense of a roof gutter (*Influencia,* p. 45).

10. See Neuvonen, *Arabismos,* esp. intro., chap. summaries, and concl., pp. 28–33, 81–83, 135–137, 257–260, 300–310. The percentage of agricultural (including irrigation) terms passing into Spanish in the three periods considered is constant: 13.2 percent (711 to mid-eleventh century), 9.6 percent (mid-eleventh to end of twelfth), and 10.1 percent (thirteenth century). Note also the lag between the generalization of a phenomenon in Al-Andalus and the diffusion of the arabism into the Romance vernaculars. Rice was grown in Spain from the eighth century on, but the use of the arabism *arroz* in Romance became commonly only after the great conquests, in the thirteenth century (*ibid.,* p. 164).

11. On *açut* see Neuvonen, *Arabismos,* pp. 132–133; Corominas, *Diccionario,* I, 352; Alcover, *Diccionari,* II, 87. Cf. the Granadan term *maglaca* (control gate?) from the Arabic *ghalaqa,* to close, cut off (Franquet y Bertrán, *Ensayo,* II, 178). *Sadd* is the common word for dam in modern Yemen (Rossi, "Irrigazione nel Yemen," p. 349).

12. On *albellon,* Neuvonen, *Arabismos,* p. 253; the Castilian form, *albollón,* passed into that language through Aragonese. Also see Diaz Cassou, *Huerta de Murcia,* p. 226 n. 1, and *Ordenanzas y costumbres,* p. 67 n. 1. For use of this word in the context of irrigation, see ARV, Gobernación, 2214, 7th hand, fol. 46v (Mar. 26, 1416): "los albellons qui de present hi son per a obs de regar les dites terres" (re. Rahana Canal, in Játiva).

13. On *alcaduf,* Corominas, *Diccionario,* I, 250–251; Neuvonen, *Arabismos,* 145–146; Alcover, *Diccionari,* I, 437; G. S. Colin, "La noria marocaine," p. 29 n. 1; Glick, "Medieval irrigation clocks," p. 426.

14. On *aljup,* Corominas, *Diccionario,* I, 137; Alcover, *Diccionari,* I, 505–506; Neuvonen, *Arabismos,* p. 150.

15. On *almenara,* Corominas, *Diccionario,* I, 148; Dozy, *Supplément,* II, 728 (citing Makkarī's use of *manāhir,* meaning "canals").

16. *Çabacequies,* Neuvonen, *Arabismos,* p. 245.

17. Cequia, Neuvonen, *Arabismos,* pp. 84–85; Corominas, *Diccionario,* I, 21; Alcover, *Diccionari,* IX, 851–852.

18. *Marjal,* Neuvonen, *Arabismos,* p. 252; Alcover, *Diccionari,* VII, 254.

19. *Noria,* Corominas, *Diccionario,* III, 522–523; Neuvonen, *Arabismos,* pp. 131–132; Steiger, *Contribución,* p. 287; R. Dozy and W. H. Englemann, *Glossaire des mots espagnols et portugais derivés de l'arabe* (Leiden, 1869), p. 195; Leopoldo de Eguílaz y Yanguas, *Glosario etimológico de las palabras españolas de origen oriental* (Granada, 1886), p. 465.

20. *Safareig,* Neuvonen, *Arabismos,* pp. 44–45; Corominas, *Diccionario,* IV, 791; Alcover, *Diccionari,* VII, 254.

21. *Sinia,* Neuvonen, *Arabismos,* pp. 34–35; Corominas, *Diccionario,* I, 21.

22. *Tanda,* Corominas, *Diccionario,* IV, 365–369. A word marked with

an asterisk means that it is a hypothetical, but probable, form which has never been documented.

23. *Tarquim, ibid.,* 387.

24. *Açarb,* Corominas, *Diccionario,* I, 348; Alcover, *Diccionari,* I, 110; Dozy, *Supplément,* I, 644; F. Valls Taberner, *Los privilegios de Alfonso X a la ciudad de Murcia* (Barcelona, 1923), p. 59 (Apr. 8, 1272). See n. 62, below.

25. *Ador,* Chabas, *Distribución de las aguas en 1244,* p. 6; Corominas, *Diccionario,* I, 41 (Aragon). In Yemen (Wadi Dahr), *dauriyah* is the modern word for turn (Rossi, "Irrigatione nel Yemen," p. 354).

26. *Albala,* Corominas, *Diccionario,* I, 81–82; Alcover, *Diccionari,* I, 412; Altamira, "Mercado de agua," p. 150. *Albalá* is more typically the Castilian form; the Catalan variant *albará* was the common medieval term for a payment voucher.

27. *Almahacen,* Musso y Fontes, *Riegos de Lorca,* p. 30; see also Corominas, *Diccionario,* I, 138–139.

28. *Almatzem,* Chabas, *Distribución de las aguas en 1244,* p. 6. The term was still in use 2 centuries later, in an inventory of divisors in Gandia (e.g., "hun almatzem o partidor"); ARV, Gobernación, 2287, 14th hand, fol. 2r, July 10, 1456. *Maqsam* is still a common term for divisor in Iran (Kerman); see Hans E. Wulff, *The Traditional Crafts of Persia* (Cambridge, Mass.: M.I.T. Press, 1966), p. 255. Compare the south Arabian expression *maqāsim al-mā',* in the sense of diversion channels (Serjeant, "Some Irrigation Systems in Ḥaḍramawt," p. 37). The other meaning of partidor — an official in charge of dividing water — is also found in southern Arabia in a word from the same root. For the *muqassim al-dayri* or "divider of the water" of Wādi Dahr, Yemen, see Thomas J. Abercrombie, "Behind the Veil of Troubled Yemen," *National Geographic Magazine,* 125 (1964), 427.

29. *Azumbre,* Neuvonen, *Arabismos,* pp. 205–206; Corominas, *Diccionario,* I, 354; Rafael Altamira, *Derecho consuetudinario y economía popular de la provincia de Alicante* (Madrid, 1905), p. 80. Compare the forms *tumen* (1435) and *tomin* (1461) from Elche (Ibarra y Ruiz, *Riego de Elche,* pp. 242, 238) and *açumen* (1173) from Veruela (Aragon) (González Palencia, "Riegos de Veruela," p. 84).

30. *Dula,* Chabas, *Distribución de las aguas en 1244,* p. 7; Corominas, *Diccionario,* II, 206–207; Ibarra y Ruiz, *Riego de Elche,* p. 206. In modern Yemen (Shoʿūb) the turn is called *daulah* (Rossi, "Irrigazione nel Yemen," pp. 352, 354), the expression used by the eleventh-century geographer al-Bakrī to describe the irrigation turn at the oasis of Touzer (now in Tunisia); *Dikr bilād Ifrīqīya w'al-Magrib,* M. de Slane, ed., 2nd ed. (Paris, 1913), p. 48. *Dula* means "turn" in the Canary Islands (Oliver Asín, *Historia del nombre "Madrid,"* p. 200 n. 1). As a synonym of *hila,* see Altamira, "Mercado de agua," p. 149. *Dula* as a measurement was also used in colonial San Antonio, Texas, whose irrigators had emigrated from the Canaries in the

eighteenth century; see *Memorandum on the Spanish and Mexican Irrigation System of San Antonio* (Austin, 1959), a pub. of the Water Division, Office of the Attorney General of Texas.

31. *Jarique,* Musso, *Riegos de Lorca,* pp. 57, 82. *Jarique* is related to the Medieval Castilian *exarich,* sharecropper (Neuvonen, *Arabismos,* pp. 128–129). For this social class see Eduardo de Hinajosa, "Mezquinos y exaricos," in *Homenaje a Don Francisco Codera* (Zaragoza, 1904), pp. 523–531, and González Palencia, "Riegos de Veruela," p. 83. Eguílaz (*Glosario,* p. 312, under *axarique*) cites an irrigation turn of the Oznar Canal (Granada) in 1575 providing that: "El tercer Domingo es de Benamohat y de los axariques. El cuarto Domingo es para Beni Muzahe y sus xariques." Benamohat and Beni Muzahe appear to be either place names or branch canals named, as in Murcia, for tribal groups. It may be, therefore, that in this case *xarique* is not a sharecropper but a sharer of water.

32. *Jarro,* Musso, *Riegos de Lorca,* p. 36; Neuvonen, *Arabismos,* p. 171. In Yecla the official who times the turns is called the *jarrero,* i.e the man in charge of the *jarro.* The *jarrero* uses a watch now, but his title recalls the original water clock. The *jarro* of Jumilla is divided into twelve *horteras* (from the Latin *fortera,* cup, bowl), a measure that also recalls the clepsydra, which usually emptied in 3–7 minutes (Ruiz-Funes, *Derecho consuetudinario,* pp. 188, 190–191). See Glick, "Medieval irrigation clocks," p. 426.

33. *Martava,* Altamira, "Mercado de riego," p. 149. *Martavero,* an official, is the practical equivalent of *acequiero;* but literally it has the same meaning as *atandador* (from tanda), one in charge of regulating turns. See also Alcover, *Diccionari,* VII, 268, and (for Novelda) Markham, *Report,* pp. 57–58. Markham observed (*ibid.,* p. 58n.) that "the word is also used in Persian and Turkish to signify a time or turn, and in India."

34. *Merancho,* Diaz Cassou, *Ordenanzas y costumbres,* p. 56 n. 3.

35. *Rafa, ibid.,* p. 64 n. 2.

36. *Sistar,* in Cueco Adrián, *La Font de la Vall de Segó,* p. 42 n. 7.

37. *Tahulla,* Musso, *Riegos de Lorca,* p. 36; Neuvonen, *Arabismos,* p. 250; Corominas, *Diccionario,* IV, 342. It was not unusual for a land measurement unit to be converted into a water unit, inasmuch as water was supposedly apportioned according to the superficies of land irrigated. Thus in Librilla (Murcia) the water unit is the *caballeria,* the equivalent of a day's water (Ruiz-Funes, *Derecho consuetudinario,* pp. 178–179); as a measure of land it was the amount of land granted to a knight (*cabellero*).

38. Neuvonen, *Arabismos,* p. 305. Measures, of which *ṭahwila* — *tahulla* — is an example, comprise 1 of the 2 categories of general exceptions to this rule (*Arabismos,* p. 306 n. 1).

39. The interior *ẓ* was, in fact, pronounced as *d* in Andalusí Arabic, but the transition *im* to *a* is hard to explain (Corominas, *Diccionario,* IV, 368).

40. A more basic problem of *tanzim* into *tanda* is a conceptual one. The only form of *n-ẓ-m* even remotely connected with irrigation is *naẓm,* a se-

ries of wells dug in a line. *Naẓama,* to put in order or arrange in a series, has a linear sense, whereas most of the other commonly used words for turn in Arabic — *daula daur, nūba* (*ʿaddān,* from *ʿadda,* "to count," is an exception) — have the sense of rotation, of returning to the same point in the cycle.

41. Markham (*Report,* p. 14) says that Makkarī called Murcia "al-Bostan." Compare references to the gardens (*basātīn*) of Valencia in the *Rawḍ al-miʿṭār,* Arabic text p. 47. See also the use of *bustān* in such expressions as *shaykh al-bāsatīn* (officer of the gardens) in the Ghūṭa of Damascus.

42. *Contribución,* p. 290; Steiger derives it, through a metathesis, from *wāqiʿ,* place (?).

43. Musso, *Riegos de Lorca,* p. 30. In Iraq the *waqt* (pl. *'awqāt*), "time," is the period from sunrise to sunset, i.e., 12 hours (Fernea, *Shaykh and Effendi,* p. 125).

44. Musso, *Riegos de Lorca,* pp. 35–36; Brunhes, *L'Irrigation,* p. 261 n. 3.

45. ARV, Procesos de Madrid, Letra S, 429.

46. Pedro de Alcalá, *Vocabulario,* ed. Paul de Lagarde (Gottingen, 1883), p. 157.

47. Marcelain Beaussier, *Dictionnaire pratique arabe-français* (Algiers, 1958), see under *khaiṭ.*

48. Martínez Morellá, *Cartas de los reyes de Castilla a Orihuela,* p. 58 (May 14, 1275). "E por fazer su oficio a todas las otras cosas que ouieron a fazer e que fagan limpiar las çequias e las filas e arrouas e los açarbes cada anyo de aquella acequia do fuere acequiero."

49. Regarding "fila de dos hulls" see ARV, Gobernación, 2297, 11th hand, fol. 18v (in the rubric it is also called "cequia de dos hulls"); also AMC, Libres de Consell, Nov. 22, 25, 1460. For "files de la doberia" see AMC, Libres de Consell, 35 (Nov. 25, 1436); and for "files de Na Orellana" AMC, Libres de Consell for Feb. 21, 1473. Fila in this usage appears to be more typical of Castellón than of Valencia.

50. Pascual Madoz é Ibañez, *Diccionario geográfico-estadístico-histórico de España,* 16 vols. (Madrid, 1845–1850), I, 535. Regarding Riu dels Ulls: *ulls,* Castilian *ojos* (eyes), is a calc on the Arabic *ʿayn,* meaning both eye and spring; see Oliver Asín, *Historia del nombre "Madrid,"* p. 172.

51. Asín Palacios, *Contribución a la toponimia árabe de España* (Madrid, 1944), p. 58.

52. With regard to the initial *f* and final *t* of *fait:* compare *riego dalffayat, dalfayt* with the forms of "tailor" in the *Repartimiento de Murcia* (e.g., Johan Perez, *alffayate,* p. 163). A derivation from *faiḍ* is less likely because Arabic final *ḍ* almost always changes to *d* (and rarely to *ç*) but never to *t* (Steiger, *Conrtibución,* p. 164).

53. Compare the toponyms, Almaguer (from *al-maghīḍ*) in Alginet, and Alcanar (Novelda), both meaning "The Canal"; Manuel Sanchis Guarner,

Introducció a la història lingüística de Valencia (Valencia, 1950), p. 90, and Asín Palacios, *Toponimia árabe*, p. 66).

54. There are no phonetic problems in this proposed etymology. See Steiger, *Contribución*, p. 272, for the loss of the interior *ha* (his illustration is from the same root, *manhar* into *almenara*). Nicolau Primitiu ("Salterio toponímico: Paiporta y Faitanar," *Almanaque "Las Provincias"* [1925], p. 201) proposed a derivation from a collective form (ending in *ar*) of *faitana* (parallel semantically to Castilian hacienda) in the sense of a country house. The derived form would be parallel to *campanar*, *quintanar*, etc. But *faitana* (which he should have written **faitana*) is apparently just a conjecture of Primitiu's (it does not appear in Alcover, *Diccionari*). The Murcian toponym Alfeitami (name of a diversion dam), could be a parallel case. It seems most likely a personal name (perhaps al-Fāṭimī, "The Fatimid," or its diminutive), but could be a form of *khaiṭ al-mi*, thread of water (cf. Pedro de Alcala, n. 46 above), which would be a close parallel to *khaiṭ al-nahr*.

55. *Repartimiento de Murcia*, pp. 55, 157.

56. Juan Torres Fontes, *Medidas de superificie y de valoración en el Repartimiento de Murcia* (Murcia, 1959), pp. 18–19.

57. "Molinería valenciana mijeval," pp. 702–704; the particle *var*, *far*, he explains, means watercourse. In reality Alfafara, which he thinks a related toponym, is an indisputable arabism meaning place where pottery is made (Asín Palacios, *Toponimia árabe*, p. 58).

58. On Favara, "Repartimiento de Valencia," p. 269 (A.D. 1240); Alfavara, ARV, Gobernación, 2241, 9th hand, fol. 34r (Aug. 11, 1430): "los cequiers de muncada de alfavara e de les altres cequies." Compare the Magribi toponym El Faouar, "The Spring"; Pierre Moreau, *Le pays des Nefzaouas* (Tunis, 1947), *pl.* 7, opp. p. 100).

59. Rascanya: *qanā* or *qanāya*, Arabic for "canal"; e.g., Pedro de Alcalá, p. 137, *canná*, pl. *canaguát*, as in *canná quibura*, "canal maestre" (apparently without the connotation of irrigation, in contradistinction to *çáguia*, pl. *çaguiát*, "regadura," p. 377). Rascayna: *Repartimiento de Valencia*, pp. 154 (A.D. 1237), 266 (A.D. 1240). Rascanya is parallel to another common toponym, *ras al-ʿayn* (fountainhead; Latin, *caput aquae*), as in Rasalany (Valencia; see Sanchis Guarner, *Introducció*, p. 90) and Ras-el-ain (Tunisia; see Moreau, *Pays des Nefzaouas*, pp. 141, 146).

60. The canal and the district it waters frequently have the same name. Algirós (in Valencia and Alcira, both) is well documented. For Algorós see Ibarra y Ruiz, *Riego de Elche*, p. 255.

61. Initial *z* into *j* is one of the few examples of this change (Steiger, *Contribución*, p. 145 n. 1). The final *b* loses its sonority and changes first to *f* (e.g., Aljarof, the form in the *Repartimiento de Valencia*, p. 480; Steiger, *Contribución*, p. 110) and then to *s*. The commentators have ignored this toponym, concentrating on the Portuguese forms *algeroz*, *algiroz*, *aljaroz*,

meaning gutter. See Eguílaz, *Glosario,* p. 180; also Dozy and Engelmann (*Glossaire,* p. 125), who remark that *mizrāb* was the current expression for gutter in North Africa and cite Freytag's definition of *zarb* as *"canalis aquae."*

62. On Ador, Jaubert de Passa, *Canales de riego,* II, 165. On *Zarahiche,* i.e., *zafariche* to *zahariche* to *zarahiche,* by metathesis, Diaz Cassou, *Ordenanzas y costumbres,* p. 58.

63. On Aljupet, Ibarra y Ruiz, *Riego de Elche,* p. 209. Cisterns were important landmarks and gave rise to numerous toponyms. See *ibid.,* p. 10 (Port d'Aljup) and Asín Palacios, *Toponimia árabe,* p. 62, Algibe. On El Albellon, Ibarra y Ruiz, *Riego de Elche,* p. 295.

64. On Calduf, *Riegos de Alcoy: ordenanzas* (Gandia, 1927), p. 2. Compare Asín Palacios, *Toponimia árabe,* p. 51, Alcadoz.

65. On Tarquin, Diaz Cassou, *Ordenanzas y costumbres,* p. 82.

66. On El Azut, Archivo de la Corona de Aragón, reg. 64, fol. 125v. On Burjassot, literally "the tower of the dam," see Asín Palacios, *Toponimia árabe,* p. 98. The *burj* was not necessarily a defense tower but, like *torre* in modern usage, also meant a country house; cf. the toponym in Islamic Córdoba, Faḥs al-Sudd, field of the dam (Torres Balbás, "Contornos de las ciudades hispanomusulmanas," pp. 453, 485). The *burj* of Burjassot may be that which gave its name to the Alborg Canal, which was apparently within the limits of the village of Burjassot (ARV, Gobernación, 2268, 4th hand, fol. 8r, May 1, 1442). The açut in question may in reality have been only a divisor or a permanent canal check, inasmuch as Burjassot is not located on the river.

67. Asín Palacios, *Toponimia árabe,* p. 111.

68. *Ibid.,* p. 103.

69. See Sanchis Guarner, *Introducció,* p. 90; Asín Palacios, *Toponimia árabe,* p. 94, Biar.

70. Julian Ribera, "Influencias berberiscas en el Reino de Valencia," *El archivo,* 1 (1886–1887), 171. Compare the name of the Moroccan town Tetuan.

CHAPTER XIII. IMAGE AND REALITY OF CULTURAL CHANGE

1. My cultural typology corresponds to Brunhes' geographical typology based on the relative scarcity of water (*L'Irrigation,* pp. 59–93). Logically, different geographical situations require the selection of different organizational models. There is general correspondence, too, to a typology based on tenure, for water is attached to the land in areas where it is plentiful ("Syrian type") and alienable where water is scarce ("Saharan type"). Compare North African tenure typology in Despois, "Development of Land Use in Northern Africa," pp. 222–223. The Yemenite legacy has been noted

by Bellver and Cacho, *Influencia,* pp. 38–39; Semple, *Geography of the Mediterranean,* p. 567; and Foster, *Geographic Structure,* p. 64. See also Stanislawski, *Portugal's Other Kingdom,* p. 10.

2. On the distinction between formal and nonformal processes of diffusion, see Foster, *Culture and Conquest,* p. 12; Glick and Pi-Sunyer, "Acculturation," pp. 151–152.

3. José M. Font Rius, "La reconquista y repoblación de Levante y Murcia," in José M. Lacarra *et al., La reconquista española y la repoblación del país* (Zaragoza, 1951), pp. 97–98; Burns, *Crusader Kingdom,* p. 7.

4. On Gandia see José Camarena Mahiques, *Historia del distrito de Gandia* (Gandia, 1965), p. 39; see also Camarena, *Colección de documentos para la historia de Gandia y su comarca* (Gandia, 1959–1961), doc. XXV (July 20, 1381), "Nombre dels fochs dels moros de la orta e contribucio de la vila de Gandia."

For Játiva see privilege of James I, Jan. 23, 1252, to the Moors of the *arrabal* of Játiva, in A. Huici Miranda, *Colección diplomática de Jaime I, el Conquistador,* 3 vols. (Valencia, 1916), I, 550–553. These Moors were granted autonomy in the administration of irrigation: "mandantes quod aliquis cequiarius Xative non intret domos vel ravallos vestros por aqua petenda vel accipienda, nisi cum uno sarraceno Ravalli predicti" (p. 553).

Regarding the *arrabal* of the Moors of Elche who irrigated from the Marchena Canal, see Ibarra y Ruiz, *Riego de Elche,* pp. 227–246.

5. Benaguasil (Beniguazir in Medieval Valencian) was one of the Pueblos Castillos irrigated from the Guadalaviar and was, in the Middle Ages, a rice-growing center largely inhabited by Muslims. The documentation is copious. See AMV, Cartas Misivas, 22, n.p. (May 17, 1454), a dispute raised by the alamí of the Moors of Benaguasil over the irrigation of vegetables.

See ARV, Gobernación, 2377, 45th hand of 1485, fol. 44r; testimony of various *moros vassals* of the lord of Callosa concerning irrigation matters.

On Chiva see ARV, Gobernación, 2227, 16th hand, fol. 27 (Nov. 12, 1421); the Moors of the *morería,* village, and barony of Chiva asserted that there was not water sufficient for irrigating in the spring from which they had been accustomed to draw water, because of drought (also, 2226, 2nd hand, fol. 40r, Feb. 27).

Torres Torres was irrigated in Muslim times (as indicated by the Arabic document of 1223 in ARV, Procesos de Madrid, S 429). After the Reconquest it continued to be irrigated by Muslim cultivators; see ARV, Gobernación, 2241, 6th hand, fol. 39r, a description of fields owned by the sons of Ali Alhais — Muslims residing in Torres Torres: "Item, un troç de vinya situat en lo dit terme sobre la cequia confronta ab la cequia e ab vinya de Alfaig moro. . . . Item, un troç de garoferal e figuerol situat en lo dit terme franch que te conffronta ab la cequia chiqua e ab vinya de Albalati."

For the Vall de Segó see ARV, Gobernación, 2210, 33rd hand, fol. 36r

(1414); an enumeration of taxes owed on certain parcels, including some fronting on canals, owned by Muslims of Quart and Quartell.

James I granted the Moors of Vall de Uxó autonomy in irrigation affairs: "et que pusquen jutgar les aygues entre si, aixi com era acostumat en temps de moros, segons ques conte en los lurs privilegis antichs" (Huici Miranda, *Colección diplomática,* I, 513).

6. For example, a dispute between Beniarbeg and Beniazmut, two villages near Ondara, in which figures the testimony of the "oldest Moors" (ARV, Gobernación, 2270, 8th hand, fol. 43).

7. *Repartimiento de Valencia,* pp. 269 (Favara), 276 (Algirós). The canal system associated with the mills of Berenguer and Michael Revell, in Boatella, is probably that of Rovella (*ibid.,* pp. 271, 277; see Nicolau Primitiu, "Molinería valenciana," pp. 700–702).

8. Font Rius, "Reconquista y repoblación," p. 100.

9. "Alfaquimus domini regi" (*alfaquim,* from the Arabic *alḥakīm,* wise man, physician; Neuvonen, *Arabismos,* p. 151. Bafiel (also Bahiel, Bahihel) was a Jew of Zaragoza who knew Arabic and served James I as interpreter (James I, *Crònica,* chaps. lxxiv, cxviii).

10. "Reallum" is possibly derived from the Arabic *rahal,* garden. See Houston, "Land Use and Society in the Plain of Valencia," p. 175.

11. For this spelling and the etymology of the toponym see Chap. XII, n. 60.

12. "In quadam magna cequia." Since this entry follows a list of donations in Campanar, the canal could be either Rascanya or Mestalla (La Petra branch), on the left bank of the river.

13. "In casalibus."

14. *Repartimiento de Valencia,* pp. 153–154.

15. *Ibid.,* pp. 364–366, lists donations in Bayren in the years 1240, 1242, and 1244.

16. That is, Upper and Lower Benieto (Benieto Sobirá and Benieto Jusá), Camarena, *Historia del distrito de Gandia,* p. 38. See map (*ibid.,* facing p. 38) for location of these settlements.

17. Peregrín de Otrosillo (or Atrosillo) was one of the knights of James I who played a prominent role in the siege of the Castle of Bayren; see the *Crònica* of James I, chaps. xv, lxvii; and Camarena, *Historia del distrito de Gandia,* p. 26. He was granted a castle and several villages in Aragon in 1252 (Huici Miranda, *Colección diplomática,* I, 559–560).

18. "Almatzem;" see Chap. XII, n. 28.

19. Two settlements on the Acequia Común of Gandia; see Fontavella, *Huerta de Gandia,* p. 82.

20. Martin d'Oblites was granted 7½ jovates of land in Bayren (Dec. 7, 1242); *Repartimiento de Valencia,* p. 365. J. de Mora was granted the *alqueria* of Benampiscar (i.e., Benipeixcar), across the Serpis from Benieto

(Sept. 20, 1240); *Repartimiento de Valencia,* p. 365. Benipeixcar is irrigated from the Vernisa Canal (Fontavella, *Huerta de Gandia,* fig. 12).

21. "Trugaman" (from the Arabic *turjamān,* interpreter, translator); see Neuvonen, *Arabismos,* p. 175.

22. The document quoted is the *Distribución de las aguas en 1244,* pp. 3–6, preserved in the Municipal Archives of Gandia (Privilegis de Gandia, fol. 73). Apparently Aragonese rather than Catalan was used in this document because Otrosillo was Aragonese.

Lorenç Ruffa (or Rufa) held land in Beniopa ("Repartimiento de Valencia," pp. 349, 468, 513), which is in the Vernisa service area, and 4 jovates in Alharrazin (Alfarrasí), one of the places that figured in the inventory of divisors; he may have been a specialist in land division if he is the same Lorenz Ruffa mentioned in the *Repartimiento de Murcia,* p. 160.

23. For example, in Granada on Mar. 2, 1498, information was taken from Mahomad ben Juncairi and Mahomad Haxu "sobre la costumbre del regar el acequia de Ynadamar" (Garrido Atienza, *Los alquezares de Santafe,* p. 7 n. 1). See also the indirect reference to an inquest for the Irués Canal in the Ebro Valley: the irrigation system was to continue "como andavan en tiempo de moros . . . que trovó verdat en antigos moros como devia andar el agua" (J. M. Lacarra, "La repoblación del valle del Ebro," in Lacarra *et al., La reconquista española,* p. 70).

24. Branchat, *Tratado de derechos,* III, 276–277:

> Per nos e per los nostres donam e atorgam per tots temps a vos ensemps e sengles habitators e pobladors de la ciutat e del regne de Valencia, e de tot lo terme de aquell regne, totes e cascunes cequies franques e liures, majors, e mijanes e menors, ab aygues e ab manaments, e ab duhiments daygues, e encara aygues de fonts, exceptat la cequia Real qui va a Puçol, de les quals cequies e fonts hajats aygua, e enduhiments e manaments daygues tots temps continuament de dia e de nit: en axí que puscats daquelles regar, e pendre aygues sen [sic] alcuna servitut e servici e tribut, e que prenats aquelles aygues segons que antiguament es e fo stablit e acostumat en temps de Sarrahins.

25. James I (July 29, 1248) conceded to the inhabitants of Morvedre all the water flowing in the Riu de Sogorb and the canal of Torres Torres for irrigation, "just as was the custom in the time of the Saracens" ("secundum pro tempore sarracenorum fuerat consuetum"); Chabret, *Sagunto,* II, 405. Alfonso X (Aug. 29, 1252) granted to Alicante certain villages with their water, "just as never they had it better in the time of the Moors" ("asi como nunca mejor las hubieron en tiempo de Moros"); Branchat, *Tratado de derechos,* III, 505. Alfonso X (July 15, 1266) granted several places to Orihuela with barren lands, springs, and rivers, "and with as many appurtenances as they ought to have, just as they had in the time of the

Moors" ("e con todas sus pertenencias quantas han de hauer assi como las ouieron en tiempo de moros"); Martínez Morellá, *Cartas de los reyes de Castilla a Oriheula,* p. 23. Privilege of Alfonso X to Orihuela (May 14, 1275) by which the cequiers were to clean the canals yearly and perform their customary duties, "and let them do this each year as they deem necessary and guide the waters in such a way that they irrigate just as was accustomed in the time of the Moors . . . and that they guide [the water] through those places where they used to go in the time of the Moors; that they make all the lands irrigate at the canal checks where they used to take their turn in the time of the Moors, and at no other place" ("e esto fagan cada año en como vieren que es menester e guien las aguas en tal manera que rieguen e ryegue tanto como solia en tiempo de moros . . . e que las guien por aquellos lugares que solian yr en tiempo de moros; que fagan que todas las tierras se rieguen por las paradas do solian tomar su tanda en tiempo de moros y no por otro lugar"); *ibid.,* pp. 58–59. Don Manuel, re. the partition of property in Elche (June 20, 1268): "Furthermore, I authorize that they ought to have the water with which were irrigated the *alquerias* of these estates just as the Moors used to have it in their time" ("Otrosi les otorgo que el agua con que se regauan las acarias do son destas heredidades que la ayan como la solien auer los moros en el so tiempo"); Ibarra y Ruiz, *Riego de Elche,* pp. 88–89.

26. See above note, privilege of Alfonso X to Orihuela, May 14, 1275.

27. "De cequiis et cequiariis" (Jan. 19, 1250), in *Aureum opus,* chap. xxxiv, fol. 11.

28. The same phenomenon is also true in the Islamic world. Compare Thoumin, *Geographie humaine,* p. 106; the stability of irrigation custom in Syria is based on an appeal to time immemorial (*qadim*).

29. *Digest,* 8.5.10; the translation is that of Charles Henry Monro, *The Digest of Justinian,* 2 vols. (Cambridge, Eng., 1909), II, 103. James I initiated a renaissance of Roman law in the lands of his crown in which many Bologna-trained jurisprudents (including Ramon de Penyafort) participated. The influence of Roman law was especially marked in Valencia; see Ferran Soldevila, *Jaume I, Pere el Gran* (Barcelona: Teide, 1955), pp. 68–69. However, it should be noted (as Robert I. Burns has pointed out to me) that Roman law was also strongly resisted, and the struggle between Roman and customary law may have given Islamic custom great impact in the thirteenth century especially.

30. ARV, Gobernación, 2253, 9th hand, fol. 24r (June 20, 1435).

31. T. F. T. Plucknett, *Legislation of Edward I* (Oxford: Clarendon Press, 1949), p. 46 n. 2: "The same date had been made the limit of writs of right by the statute of Westminster I, c. 39 (1275), i.e., a demandant cannot recover by a writ of right (or any other writ)." A statute of quo warranto of 1290 (*ibid.,* pp. 49–50) is similar in tone to the requirements for proof in the court of the governor of Valencia: "As for the writ called *quo warranto,* our

lord the king established in Whitsunday in [1290] . . . that all those who claim to have quiet possession of franchises before the time of King Richard, and can show this by a good inquest, may properly enjoy that possession." Setting limits to legal memory produced strange results in England, namely, that "an admittedly genuine deed cannot be pleaded if it was made before 1189." This did not happen in Valencia, where Arabic legal documents preceding the Reconquest could be pleaded, and were, even as late as the sixteenth century.

32. ARV, Gobernación, 2244, 15th hand, fol. 5r (Oct. 13, 1431): "segons costum antich la qual cosa es despuys que lo regne es de cristians ença que memoria de homens no es en contrari."

33. *Ibid.*, 16th hand, fol. 37v: "segons la costum antigua . . . axi com es acostumat de temps de serahins ença."

34. ARV, Gobernación, 2226, 3rd hand, fol. 26r (Mar. 29, 1421): "la manera stat e forma en que huy son los dits partidors e fila haien stat e acostumat de star per tots temps desque es la dita cequia dels archs que es dels temps de serahins e del temps molt enans ques conquistas aquest regne de Valencia per molt excellent rey en Jacme lo primer de memoria gloriosa ança continuament."

35. ARV, Real Patrimonio, Bailía, 1430, 3rd hand of 1416, fol. 22r (Aug. 5, 1416).

36. ARV, Gobernación, 2285, 14th hand, fol. 2r.

37. See Glick, "Abogados de aguas."

38. ARV, Gobernación, 2219, 4th hand, fol. 34r (Feb. 13, 1417):

> No fa res al proposit ni contra aquesta part com diu la part contraria que los hereters poden posar penes per les aygues e partir aquelles axi com los plaura; et les aygues e cequies deuen anar, star e discorrer segons foyen en temps de sarrahins . . . car cert es que los hereters ensemps ab la vila meten les penes, muden, augmenten, limiten e minuen aquelles segons los es ben vist. Et los fets ho requiren. Et aquelles ab la dita vila muden les cequies e fan discorrer les aygues per lla on los appar que stiguen e discorren mils a profit e utilitat e menys messio dels hereters.

39. Certain minor hydraulic toponyms found in medieval documentation might also have had the connotation of Islamic origin: e.g., the *cequia morisca* in the Marjals of Valencia (ARV, Gobernación, 2307, 32nd hand, fol. 18r), the *almenara morisca* of the Favara system (ARV, Gobernación, 2257, 4th hand, fol. 11r), the *azud de los moros* on the Vinalapó (Ibarra y Ruiz, *Riego de Elche,* p. 259); an *azud de los moros* also in Caspe (photo in Andrés Giménez Soler, *La edad media en la Corona de Aragón* [Barcelona, 1930], p. 334).

40. ARV, Gobernación, 2203, 28th hand, fol. 16r (May, 30, 1413): "nos demostra res picat ne obrat de nou e en altres molts e diverses lochs on es

construhida e obrada dargamassa molt fort e antigua e en altres parts on es feta de pedra mamposta fort antiga e la qual obra se mostra esser tota morisqua e obrada de temps de moros e la qual dita cequia del dit açut vell tro a hun moli destrohit que es prop de hun loch despoblat appellat facuquer es feta e construida segons dit es antiga e morisqua."

41. ARV, Gobernación, 2282, 12th hand, fol. 12v (Apr. 21, 1452):

> E sis volia dir als en contrari ço que nos creu diu . . . que los molins dels dits mossen pere dodena e en marti scola son molins molt antichs de temps de moros e re ipsa se mostren en cascu de aquells edifficis morischs los quals per qualsevol vehent aquells seran haguts per hedificis morischs e construhits en temps de moros ans que Valencia fos de cristians e per hedificacio morischs e entichs e fets en temps que Valencia era de moros los dits hedifficis son hauts per los vehents aquells.
>
> Resulta donchs ut concludit necessarie que sils dits molins son a temps de moros ans que Valencia fos de cristians que en aquell temps que foren construhits los fon donada laygua e es propria llur e son en possessio pacifica vel quasi de haure e usar de les dites parts de la aygua de la dita cequir de algiroç. . . . diu . . . que lo moli pilater del qual la part altra fa gran festa appellat den ferrada es moli nou e lo casal de aquell e totes les obres de aquell son noues ço es ques mostren fetes de cent anys ença e noy ha res de antiguor ne hedifici algu de temps de moros ans cascun entenent de obres haura e ha la construcio e fabrica del dit moli pilater per obra e fabrica costumada de temps de cristians e de cent anys ença e menys o poch mes.

42. AHN, Ordenes Militares, Orden de Montesa (documentos particulares), pergamino no. 653 (Sept. 27, 1310): "ells justament ab raho tenen dins los dits azuts antichs de temps de sarayns pertayens a la dita vila de Tales laygua que passa lo terme de Tales e discorrent en los dits açuts."

43. AMV, Cartas Reales, 8, fol. 181v (July 30, 1609): "En lo terme de la ciutat de Murcia ques en Castella en lo nostre riu de Segura y a edificat un açut en dit riu fet de temps de moros ab tal industria y art que ab laygua quey de damunt de aquell pasa se sustenta esta ciutat y totes les viles y pobles desa contribucio esta fet a modo de cuchara." It is not clear why this document, pointed out to me by L. P. Harvey, should have been written in Catalan and included in the papers of the city of Valencia.

44. *Fori regni valentiae*, fols. 90v–91r (1342):

> Item com segons fur de Valencia posat en rubrica de servitut de aygua, e de altres coses, lo capitol qui comença, l'aygua del flum publich [i.e., no. 38, fol. 90v] etc. deu esser partida segons la manera, e la granea de les possessions a regar, e que aço sia feyt sens injuria del altre. E senyor apres la confeccio del dit fur alcuns hajen staliades, e fetes staliar moltes terres, e diverses: e les quals a regar prenen, e reben del dit flum gran partida de l'aygua en gran dan, injuria, e perjudici de les terres de la

orta de Valencia: les quals ja en temps de Sarrahins eren laurades, e hauien compliment d'aygua del dit flum a regar aquelles ço que alcunes de vegades entre l'any n'an gran minua e fretura per raho de les aygues que alcuns prenen a ops de regar les terres que apres son staliades. E com no sia cosa justa ne consonant a raho que les terres que no hauien acostumat de regar hajen gran abundancia d'aygues, axi que quasi neguen que les terres antigament laurades en la dita orta n'ajen gran minua, e fretura per la qual los splets se perden.

45. Branchat, *Tratado de derechos,* III, 303–306 (Oct. 5, 1368).

46. See Chap. VII, n. 25

47. J. C. Russell, "The Medieval Monedatge of Aragon and Valencia," *Proceedings of the American Philosophical Society,* 106 (1962), 496. In 1361 there were in addition some 2,000 hearths in the huerta; no figures were given for 1418.

APPENDIXES

APPENDIX 1. MUNICIPAL PROVISIONS FOR ROAD MAINTENANCE, 1396

1. AMV, Manuals de Consell, 21, fol. 40r (Aug. 14, 1396).
2. Officials of the Fabrica de Murs i Valls (Lop, *Murs e Valls,* pp. 23, 69).

APPENDIX 2. IRRIGATION OF VINEYARDS

1. ARV, Gobernación, 2262, 3rd hand, fol. 23r (Aug. 28, 1438).
2. By my calculation the total is 31 *lliures* 8 *sous.*

APPENDIX 3. CROP PRIORITIES IN TIME OF DROUGHT, 1376

1. AMV, Manuals de Consell, 17 fol. 62r (July 18, 1376). "De forma de regar los esplets."

APPENDIX 4. REGULATIONS OF THE WATER COMPANY OF BENIDORM

1. ARV, Varia, 126, fols. 81r–82v. "Capitols de que deuen esser fets e ordenats entre los campanyons de la aygua de benidorm."

APPENDIX 7. PLANS FOR HYDRAULIC WHEEL IN VALENCIA TO IRRIGATE GARDEN OF THE LONJA, 1529

1. AMV, Manuals de Consell, 63, n.p. (Nov. 27, 1529).

2. For this bridge (on the Quart road) see Lop, *Murs e valls*, pp. 490–491.

3. *Carrero açucach: açucach* (or *atzucach*) is derived from the Arabic *zanaqa* (narrow street). See Arnold Steiger, *Origin and Spread of Oriental Words in European Languages* (New York: S. F. Vanni, 1963), pp. 35–36.

APPENDIX 8. DIVISION OF THE BARADA RIVER IN 125 H. (742/743 A.D.)

1. Ibn ʿAsākir, *Tarīkh Madīnat Dimashq*, ed. Ṣ. Munajjid, vol II (Damascus, 1951), p. 151. A less complete version is available (*ibid.*, p. 147), and a later version, almost identical to the one translated here, is in Ibn Shaddād, *Al-Aʿlāq al-Khaṭīra*, ed. Sami Dahan (Damascus, 1963), pp. 15–17.

2. Umayyad Caliph, reigned A.D. 724–743.

3. Munajjid says he was a surveyor (*ṣāḥib al-misāḥa*), Ibn ʿAsākir, p. 147 n. 2.

4. *Yamāzu* — the root *m-y-z* means "to separate." A substantive, *tamāz*, (apportioning) appears on p. 152. The current word for "water divisor" in the Ghūṭa, *mezzaz*, is more probably related to this root than to *mazza*, "to suck" (Tresse, "L'Irrigation," p. 475); this is more in line with the literalism in irrigation terminology of other Islamic systems.

5. *Maskaba* (alternating with *masākib*, another plural of *maskab*) — the root *s-k-b* means "to pour forth." E. W. Lane defines *maskaba* as "the higher, or highest, rivulet, or small channel for irrigation, from which other rivulets, or small channels, are supplied with water"; *Arabic-English Lexicon*, 8 vols. (New York: Ungar, 1955), I, part 4, 1389. This suggests a connotation similar to that of the Valencian fila, a continuously flowing proportional part of the river's capacity.

6. An imam of the mosque of Damascus (Ibn ʿAsākir, p. 147 n. 6).

7. The 3 versions differ somewhat, but the sense in all is that these 4 *masākib* were part of the Bānās's 30.

8. The *masākib* total 184, suggesting a 2-week turn on the river if one *maskaba* was then (or was originally) equal to 1 hour (184 hours = 15 days 4 hours).

9. Ibn ʿAsākir, p. 151, and ibn Shaddād read *al-qanā*. But this reading, Ibn ʿAsākir, p. 147, is clearly preferable because al-Qanawāt is the famous canal serving the city of Damascus.

Index

Major variant spellings of place names are in parentheses following the initial entry.

Bei Fragen zur Produktsicherheit wenden Sie sich bitte an:
If you have any questions regarding product safety,
please contact:

Walter de Gruyter GmbH
Genthiner Straße 13
10785 Berlin
productsafety@degruyterbrill.com